W9-ACI-340

Biomechanical Analysis of Fundamental Human Movements

Arthur E. Chapman, PhD, DLC

Professor Emeritus

Simon Fraser University

Burnaby, Canada

Human Kinetics

Library of Congress Cataloging-in-Publication Data

Chapman, Arthur E., 1941-
    Biomechanical analysis of fundamental human movements / Arthur E. Chapman.
        p. ; cm.
    Includes bibliographical references and index.
    ISBN-13: 978-0-7360-6402-6 (hard cover)
    ISBN-10: 0-7360-6402-8 (hard cover)
    I. Human mechanics. I. Title.
    [DNLM: 1. Movement--physiology. 2. Biomechanics. WE 103 C466b 2008]
    QP303.C446 2008

    612.7'6--dc22

                            2007046933

ISBN-10: 0-7360-6402-8
ISBN-13: 978-0-7360-6402-6

Acquisitions Editor: Loarn D. Robertson, PhD; **Developmental Editor:** Judy Park; **Assistant Editors:** Lee Alexander and Jillian Evans; **Copyeditor:** Joyce Sexton; **Proofreader:** Pam Johnson; **Indexer:** Bobbi Swanson; **Permission Manager:** Carly Breeding; **Graphic Designer:** Bob Reuther; **Graphic Artist:** Kathleen Boudreau-Fuoss; **Cover Designer:** Nancy Rasmus; **Photo Asset Manager:** Laura Fitch; **Photo Office Assistant:** Jason Allen; **Art Manager:** Kelly Hendren; **Illustrators:** Arthur E. Chapman and Alan L. Wilborn; **Printer:** Edwards Brothers

Printed in the United States of America   10   9   8   7   6   5   4   3   2   1

Human Kinetics
Web site: www.HumanKinetics.com

United States: Human Kinetics
P.O. Box 5076, Champaign, IL 61825-5076
800-747-4457
e-mail: humank@hkusa.com

Canada: Human Kinetics
475 Devonshire Road Unit 100, Windsor, ON N8Y 2L5
800-465-7301 (in Canada only)
e-mail: info@hkcanada.com

Europe: Human Kinetics
107 Bradford Road, Stanningley, Leeds LS28 6AT, United Kingdom
+44 (0) 113 255 5665
e-mail: hk@hkeurope.com

Australia: Human Kinetics
57A Price Avenue, Lower Mitcham, South Australia 5062
08 8372 0999
e-mail: info@hkaustralia.com

New Zealand: Human Kinetics
Division of Sports Distributors NZ Ltd.
P.O. Box 300 226 Albany, North Shore City, Auckland
0064 9 448 1207
e-mail: info@humankinetics.co.nz

This book is dedicated to my wife,
who raised my children,
who in turn participated in many casual experiments
without their knowledge or permission,
and further to my daughter,
who has provided two more experimental subjects.

I cannot sufficiently express my gratitude
to all my graduate students
who I believe are and will continue to be friends.

Drs. J.D.G. Troup and D.W. Grieve
gave me both intellectual and scientific grounding
in this area of lifelong interest.

Last of all,
I should like to dedicate this book to my parents,
both of whom knew how to kick a ball
but didn't know why scientifically
and sadly will never know.

Contents

Preface **vii** ◇ Acknowledgments **xiii**

Preface

Have you ever asked yourself, Why can't sprinters accelerate forever? Why can you land without injury from a height that is much greater than the height you can jump up to? Why is running a much more costly activity per distance traveled than walking? Why can't a pole-vaulter increase the height of the vault continuously by using poles of ever-increasing length? Why do bicycles have a range of gears? Why is it ungainly to walk without free movement of the arms? What are the energy and safety requirements of lifting a large load one time in comparison with lifting 1/10 of the total load in 10 repetitions? The answers are to be found in your mechanical properties, working according to the laws of mechanics in our gravitational field.

Human movement has existed for much longer than we have been able to talk about it. Despite the development of language, it was only recently in human existence that laws of movement of bodies, human and otherwise, were formalized. Now we are in a position to be able to take a scientific approach to the understanding of human movement. This area of study is biomechanics—a name that indicates the application of scientific principles and laws toward an understanding of biological systems. Specifically we are dealing with biomechanics of human movement, which in itself is a small part of the whole range of investigation in human biomechanics.

This book adheres to rigid mechanical and biological definitions and concepts. For example, force is recognized as that influence which accelerates a mass. Should there be no acceleration, either the force is zero or there must exist an equal and oppositely directed force (Newton's third law). It is true that application of a force increases the velocity of a thrown object. Yet this statement is inaccurate in the sense that we require more knowledge about the force in order to calculate velocity. In fact, it is either force multiplied by time (impulse) or force multiplied by the displacement of the force (work) that is required. Correct terminology must also be applied to our description of human motion. The term "hip flexion" has a specific meaning and refers to the movement of a joint. The phrase "flexing one's muscles" has no meaning in the biomechanical world.

The ability to determine the effect of some influence on motion or to calculate the speed of an action requires the language of a science known as mathematics. Without mathematics there can be no accurate numerical description of motion and no ability to relate biomechanical concepts. Numerical calculus is particularly important as a means of describing human motion. For example, force produces acceleration; and the relationship between acceleration, velocity, and displacement can be understood only if the concepts of differential and integral calculus are mastered.

No engineer would design a machine without an understanding of the mechanical properties of the motor driving it. Although we are not designing a human machine, knowledge of the mechanical properties of muscle enables us to understand why we do things in the way we do and what modifications we might make in the way we do something to enhance performance. Since muscle shows force–velocity and force–length relationships, its ability to produce acceleration depends on the instantaneous kinematic state (position and velocity) of the body.

An understanding of the mechanical properties of ligament, tendon, cartilage, and joint structure and articulation is also necessary. For example, cartilage can dissipate dangerous amounts of energy, and ligament and tendon can provide a temporary storage of energy that can be returned to the system when appropriate.

Audience and Approach

This book is written for those students who wish to understand human movement by means of scientific, quantitative biomechanical analysis. The information contained here should satisfy this wish as it provides an approach that is uncompromisingly mathematical and mechanical.

Earlier books in this area explained human movement by recourse to well-established mechanical laws. Such explanations were descriptions of how a variety of mechanical phenomena were exhibited in a variety of activities. The aim was to educate readers about the reasons we do things the way we do. This approach may be criticized as merely an academic exercise since we have been performing these movements without description for millennia. A further criticism is that these early books described the status quo of current human activities without facilitating an understanding of how movement may be modified. Yet such a development was a necessary precursor to more detailed study, as numerous important mechanical reasons for a particular type of movement were illustrated.

Many books on biomechanics tend to use examples of research that can appear to be rather disjointed. The problem with this type of compilation of information is that no formal approach to the study of human movement is either used or proposed. This book takes a systematic approach by examining the mechanical essentials of a task and then explaining how these essentials can be achieved by use of the biomechanical properties of the body. The purpose is to provide strategies and techniques for human performance and also biomechanical analysis of human movement from mathematical and mechanical viewpoints. Consequently this book is written for those individuals who either have a good grounding in mechanics or are prepared to learn mechanics. The information will be of use to teachers of human movement, to sport apparel and safety equipment designers, and to rehabilitation specialists. The current information will be useful to any person who wishes to understand the mechanical interaction of the human with the environment and the force, work, and energy either induced or required in such interaction. This material is essential for those who wish to perform advanced research in the area of human biomechanics. The material that is presented is necessary to allow the reader to advance from qualitative to quantitative understanding of human biomechanics; and it is the basis for calculation, or at the very least estimation, of the effects of adopting a change in technique in a particular activity. The mechanics, in conjunction with the information on properties of body structures, will enhance the reader's ability to estimate or calculate loads either applied to the body as a whole or induced in individual structures.

Since the early stage of this discipline, a great deal of research has been conducted for a wide range of reasons. Some of the reasons have been specific—for example, to identify the pattern of forces at the shoe-ground interface in running. Some implications of this work are in the design of running shoes and the avoidance of repetitive strain injuries in long distance running. Yet whatever the intention of the researcher, all of these investigations involve mechanics and mathematics.

Human biomechanics is a cross-disciplinary approach to understanding our biomechanical system through the laws of mechanics. Therefore information is provided on the structures involved in human movement such as bones, ligaments, cartilage, and muscles. The important properties of these structures are necessarily mechanical and include their viscoelastic characteristics and their properties of energy storage, release, and dissipation.

Muscle is given special attention as the force and work producer that is under our volition, as opposed to gravity, about which we can do nothing. Many of the mathematical techniques necessary to perform mechanical calculations are included in appendix B, but some are incorporated into the text in cases in which the material requires understanding of additional mechanics.

As in all textbooks, the question arises as to the depth of information that can be expected. This book is designed for upper-level undergraduate and graduate students of human movement. It should also be useful for those graduate students whose specialization is not biomechanics but who need to analyze human movement for other reasons; motor behaviorists are a good example.

How This Book Is Organized

The text is organized into two parts comprising 11 chapters. Part I includes three chapters covering the principles of biomechanics. Chapter 1, "Biomechanical Structures of the Body," deals with structures of the body including the skeletal framework and the properties of other tissues. The aim is to give the reader an appreciation of the biomechanical properties of the structures that either move the human "machine" or restrict its movement. Chapter 2, "Essential Mechanics and Mathematics," deals with the mechanics and mathematics necessary for our purpose. This order of presentation of the mechanical concepts differs from that of most mechanics textbooks, and it includes only information pertinent to our study. The aim is to develop mechanical concepts by means of a series of equations. Furthermore, readers will see that this material not only is mechanical but also includes the mathematics needed to manipulate the equations. Without mathematics, mechanics either makes little sense or is at best vague. Books that purport to teach biomechanics nonmathematically usually provide no more than a description of motion using mechanical terminology. To keep this chapter from becoming a lengthy mechanics primer,

some mechanical concepts are dealt with in the chapters where they naturally arise.

Chapter 3, "Foundations of Movement," deals with those factors that move us, such as muscular force and gravity; those factors that enable us to move, such as friction; and the manner in which intersegmental motion occurs. Introductory mechanics usually deals with single rigid bodies. In our case we have a multisegmental body comprising head, upper arms, forearms, trunk, thigh, and so on. Obviously these segments move in relation to each other, and chapter 3 describes the manner in which this is accomplished. This relative motion, known as articulation, is one reason we can perform a wide range of activities, but also why we are sometimes in danger of injury. In other words, a machine with many moving parts is in greater danger of breaking down than is a single lump of matter.

Part II, "Fundamental Human Movements," is an examination of the mechanics of a wide range of fundamental human activities. While this part can be seen as the body of the book, readers will not be able to understand all the material unless they have completely mastered the preceding chapters. However, readers are encouraged to preview any one of the fundamental movements in order to appreciate what type of scientific information is necessary.

Devoting part II to the biomechanical analysis of fundamental human movements was an alternative to dealing with the intricacies of a variety of movements in a particular sport. The reason for this approach is that most, if not all, of our special activities (e.g., sporting and work) involve aspects of the fundamental movements. For example, running is a part of almost all of our sporting activities. A further reason is that many of our apparently simple, well-learned, and mundane movements often produce injury, as a result of either a single traumatic event or excessive repetition. The format for analysis of each fundamental activity involves a formal process in which the first step is to describe the primary *aim* of the activity. The next step is to convert the verbal description into a *mechanical* aim such as maximization

---

of work or maximization of potential energy. Structures in the human body that can contribute to achievement of the mechanical aim are then identified, and the manner in which they can or should be used is determined based on knowledge of their properties. At this stage the *biomechanical* analysis of the movement is complete, and what follows is an examination of *variations* of the primary aim. For jumping, for example, the high jump, long jump, triple jump, and pole vault are examined for their particular requirements. Means of *enhancing* performance are examined with respect to changing the properties of body structures involved. A section on *safety* identifies potential injuries to body structures and then identifies strategies for avoidance of injury based on the manner in which the structures are used in the activity. Some worked *practical examples* are given to reinforce both the mathematical manipulation of biomechanical data and the process of solving a biomechanical problem. Finally, a chapter summary refers to the major points using common language rather than the technical language that predominates in the body of the chapter.

Chapters adhere to the following scheme for each of the fundamental movements:

1. *The aim:* This section identifies the primary mechanical aim of a given human activity from its initial verbal description.

2. *Mechanics:* In this section we identify the mechanical formulas which best describe how to achieve the mechanical aim. This second stage is both something of an art and a matter of convenience, since all mechanical formulas are related. For example, if we know force, time, mass, and displacement, we can deduce both momentum and energy. In some cases momentum rather than energy (or vice versa) may be the best way to capture the nature of the activity in question.

3. *Biomechanics:* This discussion identifies structures in the human body that can contribute to achievement of the mechanical aim and explains how these structures should or can be used based on knowledge of their properties.

4. *Variations:* This section examines variations of the primary aim. For example, for jumping, the particular requirements of the high jump, long jump, triple jump, and pole vault are identified.

5. *Enhancement:* This discussion focuses on ways in which the structures involved may be modified to enhance performance of the aim.

6. *Safety:* This section deals with how potential injuries to the structures may be identified and avoided. In some cases the questions of enhancement and safety are somewhat synonymous. In those cases safety is dealt with under the heading "Enhancement and Safety."

7. *Practical examples:* Worked examples are presented to demonstrate the choice of mechanical approach and the mathematical techniques that can be used to solve biomechanical problems, as well as to illustrate how biomechanics facilitates understanding of the nature of a given movement and its limitations. In many cases the examples can only approximate the biomechanics of the real situation because the muscle force produced depends upon the kinematics that the muscle itself produces. This leads to a complex situation that cannot be solved through application of the principles of calculus to an equation of the motion. This problem has been solved in the text through use of simulation of an activity whose solution requires numerical analysis. However, instruction on any other than basic numerical analysis is beyond the scope of this book.

The chosen format for this book is analysis of fundamental human movements because these movements have their application in a variety of everyday human situations. While the analysis uses examples from a variety of human situations, this is not a textbook on how to play a given game or how to work in a specific occupation. Many of these movements are combined and incorporated into a human situation, whether it is sporting or occupational or whether it has to do simply with survival. It is hoped that mastery of the fundamentals will provide a body of information and a method of analysis that can be applied to human tasks of interest to the reader.

The overall aim is to impart accurate scientific knowledge of a biomechanical nature in order to counter the misinformation, obfuscation, and ambiguity that surround this area of the study of human movement.

Special Features

The text is supplemented by several special elements. Within each chapter the reader will find Key Points, which summarize critical information; Recommended Readings; and Practical Examples as described for part II.

The Recommended Readings include sources of basic reference material on which this book is based, categorized according to discipline and application of biomechanics. The bibliography includes journal titles for those interested in delving deeper into a specific application, and conference proceedings that are accessible by modern electronic means.

In addition to a glossary, a bibliography, and a detailed index, several unique elements are included at the back of the book. Appendix A presents key symbols for mechanical variables, and appendix B presents mechanical formulas. Appendix C contains 68 problems that readers can tackle to test their grasp of biomechanical analysis; these are categorized by mechanical concept rather than by fundamental human

movement. The topics include kinematics, forces and moments, impulse–momentum, and work–energy. The answers to the problems are provided in appendix D. Readers will have to develop the diagrams for each problem in the manner done for the worked examples in the text.

The author hopes that upon completion of this material, the reader will be armed with a full understanding of this branch of human biomechanics. Readers should then be able to identify the appropriate approach to investigating activities they are interested in. They should also be better armed with the ability to suggest novel ways of performing an activity. Last but not least, readers should be able to describe to others what is happening in a human movement, why it is happening, and what makes it happen. Readers' descriptions will employ correct usage of mechanical terms, and readers will appreciate that much popular use of mechanical terms is misleading at best and at worst incorrect.

Many books on human biomechanics have taken a mechanical concept and identified activities in which the concept is implicated. This approach is useful for learning mechanics through experience of well-recognized activities. The unique feature of this book is that the fundamental activity is of prime importance and the biomechanical concepts that best facilitate its understanding are identified.

Acknowledgments

I should like to acknowledge the assistance of Human Kinetics personnel: Loarn Robertson, and two individuals called Melissa and Joyce whose names kept appearing on my computer screen. I should like to acknowledge Judy Park in particular, whose unfailing cheerfulness and optimism helped a great deal. These individuals were instrumental in the origin and completion of this book. I also owe a debt of gratitude to those many students and others who were willing participants and coworkers in my biomechanical experiments, as well as colleagues who kept me thinking about human biomechanics.

PART I

Introduction

When making sense of the biomechanics of human movement, we need three basic subsets of information. Chapter 1 concerns the manner in which the human system is constructed. Within this area we require information on the framework of the body, how parts of the framework are linked together, which structures aid the linkage, and which produce relative motion of the parts. If our interest is in safety, we also need to know the strength of the linking tissues and the mechanical conditions predisposing them to injury. The second subset of information, in chapter 2, deals with the conceptual and practical tools that facilitate biomechanical analysis. The mechanics give us an understanding of the way bodies move in our gravitational world, and the mathematics allow us to use mechanical equations to produce numbers representing movement. Chapter 3 concerns the foundations of movement. External agencies such as gravity and friction are considered. Mechanical properties of muscles are dealt with from the point of view of the internal forces that they create, how they predispose us to certain optimal patterns of movement, how they are used to generate maximal momentum and maximal kinetic energy, and how they facilitate redistribution of both momentum and energy among body parts. Chapter 3 ends with a discussion of the forces occurring between body parts and the analytical framework for their determination. Successful understanding of part I will allow the reader to tackle the analyses of fundamental human movements in part II.

Biomechanical Structures of the Body

S tudy of the mechanics of a system must begin with knowledge of the size, shape, and inertial properties of the parts that compose it. Further knowledge is required on the kinematics and frictional characteristics of the mechanical linkage of connected parts. The study is completed by knowledge of the properties of the motive forces driving the system. This outline relates particularly to a system that does not change once constructed. Alternatively, the biomechanical system is subject to change. Its individual parts or segments, largely constructed of bone, muscle, and fat, are nonuniform and can change their size, shape, and inertial characteristics with age, use, and nutrition. The linkages or joints between segments have some interesting properties that can also change with these factors. Lastly, the motors or muscles driving the segments have complex properties that are subject to change. It is these animal motors that not only allow us to move, but also predispose us to certain patterns of movement depending upon the goals of our movement. This chapter concerns the structure of the human body primarily from the point of view that we are a moving machine capable of an enormous range of skills, but also capable of breakdown. The underlying theme is the biomechanics of the form and properties of the constituent parts of the human body. To this end we will confine ourselves to things biomechanical and leave the other wonderful features of human beings to others.

A question arises: How much do we need to know about the form and function of the body in order to perform biomechanical analysis? The answer is, "What are the goals of our investigation?" What we require in order to understand the biomechanical causes of lateral epicondylitis or "tennis elbow" is quite different from the knowledge we need to understand the biomechanical process of performing a handstand in gymnastics. Another question arises: What depth of knowledge is required? The answer in this case is, "How accurate do you want the results of your investigation to be?" In this chapter the aim is to present a body of biomechanical knowledge of the structure that is sufficient for understanding and analyzing a wide variety of fundamental human movements. A simpler statement might be that we need to know what to plug into our equations of motion so that we can tell a true story about human movement.

Not all of the material in this chapter is directly pertinent to the biomechanics of a single, fundamental human movement. For example, it is not necessary to know the ultimate strength of bones in order to understand how we jump. Such information is included because knowledge of the mechanical properties of biomechanical structures is important if we are to understand the effects of loads induced during fundamental activities. An exhaustive study of this area has been published by Nordin and Frankel (2001). The presentation in this chapter assumes that readers are familiar with the anatomical terminology for principal axes and planes; joint motion; and names of muscles, their locations, and points of origin and insertion.

The Frame

The frame or endoskeleton is the most rigid supporting structure in the body. This endoskeleton is the basic structure upon which the rest hangs. Long bones give us reach and leverage so that we can transport ourselves and manipulate objects that are at some distance from the mass of the body. Such aims are achieved by the long bones comprising the upper and lower limbs by means of enlargements at each end where they contact other bones to form an articulation, and at various other places along their length where tendons insert. The purpose of such enlargements is dealt with later in relation to their mechanical significance. Flat bones such as the iliac bone and the scapula have large surfaces where large areas of muscles can be inserted. Although the leverage of these muscles may be small, their size is an important part of their biomechanical energy-generation mechanism. This avoids the need for us to carry large muscles at the periphery. Such an arrangement would greatly increase the inertial properties of the body, and result in a concomitant increase in energy cost to the whole system during movement. The bones of the cranium are flat bones which protect the brain, yet they also perform the biomechanical function of providing a large surface of attachment for some muscles involved in feeding. Many other oddly shaped bones, such as the carpal bones of the wrist and their counterpart tarsal bones of the ankle, have important functions during our forceful mechanical interaction with the external world.

The bones of the frame are effectively rigid in comparison with the softer mechanical tissues. In reality, bones have elastic and other properties that are implicated in movements involving high-impact forces. These properties allow bending rather than breaking of bones, and as such contribute to energy dissipation when bones are loaded, as in falling. Unfortunately, when forces are too large or there is too much energy to absorb, bones can break. A further property of bone is that as a living tissue its cells are constantly being destroyed and created. This property is known as bone remodeling, in which both nutrition and applied forces play an important part. Without this feature any bone would wear out and lose its characteristic shape (see Currey, 1984).

■ **Key Point**

Bone shapes are implicated in joint leverage, the amount of muscle that can be attached, and the size of the articulation surface.

Articulations

Movement of one bone relative to another is known as articulation, so the joints are called articulations. Figure 1.1 shows a very rough approximation of the human knee joint linking the femur and tibia.

The knee joint provides a general example, as it displays most of the features of joints in general and is particularly subject to injury in many activities. Articulation occurs smoothly because the ends of the bones are covered in an especially smooth hyaline cartilage with low friction characteristics. You will be able to see this smooth surface on the ends of the bones of the next chicken leg you eat. In some joints there is also fibrocartilage, which not only aids in reducing joint friction but also modifies the shape of the articular surfaces. The knee joint contains two fibrocartilages; these are the structures removed when you have your

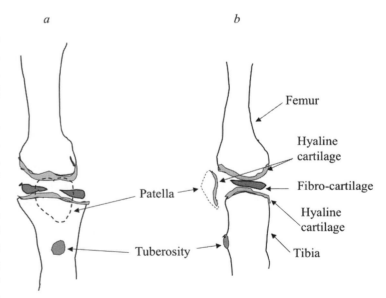

FIGURE 1.1 Bones and cartilages of the knee articulation, seen from *(a)* the front and *(b)* the side. The patella articulates upon the anterior aspects of the femoral condyles during movement.

"cartilage" removed. With advances in surgical techniques, cartilage removal is less common than previously. This is a good thing because cartilage removal severely alters the mechanics of the articulation. Within the knee joint are also the patella (commonly known as the kneecap) and a raised portion on the front of the tibia known as the tibial tuberosity. These extra parts contribute to the effectiveness of muscular force and are dealt with later. All of the joints of the body exhibit some or all of the structures described depending on the required range of movement and some other factors. Cartilage is one of many connective tissues that are involved in biomechanics (see Armstrong and Mow, 1980).

Cartilage displays the properties of elasticity and viscosity. Elasticity, or spring-like behavior, is simply a change in shape in response to an applied force. A perfectly elastic material stores strain energy equal to the force times the amount of deformation. Upon recoil, the elastic material gives up its strain energy to the external environment, appearing as work equal to that which originally deformed it. While the common understanding of viscosity concerns fluid flow, viscosity also represents a property in which the resistance to deformation is proportional to the rate of deformation. Unlike the spring, the viscous component does not return by itself to its undeformed state following deformation. Therefore the energy required to deform it is not returned but is dissipated as heat. In fact, a force in the opposite direction is required to return it to its original state. Elasticity and viscosity are shown by the springs and shock absorbers (dampers) of your automobile, respectively (see Ozkaya and Nordin, 1991).

Figure 1.2 shows the behavior of springs and dampers; S and S-D represent, respectively, a spring alone and a parallel spring and damper. The area under the graph, the product of force and deformation, represents work done or energy stored. The arrows represent the direction of deformation and relaxation in response to the applied force, which increases and subsequently decreases. In the spring model, deformation and relaxation take place along the same path, so the energy returned is equal to the energy used to deform. In the S-D model the paths are different, the difference representing energy or work lost. These characteristics are amply suited to shock attenuation and energy dissipation in running and jumping. Many model combinations of springs and dampers are used to represent the behavior of biological and other physical systems (see Nordin and Frankel, 2001).

Ligaments

Ligaments generally link bone to bone across the articulation as seen in figure 1.3. The collateral ligaments are on either side of the knee joint and are subject to significant stretch when the knee experiences an applied force to either side. The cruciate ligaments are within the interior of the joint, and each is subject to

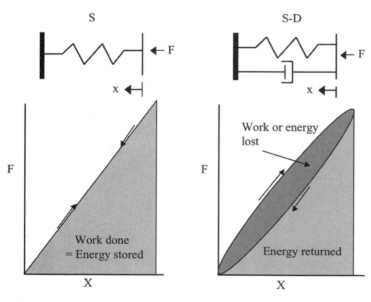

FIGURE 1.2 The relationship between force and extension of two models showing how energy is dissipated by a viscous component but not by a pure spring.

disruption due to different directions of applied force. The patellar ligament is unusual in that it is an extension of the quadriceps tendon and serves to apply the force generated by the muscles to the tibia.

It is popularly held that the role of ligaments is to arrest articulation at the extremes of joint motion. However, muscles have much better properties and are positioned most effectively to serve this function. Ligaments have a very important part to play in controlling the relative motion of the articular surfaces, even within the midrange of joint motion. This is one reason why knee and ankle joints feel unstable when one of their ligaments is ruptured, even in tasks with limited range of motion such as walking (see Daniel et al., 1990). Ligaments also display the characteristics of elasticity and viscosity as described previously for cartilage.

Ligaments transmit forces of unknown magnitudes. It is possible to calculate the net, single value of force between two connected bones if the segmental dimensions and inertial values are available. However, the net force is the result of forces in all structures including muscles and ligaments. Unfortunately, there are too few equations to allow identification of the multiple sources of this net force.

Joint Lubrications

Most movable joints are surrounded by a connective tissue capsule that retains synovial fluid. This specialized fluid provides nutriment to those areas where it would be inconvenient to have many blood vessels, which could rupture during joint motion. Its biomechanical significance is to provide lubrication to the articular surfaces as is common in other machines. Lubrication reduces friction between articular surfaces, which reduces both wear and energy cost. In the lubrication process in general, synovial fluid is squeezed in and out of microchannels in the hyaline cartilage due to pressure between surfaces in contact. Such a process is enhanced by the surprising quality of elasticity that synovial fluid possesses. This allows areas of cartilage not in contact to be bathed in synovial fluid in

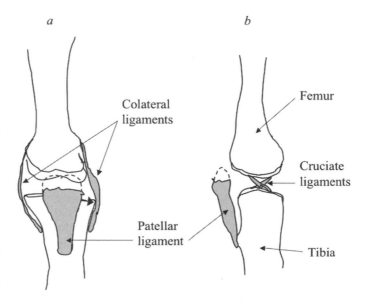

FIGURE 1.3 Cruciate and collateral ligaments of the knee joint.

■ **Key Point**

Ligaments are tough elastic tissues that provide stability of a joint and control the relative motion in articulation.

The skeleton is the support structure that provides areas of muscular attachment and articulation of connected segments.

■ **Key Point**

Joints are enclosed in a membranous capsule that retains synovial fluid, which in turn fills microchannels in cartilage and reduces joint friction.

anticipation of future contact during joint movement. One further property of synovial fluid is "thixotrophy," a decreasing resistance to motion as motion speed increases. This quality is replicated by a bucket of wet sand, which is difficult to stir initially but easier to move once stirred up. When we are in a fixed posture such as sitting for an extended period, our initial movements feel "stiff," more so in people who are aged and often with pain. With movement the effect goes away as the synovial fluid is redistributed. The effect of age is due to reduced ability to retain water, an effect that reduces the amount of synovial fluid and changes the characteristics of both synovial fluid and connective tissue in general.

Muscles and Tendons

Muscle is what we know as flesh or meat, most commonly encountered in the butcher shop or the grocery store. In addition to being good to eat (at least in the opinion of some), muscles (in this case skeletal muscles) are wonderful engines (see MacIntosh et al., 2006). They produce force over a range of lengths. The force occurs only as a tendency to pull the ends of the muscle toward each other. Unfortunately, this engine cannot be put in reverse, but there are muscles on the other side of the joint to perform this function. Simultaneous production of force and muscle shortening represent work done. Force accompanying muscle lengthening due to an external load represents work against the load and therefore dissipation of energy. Muscle is composed of fibers of various lengths depending upon its architecture. The fibers are grouped together in bundles of ever-increasing size until we see a whole muscle surrounded by connective tissue. A cross-sectional cut of meat shows the arrangement of the connective tissue bundles.

■ **Key Point**

Muscle, the single force producer under conscious control, develops tension that is transmitted to the bone by tendons of various lengths.

Muscle is generally attached to bone by a tendon as seen in figure 1.4. In the case of the knee joint, the continuous connection between the quadriceps tendon and the patellar ligament envelops the patella. The quadriceps muscle is actually four muscles, although only three can be seen (the fourth lies under the others). Figure 1.4b shows how the quadriceps muscle is well placed to produce extension of the knee joint. Bones such as the patella and bony prominences such as the tibial tuberosity increase the lever arm of muscle, thus increasing the moment of force.

Tissue Strength

There are limits to the strength of tissue. One important characteristic is known in engineering as "toughness." This is simply the amount of energy the tissue can absorb before it ruptures. But even before the toughness level is reached, most tissues will show indications of beginning failure. For example, a tendon that is overstressed will experience rupture of some of its constituent fibers. Removal of the rupturing load will arrest further rupture and allow healing to take place. Continued load application will end with complete

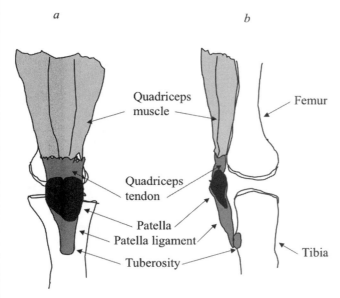

a *b*

Quadriceps muscle

Femur

Quadriceps tendon

Patella

Patella ligament

Tuberosity

Tibia

FIGURE 1.4 Extensor muscle force of the knee is transmitted via the quadriceps tendon and the patellar ligament to the tibial tuberosity.

rupture or separation and lack of any mechanical connection of the muscle to the bone. Repair of this condition is not the subject of this book, but it is not difficult to appreciate that such repair is painful, difficult, and long-term.

While toughness is an important single value predicting rupture, the way in which the rupturing load is applied is also important. Figure 1.5 illustrates the relationship between stress and strain when ligament and other connective tissues are subject to increasing load with different rates of application. The curved shape of the relationships is due partly to viscosity in the stretched tissues and partly to the construction of the ligament. Ligaments comprise bundles of collagen fibers that are essentially parallel but are cross-linked by other fibers. Application of a small load begins to stretch out some of the collagen fibers. Further increase in load brings successively more fibers and cross-links into play in the manner of adding more springs in parallel. Therefore the stiffness of the ligament increases as indicated by the increase in slope of the tangent to the stress–strain curve. The more rapid the rate of increase of stress, the greater the stiffness of the ligament due to the viscosity in the ligamentous system. When a stress is applied rapidly, the tissue tends to rupture with a higher stress, but a smaller amount of strain. Conversely, slower application of load induces rupture at smaller loads, but not before a larger amount of stretch has been achieved.

These properties of tissues often explain why serious injury occurs in circumstances in which there might appear to be no danger and vice versa. The mechanical properties of tissues are complex, particularly if we include the effect of aging, nutrition, and repetition of the same movement. For example, what could be lower in force than the tapping of a keyboard or a checkout machine in a store? But repetition of this low-force activity only too often results in the painful condition of carpal tunnel syndrome. This condition is a consequence of repeated frictional force on the finger tendons that rub on the ligament surrounding the wrist.

Tissue Injury

Despite the wonderful construction of bodily structures, there is an ever-present danger of injury. The mechanics of tissue injury is a wide field of study. This section is not intended to provide a comprehensive treatise on injuries, but rather to give a brief account of how injury is sustained in fundamental human movements.

Injury can occur from a single event or from repetition of an action over a period of time. Single trauma involves the inability of a given structure to dissipate a given amount of energy. A broken tibia in soccer usually involves direct contact with an opponent which induces bending stress that is too great for the tibia to sustain. In other words, the ability to dissipate energy in bending is insufficient. A load dropped by a manual laborer

■ **Key Point**

The viscoelastic properties and internal structure of connective tissue make their strength, ultimate stress, and ultimate strain dependent on their rate of loading.

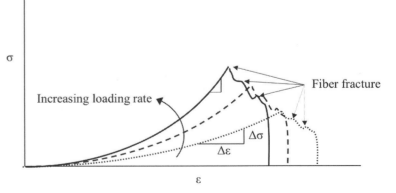

Stress = Force / cross-sectional area; $\sigma = F/A$
Strain = change in length / initial length; $\varepsilon = \Delta l / l$
Stiffness = Young's modulus; $Y = \Delta\sigma / \Delta\varepsilon$
Toughness = energy to fail = $\int \sigma \, d\varepsilon$

σ

Increasing loading rate

Fiber fracture

$\Delta\sigma$

$\Delta\varepsilon$

ε

FIGURE 1.5 Stress–strain diagrams show increasing stiffness with increasing rates of loading.

can crush the toe for the same reason. The problem in these cases is that there are no intervening tissues between the applied force and the bone that could dissipate some of the energy. Shin pads protect the soccer player from injury only slightly because their deformation and resulting energy-dissipating characteristics are small. In contrast, the steel-toed boots worn by the manual worker can prevent the application of any force to the toes. In this case the energy dissipation leaves the steel toe permanently deformed, but the cost of a new pair of boots is less in many ways than the cost of a crushed toe.

Another source of injury is repetition of small forces that are not singly injurious. The fact is that all materials, whether they are bones, ligaments, or steel girders, have a microstructure. Not all of the elements in these structures have the same mechanical characteristics. A single small force may injure a small element in the structure and go unnoticed. In living tissues of the body, the structure will either mend or be replaced over time. But immediate repetition will induce more damage and contribute to the mechanical fatigue of the structure. This is a common scenario in structural engineering, but it also applies in humans. It results in the condition known as shinsplints that is experienced as severe soreness on the anterior aspect of the tibia. Numerous biomechanical structures may be involved. These include the short tendinous material at the origin of the tibialis anterior muscle on the tibia and the periosteum covering the bone. There may even be evidence of microstructural breakdown of the bone underlying the periosteum. It is clear that small forces are no guarantee of the avoidance of injury. The only protection from repetitive injuries is to reduce the number of repetitions before injury occurs. While it is easy to avoid a single trauma by wearing steel-toed boots

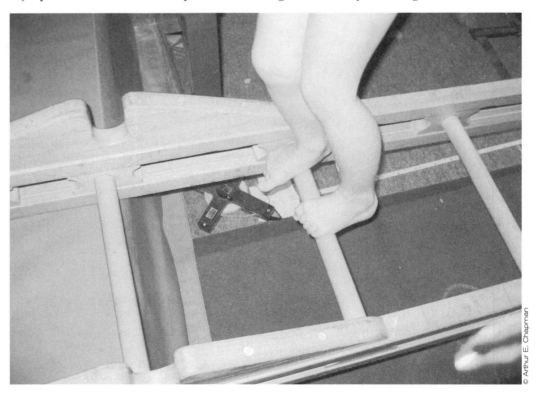

The knee and ankle joints, which support the greater part of the body mass against gravity, are particularly susceptible to injury when walking on irregular surfaces.

before the fact, the number of repetitions leading to repetitive strain injuries is known only after the fact. The latter is the case even with the best running surface and the best running shoes (see Whiting and Zernike, 1998).

Summary

The body is composed of a series of complex structures that are admirably designed for their functional roles and that generally exhibit nonlinear biomechanical properties. Safety from injury is partially ensured by the increasing stiffness that accompanies increasing deformation of biological structures. Ease of movement is partially ensured by the property of synovial fluid that is manifest as decreased friction with increased movement speed. Movement is enhanced by the shapes of bones, which enhance leverage. Unwanted amounts of energy are dissipated by the viscoelastic properties of tissues, but eventually there may be too much energy for the tissues to dissipate and injury will occur. As with all materials, there is a general inverse relationship between the magnitude of an external load and the number of repetitions of that load that will lead to injury. However, the cumulative fatigue shown by man-made materials in repetitive loading can be offset in biological materials by the processes of remodeling and repair.

RECOMMENDED READINGS

Armstrong, C.G., and Mow, V.C. (1980). Friction, lubrication and wear of synovial joints. In *Scientific foundations of orthopaedics and traumatology,* ed. R. Owen, J. Goodfellow, and P. Bullough. London: William Heinemann Medical Books.

Bartel, D.L., Davy, D.T., and Keaveny, T.M. (2006). *Orthopaedic biomechanics: Mechanics and design in musculoskeletal systems.* Upper Saddle River, NJ: Pearson/Prentice Hall.

Currey, J.D. (1984). *The mechanical adaptations of bones.* Princeton, NJ: Princeton University Press.

Daniel, D.D., Akeson, W.H., and O'Connor, J.J. (1990). *Knee ligaments: Structure, function, injury and repair.* New York: Raven Press.

MacIntosh, B.R., Gardiner, P.F., and McComas, A.J. (2006). *Skeletal muscle: Form and function.* Champaign, IL: Human Kinetics.

Nigg, B.M., and Herzog, W. (Eds.) (1994). *Biomechanics of the musculo-skeletal system.* New York: Wiley.

Nordin, M.H., and Frankel, V.H. (2001). *Basic biomechanics of the musculoskeletal system.* Philadelphia: Lippincott, Williams & Wilkins.

Ozkaya, N., and Nordin, M. (1991). *Fundamentals of biomechanics: Equilibrium, motion, and deformation.* New York: Van Nostrand Reinhold.

Whiting, W.C., and Zernicke, R.F. (1998) *Biomechanics of musculoskeletal injury.* Champaign, IL: Human Kinetics.

CHAPTER 2

Essential Mechanics and Mathematics

The purpose of this chapter is to provide sufficient mechanics and mathematics to facilitate understanding of the remainder of this book. For many people, mathematics has been a frightening area of their school curriculum. Additionally there is a belief that one can or cannot do mathematics and that there are no half measures. Yet many years of teaching biomechanics have led me to the belief that anyone can learn almost anything if such learning affords further understanding of an area of present interest. If your interest is the fundamentals of human movement (or performance), it is essential that you understand some mechanical laws. These laws can be discussed sensibly only if we all agree on the definitions of words used to formulate the laws. This is the mechanical area of our study. The mathematical area contains techniques that facilitate comparison of mechanical terms. In a historical context, understanding of the movement of bodies, be they humans or planets, stagnated until the techniques of calculus were elucidated. While this chapter covers mechanical and mathematical concepts by linking them to human movement, it is far from a comprehensive study of mechanics and mathematics.

The plain fact is that the laws of mechanics govern our movement, and mechanical techniques facilitate understanding of the relationships between mechanical concepts and allow us to put numbers to them. In this chapter we begin with the useful technique of dimensional analysis to avoid making mistakes with our equations of motion. Next we look at kinematics, which is the study of the motion of bodies, be they the whole body or interacting segments such as the upper arm and forearm. Kinematics therefore includes displacement, velocity, and acceleration and their rotational or angular equivalents. It is this area in which calculus is so important. The many mechanical variables that have both magnitude and direction are known as vectors. Knowledge of vectors is necessary because we need to know where, in which direction, and by how much our force is applied, for example (or how big our velocity is and which way it is going). Therefore we deal with their addition and multiplication because these operations do not necessarily follow ordinary arithmetic rules. After kinematics we deal with kinetics, the study of the force and work that produce our motion. As part of this discussion we calculate the impulse applied by a force acting over time or the work done by a force acting over a displacement. These two kinetic concepts lead us to momentum and kinetic energy, respectively, both in linear and in rotational terms. Lastly, we deal with the much-used and much-misunderstood concept of power, which is the rate at which work is changing over time.

The mechanical concepts appearing here can be represented by various combinations of the dimensions mass (m), length (l), and time (t). Familiarity with dimensional analysis is useful to students because the combinations of dimensions on each side of the equals sign in an equation must match. Not only is this process used for verification of relationships; it is also helpful as a means of jogging the memory should the student not be quite sure of the relationship. As an example, consider what might be the effect of a force F applied to a mass over a certain displacement s.

$$Fs = ?$$

We know that force equals mass multiplied by acceleration, or mass × length/time², so

$$F = ma = m \times l/t^2,$$

and because s is represented by the dimension L, the product Fs is

$$Fs = m \times l/t^2, \times L = m \times l^2/t^2.$$

We also know that l/t is velocity or v, so l^2/t^2 is v^2. Therefore Fs, which is work, has the same dimensions as mv^2 and kinetic energy (KE). Throughout this book mechanical variables are reported according to the International System (SI; Système International) of units. This system and the symbols used for naming variables are shown in Appendix A of Robertson and colleagues, 2004. Occasionally values expressed in SI units (e.g., kilograms) are converted into values in the Imperial System (e.g., pounds) since some readers may be more familiar with the latter.

Kinematics

If we are to understand the sources of motion of a runner, a diver, the foot in kicking, or any other mechanical activity, we require a method for describing motion. The three interrelated kinematic expressions of motion that we use are displacement, velocity, and acceleration. The term kinematic is used to describe the study of motion with no regard to its cause. The mathematical process by which these variables are related is known as calculus, both differential and integral. The development of calculus represented a great leap forward in the understanding of motion of bodies. Learning calculus is not simply a process of learning formulas for differentiation and integration. It involves understanding what calculus does in terms of relating first, second, and third differentials and their integral counterparts. This is the conceptual part of calculus, and no learning of formulas will achieve this conceptual understanding.

In mechanics, displacement *(s)* of a runner, for example, represents the straight-line distance between two points, irrespective of the path taken between the points. Displacement has the dimension of length *(l)* and is measured in convenient units such as meters, yards, kilometers, and miles. Velocity *(v)* of the runner is the rate at which displacement is occurring with respect to time. Velocity therefore has the dimensions of length divided by time (l/t) and is measured in meters per second, miles per hour, and so on. Finally, acceleration *(a)* is the rate at which velocity is changing with respect to time. Acceleration is therefore v/t. But since v is l/t, acceleration has the dimensions of $(l/t)/t$ and is measured in meters per second per second and so on. Table 2.1 shows these relationships.

As we have stepped successively from displacement to acceleration we have divided successively by time. Were we to step from acceleration through velocity to displacement we would multiply by time. It should be noted that the words and expressions presented in the previous paragraph are strict definitions of kinematic variables. These words are used frequently in common speech, but they are often used loosely and certainly incorrectly if they do not have the meanings specified here.

■ Key Point

If in doubt, one must verify that the dimensions on the two sides of an equation are equal.

TABLE 2.1

Translational Kinematic Variables

Translational kinematic variable	Dimensions	Units	Symbol
Displacement	l	Meters	s
Velocity	$v = l\,/\,t$	Meters per second	v
Acceleration	$a = v\,/\,t$ $a = (l\,/\,t)\,/\,t$ $a = l\,/\,t^2$	Meters per second squared	a

As a simple practical example we will consider the motion of a person falling from rest. Figure 2.1 shows acceleration and velocity. A property of our earth-bound system is that all bodies fall with a constant acceleration of 9.81 meters per second per second (m/s^2) as represented by the broken horizontal line. The inclined solid line represents velocity, which is increasing and which is obtained by integration as follows. The velocity change from 0 to 1 s is represented by the rectangle of area a1. This area is acceleration (9.81 m/s^2) multiplied by time (1.0 s), which gives a change of velocity equal to 9.81 m/s. The velocity at the beginning of the 2nd second is therefore 9.81 m/s. Then area a2 gives a change in velocity of another 9.81 m/s between 1 and 2 s, which when added to the existing velocity of 9.81 m/s gives a velocity equal to a1 plus a2 or 19.62 m/s at the end of 2 s. This continues for as long as the body is falling, so that after 5 s the total change in velocity is equal to the sum of the five areas and is 49.05 m/s. This process is known as integration by summation and shows that velocity continues to change as long as acceleration is present. If we were to reverse the process and calculate acceleration from velocity, we would observe that the slope of the solid velocity

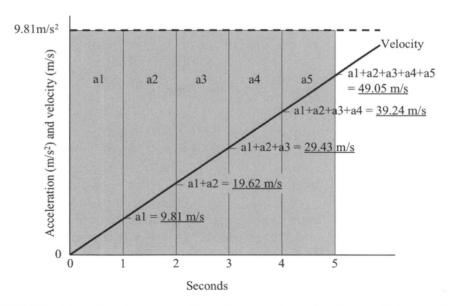

FIGURE 2.1 Integration by summation of areas of acceleration multiplied by time to produce velocity.

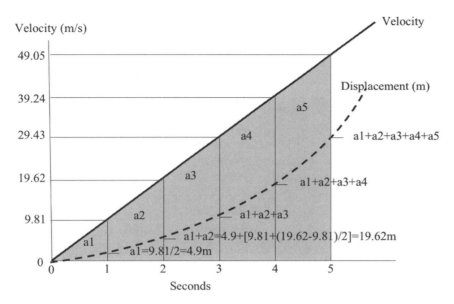

FIGURE 2.2 Summation of the velocity–time areas to produce displacement.

line is constant and is 49.05 m/s divided by 5 s (or 29.43 m/s divided by 3 s or 19.62 m/s divided by 2 s) and equal to 9.81 m/s².

The process of calculating displacement (m) from velocity (m/s) is shown in figure 2.2. The displacement from 0 to 1 s is represented by the triangular area a1. This area of this triangle represents velocity (9.81 m/s) multiplied by time (1.0 s) divided by 2, which gives 4.9 m as shown by the broken line. Area a2 represents the additional displacement from 1 to 2 s. It is equal to the area of a rectangle 9.81 m/s × 1.0 s plus the area of the triangle above it, which is (19.62 − 9.81)/2. When this calculation is added to the displacement at 1 s, a displacement of 19.62 m is seen at 2 s. It will be noticed that displacement at each successive second is increasing by increasing amounts as the successive areas (a1 to a5) are increasing. After 5 s the total displacement is equal to the sum of the five areas or the area of the large triangle (49.05 × 5/2) and is 122.62 m. Do not fall for 5 s. It must be emphasized that when we integrate acceleration to obtain velocity (or velocity to obtain displacement), the final velocity is the initial velocity plus the change in velocity over a given time period *(t)*. The equation encapsulating this process is as follows:

$$v_f = v_i + at \qquad \text{(Equation 2.1)}$$

where v_f = final velocity; v_i = initial velocity; a = acceleration; and t = duration of the time period.

Equation 2.1 can be expanded to obtain displacement *(s)* from velocity as follows:

$$s_f = s_i + v_i t + at^2 / 2 \qquad \text{(Equation 2.2)}$$

where s = displacement over the time period.

These equations are special cases and apply only to uniformly accelerated motion or constant acceleration, as we began by considering gravity as our accelerating effect. A third equation completes the equations of uniform motion as follows:

$$v_f^2 = v_i^2 + 2as \qquad \text{(Equation 2.3)}$$

It must be stressed again that these equations can be used only for motion that has constant acceleration; this includes zero acceleration. Their application therefore is confined to humans in freefall over displacements and with velocities that are small. At greater velocities, air resistance provides an opposing force that increases with increasing velocity and changes acceleration as a consequence. Therefore the equations that have been presented are subject to insignificant error for human jumping, but must be modified to predict the motion of missiles such as the discus and javelin. Equations 2.1 through 2.3 have little application if the accelerating force is due to muscular contraction because muscular force varies throughout a movement. The major reason for presenting these equations has been to show how displacement, velocity, acceleration, and time are related.

As an alternative to using a given formula to solve a problem without understanding the process involved, we will use a fundamental method of differentiation to obtain velocity from the displacement–time graph shown in figure 2.2. Displacement and time are related by Equation 2.2. The term s_i can be set to zero at zero time since we are interested only in obtaining velocity after zero time. This leaves us with the following:

$$s_f = v_i t + at^2 / 2,$$

which is the displacement at time t. After a short period Δt, a displacement of Δs has occurred, so the equation becomes

$$s_f + \Delta s = v_i(t + \Delta t) + a(t + \Delta t)^2 / 2 = v_i(t + \Delta t) + a(t^2 + \Delta t^2 + 2t\Delta t) / 2.$$

To obtain Δs we subtract the second expression from the first to give

$$\Delta s = v_i \Delta t + a(\Delta t^2 + 2t\Delta t) / 2.$$

The velocity $\Delta s / \Delta t$ is then obtained by dividing throughout by Δt to yield

$$\Delta s / \Delta t = v_i + a(\Delta t + 2t) / 2.$$

The instantaneous velocity $v_t = ds / dt$ is obtained as the time difference tends to zero, so as

$$\Delta t \rightarrow 0, \Delta s / \Delta t \rightarrow ds / dt; v_t = v_i + at,$$

which is the expression for velocity seen in Equation 2.1. This leads to the general expression for differentiating a power series as follows. Consider a relationship between y (the dependent variable) and x (the independent variable) such that

$$\text{if } y = x^n, dy / dx = nx^{(n-1)}. \tag{Equation 2.4}$$

Also note that integration of a power series involves the reverse operations plus the initial value of the constant of integration (C). Integration produces the following:

$$\text{If } y = x^n, \int y dx = \int x^n dx = x^{(n+1)} / (n + 1) + C. \tag{Equation 2.5}$$

Practice in verifying Equations 2.1 through 2.3 by means of these operations should establish these concepts of calculus. One pragmatic reason for such understanding is that profiles of kinematic variables can often be represented as power series; only with such understanding can computer programs be written to derive one kinematic variable from another when they are expressed relative to time. The preceding equation is a general expression for the integral of y with respect to x. A more specific form occurs when the limits or boundaries of the integration are specified. For example, if we wish to know the integral of y with respect to x over the range from $x = 1$ to $x = 3$, the equation would appear in the following form:

$$\int_{x=1}^{x=3} y\mathrm{d}x = \int_{x=1}^{x=3} x^n\mathrm{d}x = [x^{(n+1)} / (n+1)]_1^3$$

The solution involves substituting 1 for x in the square brackets and subtracting the result from that obtained when 3 is substituted for x as follows:

$$\int_{x=1}^{x=3} y\mathrm{d}x = \int_{x=1}^{x=3} x^n\mathrm{d}x = [(3^{(n+1)} / (n+1)) - (1^{(n+1)} / (n+1))]$$

$$\int_{x=1}^{x=3} y\mathrm{d}x = \int_{x=1}^{x=3} x^n\mathrm{d}x = (3^{(n+1)} - 1^{(n+1)}) / (n+1)$$

Suppose that $n = 2$, such that $y = x^2$. The preceding result is now the following:

$$\int_{x=1}^{x=3} y\mathrm{d}x = \int_{x=1}^{x=3} x^2\mathrm{d}x = (3^{(3)} - 1^{(3)}) / 3 = 26 / 3 = 8.67$$

This is rather abstract in its formality, but we find application in the forthcoming biomechanics when x (the independent variable) is substituted by t for time, or θ for angle.

Frequently some function of an angle θ requires differentiation and integration with respect to time, and the following technique of calculus is useful for this purpose. Suppose that $x = r\sin\theta$ where r is a constant (the use of this will become apparent in later chapters). If we wish to obtain $\mathrm{d}x/\mathrm{d}t$ as the velocity of x, we require $\mathrm{d}(r\sin\theta)/\mathrm{d}t$. There is no fixed solution to this function, so the following is used:

$$\mathrm{d}(r\sin\theta) / \mathrm{d}t = r\mathrm{d}(\sin\theta) / \mathrm{d}t = r[\mathrm{d}(\sin\theta) / \mathrm{d}\theta][\mathrm{d}\theta / \mathrm{d}t]$$

Since $\mathrm{d}(\sin\theta) / \mathrm{d}\theta = \cos\theta$, and $\mathrm{d}\theta/\mathrm{d}t = \omega$ we obtain the following result:

$$\mathrm{d}(r\sin\theta) / \mathrm{d}t = r\cos\theta(\mathrm{d}\theta / \mathrm{d}t) = r\omega\cos\theta \qquad \text{(Equation 2.6a)}$$

A similar result for $\cos\theta$ is

$$\mathrm{d}(r\cos\theta) / \mathrm{d}t = -r\sin\theta(\mathrm{d}\theta / \mathrm{d}t) = -r\omega\sin\theta. \qquad \text{(Equation 2.6b)}$$

One further technique of importance is used when the need arises to differentiate the product of two or more variables that both vary with respect to the same third variable. For example, if x and y both vary with respect to time, the following is obtained:

$$\mathrm{d}(xy) / \mathrm{d}t = x\mathrm{d}y / \mathrm{d}t + y\mathrm{d}x / \mathrm{d}t \qquad \text{(Equation 2.7)}$$

These techniques are essential if we are to obtain relationships between translational and rotational velocities and also accelerations. The information just presented is somewhat difficult to absorb all at one time. Readers should review

■ **Key Point**

Integration only produces the change between the limits of integration; the final result requires the addition of the value at the beginning of integration (C).

the equations when they are referred to in subsequent chapters. One final point involves the shorthand notation for differentials. So far we have represented differentiation of x with respect to time as dx/dt. The second differential of x is represented by $d(dx/dt)/dt$ or d^2x/dt^2. It is common shorthand to indicate that $dx / dt = \dot{x}$ and $d^2x / dt^2 = \ddot{x}$. This is done only when we are differentiating with respect to time and not with respect to some other variable.

So far we have considered only translational motion and some calculus techniques applied to angular motion. However, there are equivalent terms in rotational or angular motion. Table 2.2 shows the angular equivalents.

Note that the radian as a measurement of angular displacement is dimensionless. It is the angle subtended by the arc of a circle divided by the radius of the circle as shown in figure 2.3.

The arc represented by the broken line is equal in length to the radius, so the angle is r/r, which equals 1 rad. The arc represented by the solid line is of unknown length l, so the angle is l/r radians. The common measure of angle is degrees, but this is just a convenient name with a convenient number of degrees in a circle (360°). Historically there have been various numbers of divisions of the circle (e.g., 400). The radian has a fundamental meaning and is dimensionless—a length divided by a length. Although not apparent as yet, use of the radian as an angular measure is the only manner in which translational and rotation energy can be equated numerically. The symbolic conventions for angle, angular velocity, and angular acceleration are generally the Greek letters theta (θ), omega (ω), and alpha (α).

Circular Motion

A particularly important kinematic phenomenon is that relating linear and angular motion. An object may be traveling in a circular path at a constant speed in a tangential direction at any instant. However, the velocity is changing. The reason is that the direction of the velocity vector is changing although the magnitude is constant. In this case there must be acceleration a occurring in a radial direction (i.e., toward the center of the

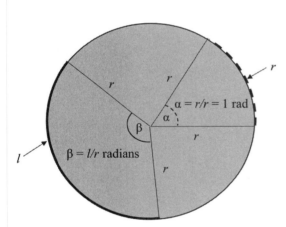

1 rad = 57.3 deg
2π rad = 360 deg
    π = 3.14159

$\alpha = r/r = 1$ rad

β

$\beta = l/r$ radians

FIGURE 2.3 Calculation of angular displacement in radians.

TABLE 2.2

Angular Kinematic Variables

Angular kinematic variable	Dimensions	Units	Symbol
Displacement	θ (dimensionless)	Radians	θ
Angular velocity	$\omega = \theta / t$ $\omega = 1 / t$	Radians per second	ω
Angular acceleration	$\alpha = \theta / t$ $\alpha = (\theta / t) / t$ $\alpha = 1 / t^2$	Radians per second squared	α

circle of motion). If v is the magnitude of the tangential velocity, r is the radius, and ω is the angular velocity,

$$v = \omega \times r. \qquad \qquad \text{(Equation 2.8)}$$

The radial acceleration a is

$$a = v^2 / r = \omega^2 r. \qquad \qquad \text{(Equation 2.9)}$$

If the accelerating object has a mass of m, the force producing the circular motion is

$$ma = mv^2 / r = m\,\omega^2 r. \qquad \qquad \text{(Equation 2.10)}$$

■ **Key Point**

Velocity changes when speed, direction or both change.

Vectors

Most kinematic and kinetic variables are vector quantities; they possess both magnitude and direction. One must take both qualities into account when applying the rules of addition and multiplication of vectors. Fortunately, addition and multiplication of vectors can be done graphically as well as numerically. Figure 2.4 shows the process of addition of vectors V1 and V2.

The mathematical solution to vector addition is shown in the left panel under Addition, while the graphical process of adding V1 and V2 in a head-to-tail manner appears on the right. These two vectors might be two forces applied to a single object in the directions shown. Any number of vectors can be added in this manner providing that they have the same dimensions. Subtraction of these vectors will be taken care of by their directions. Students should verify the smaller value of the resultant (V1 minus V2) by drawing V2 in the opposite direction. There are two results of vector multiplication, namely vector (cross) product and scalar (dot) product. Unlike vector addition, vectors of different dimensions can be multiplied. The terms "cross" and "dot" result from the manner in which the angle is treated. On the left, a moment of force (M) results from the application of a force (F) at some distance (d) from the origin (O) in the y-z plane. The vector product is a vector of magnitude M directed along the x-axis and positive in the direction determined by the right-hand rule. Briefly this means that if the flexed fingers of the right hand curve in the direction of the turning effect of the moment, the direction of M is positive in the direction in which the thumb is pointing. Alternatively, the scalar product is not a vector and therefore has no direction, although it is calculated by multiplying the magnitudes of the components of the original vectors in the same direction. If we choose the direction along F, the result is F multiplied by $s\cos\theta$. If we choose the direction along s, the result is s multiplied by $F\cos\theta$. Since these two calculations have an identical result, it is clear that we can choose any direction we wish in order to calculate work. This is how work is calculated in

Addition

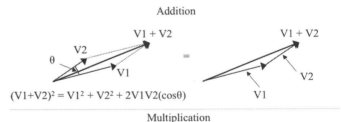

$$(V1+V2)^2 = V1^2 + V2^2 + 2V1V2(\cos\theta)$$

Multiplication

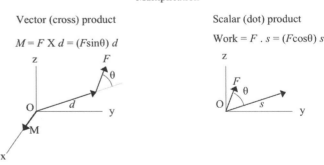

FIGURE 2.4 Vector addition and multiplication by graphical and mathematical methods.

the bottom right panel of figure 2.4, and work *(Fs)* results in energy that does not have direction.

Kinetics

Kinetics is the study of motion with regard to the sources of the motion. Terms such as force, momentum, and energy refer to kinetic variables. Newton's first law states that every body continues in a state of rest or uniform motion in a straight line unless acted upon by an external force. We can expand this law a little by replacing "external force" with "a net external force." This expansion obviates the confusion of having external forces acting that do not change the state of rest or uniform motion because their net vector sum is zero. This law is the foundation of the most fundamental equation of mechanics:

$F = ma,$

which tells us that instantaneous acceleration *(a)* of a mass *(m)* is proportional to the instantaneous net external force *(F)*. However, instantaneous force tells us little about the kinetics of motion through time and space. We need to observe or calculate the force over a period of time or a displacement to gain information about the nature of human motion. By observing force with respect to time, otherwise known as mechanical impulse, we can see how our momentum is changed. Furthermore, by observing force with respect to displacement (or the distance through which it acts), we can calculate work (W) and relate it to kinetic energy (KE). These are the two major kinetic processes by which we have come to understand the biomechanical functioning of the body.

Momentum

Sir Isaac Newton penned the relationship between force and acceleration in his second law, which states, "The rate of change in momentum is proportional to the applied force and occurs in the direction of the force." Therefore, whatever the force is at any instant, the rate of change of momentum will have a single value at that instant. Of course force has to be applied for a period of time in order that an object moves. Here we have introduced the concept "momentum," which can be considered the amount of motion of a body. The meaning of momentum can be developed as follows from Equation 2.1:

$v_f = v_i + at$

Multiplying both sides of this equation by mass gives us

$mv_f = mv_i + mat$ (Equation 2.11)

where *mv* is momentum or mass multiplied by velocity. Using the fundamental equation showing the relationship between force and acceleration of a mass,

$F = ma,$

where *F* is force, we obtain

$$mv_f = mv_i + Ft \text{ or } Ft = mv_f - mv_i. \qquad \text{(Equation 2.12)}$$

So the right side represents change in momentum, and the left side represents a new term known as mechanical impulse. This is the impulse–momentum relationship. In summary, we must apply a force for a period of time in order to change the momentum of a body.

Impulse–Momentum

Generally force varies during movement, so we resort to integral calculus to determine the change in momentum as follows:

$$\int_{t_i}^{t_f} F dt + mv_i = mv_f \qquad \text{(Equation 2.13)}$$

The integral sign indicates the area of a graph relating force to time. When this area is added to the initial momentum, the final momentum is obtained. Figure 2.5 shows this effect graphically in the case of kicking a stationary ball horizontally.

In this case the initial momentum (mv_i) is zero, so the total change in momentum is due to the shaded area $\int F dt$ under the force–time curve. Notice that the momentum changes at an ever-increasing rate as the force increases. As the force decreases, the momentum continues to increase because an accelerating force remains. However, the change in momentum occurs at an ever-decreasing rate. As the force reaches zero and the ball leaves the foot, the velocity remains constant. It is clear that greater momentum will result from greater force, or greater time of application of the force, or both, as illustrated by Equations 2.12 and 2.13. This fact should put paid to the assumption made by some sportscasters who insist that a ball picks up speed after leaving the foot. This assumption is based upon an illusion; the ball can only slow down in the horizontal direction after leaving the foot.

Center of Mass

The impulse–momentum relationship has been established for linear or straight-line motion of a body. By "body" we mean any rigid mass of material. The mass represents the inertia of the body. In a mechanical sense, inertia represents the tendency of a body to continue what it is doing at any time unless it is acted upon by an external force. An important property of a body is its center of mass (henceforth represented by CM). This is a theoretical point about which the body's mass can be considered to be equally distributed. Therefore the body will remain in balance if it is supported by force acting vertically through the CM. You can perform a simple test of the position of the CM by supporting a mass, a golf club, for example, on one finger and moving the position of

■ Key Point

Whereas force indicates only the instantaneous acceleration, impulse gives the change of momentum due to force integrated with respect to time.

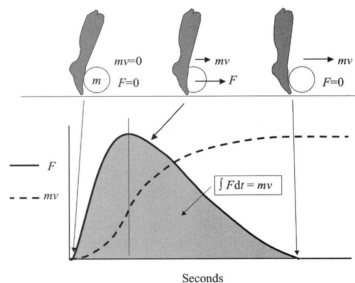

FIGURE 2.5 Application of an impulse in kicking where the area under the force–time curve (impulse) equals the change of momentum of the ball.

Velocity of a kick is maximized by maximizing the force–time integral, or impulse.

the finger until the object has no tendency to rotate in any direction. In this case the CM lies directly above the finger. Figure 2.6 shows how the position of the CM depends upon the orientation of the separate masses.

When the body parts are reoriented from figure 2.6a to figure 2.6b, the CM moves anterior to the pelvis in the direction of movement of the centers of mass of the limbs. With reorientation of parts from figure 2.6a to figure 2.6c, the CM moves in the upward and lateral directions of movement of the centers of mass of the right limbs. The position of the CM always moves in the direction of the displaced masses, and it does not have to lie within the physical limits of the body. When a force is applied to a body, it is the position of the CM that accelerates in the direction of the force. If the configuration of the body is fixed, all of its parts accelerate following parallel lines. Should the configuration change, as in the case of jumping, the CM will accelerate in the direction of the force, but other parts of the body can accelerate along different lines in different directions. So the human body must be considered to be a set of connected rigid segments that can change their positional relationships relative to each other, each having its own CM where its inertia is considered to be located.

Another important property related to the CM is that an unsupported body that is free to rotate will do so about an axis through the CM.

■ **Key Point**

Momenta of the CM of individual segments add vectorially to equal the momentum of the whole-body CM.

Moment of Inertia

A further property of the body is similar to mass and is pertinent to rotational or angular motion about a specified axis through the CM. This is moment of inertia (henceforth represented by MI, symbol I or J), which is a measure of a combination of mass and distribution of the mass away from the CM. Moment of inertia is in fact calculated as the product of mass and the square of the displace-

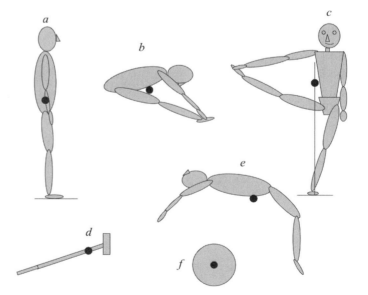

FIGURE 2.6 Distribution of segmental masses determines the position of the whole-body CM.

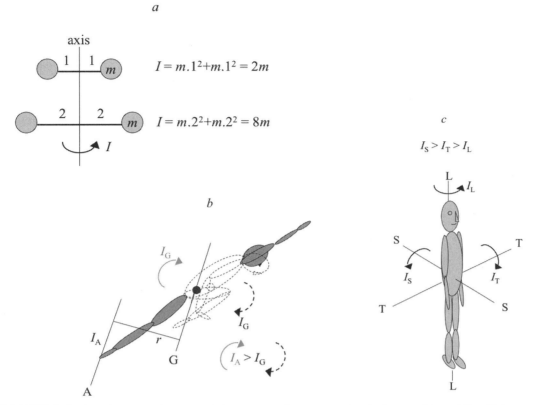

$$I = m.1^2 + m.1^2 = 2m$$

$$I = m.2^2 + m.2^2 = 8m$$

FIGURE 2.7 Distribution of segmental masses changes not only the position of the CM, but also the magnitude of MI by a squared factor of the distribution.

ment of the mass from the CM, and has the dimensions of ML^2. Figure 2.7 gives a diagrammatic meaning to MI.

Figure 2.7a shows how the square of the distribution of the mass contributes to the calculation of MI. Figure 2.7b shows how the MI in the extended posture is greater than that when the body is tucked because in the latter case, the body parts are brought closer to the CM. Note also the change in the relative position of the CM between postures. In figure 2.7c the three principal, mutually perpendicular axes are depicted (L = longitudinal, S = sagittal, and T = transverse). Note the difference in moments of inertia about each axis due to the shape of the human body. This indicates that the moment of inertia is greatest about the sagittal axis and least about the longitudinal axis even though the body has not changed shape. Were this human form to be unsupported and falling, it could have rotation about each of these principal axes as is exhibited in numerous airborne gymnastic maneuvers.

Frequently moment of inertia is represented as mk^2, which is the product of the mass and the square of radius of gyration about the chosen axis. Radius of gyration k can be explained as the radius of an infinitely thin circular ring with all of the mass lying at the same distance k from the center of the ring. Since all particles have the same distribution from the axis, the total moment of inertia is mk^2 according to the definition of moment of inertia.

One final concept concerning moment of inertia is the placement of the axis about which it is calculated. In figure 2.7b, two axes are shown as perpendicular to the plane of the figure and labeled G and A. The former is through the CM, and the latter is parallel to G and displaced at a distance *r*. The moments of inertia are different when calculated about the two axes. Whereas the centers of mass of the lower limbs are oriented at a similar distance from both G and A, the upper body is oriented at a much greater distance from A than from G. Therefore it is to be expected that I_A is greater than I_G. The calculation to convert moment of inertia between parallel axes is known as the parallel axis theorem.

$$I_A = I_G + mr^2 \qquad \text{(Equation 2.14)}$$

■ **Key Point**

Distribution of segmental masses determines the MI about the whole-body CM, and the parallel axis theorem relates MI calculated about one axis to that calculated about a parallel axis.

In gymnastics activities, it is clear that the moment of inertia of the body is smaller during rotation about an axis through the CM than during rotation in an extended form about the horizontal bar.

Angular Impulse–Angular Momentum Relationship

The previous information on rotational inertia allows us to develop the rotational equivalents of the prior equations that apply to linear motion. In rotational or angular terms, the translational equation $F = ma$ becomes

$$Fd = I\alpha \qquad \text{(Equation 2.15)}$$

where *Fd* is the force multiplied by the perpendicular distance between the line of action of the force and the axis of rotation. It is measured in units of newton. meters (N.m). α is angular acceleration measured in radians per second squared (rad/s²).

The term *Fd* is known as moment of force (*M;* note the equivalence with moment of inertia) and is sometimes known as torque (τ). We shall use the word "moment" for the purpose of economy and use the symbol *M* to denote it. So Equation 2.15, which is the rotational equivalent of $F = ma$, becomes

$$M = I\alpha. \qquad \text{(Equation 2.16)}$$

The total effect of an off-center or eccentric force on the linear and angular acceleration of a body is shown in figure 2.8.

In this figure we are looking down on a foot kicking a soccer ball with the intention of imparting a sidespin. Force *(F)* is seen to accelerate *(a)* the mass *(m)* linearly, and moment *(M* or *Fd)* is seen to produce angular acceleration (α) of the moment of inertia *(I)*. Should *d* equal zero, by which we mean force is acting though the CM, only linear acceleration will be present. If the force lasts for a period of time, linear and angular velocity will be produced. The ball will then gain both linear *(mv)* and angular momentum. The latter results from the angular equivalent of the linear Equation 2.13, which becomes

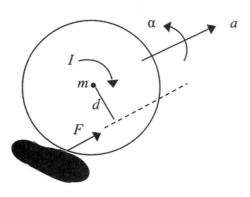

Linear $F = ma$ Angular $Fd = M = I\alpha$

FIGURE 2.8 Imparting sidespin to a ball by means of a force eccentric to the CM.

$$\int M dt + I\omega_i = I\omega_f \qquad \text{(Equation 2.17)}$$

where ω_i = initial angular velocity; ω_f = final angular velocity; I = moment of inertia; and M = moment of force.

Angular momentum is equal to the product of moment of inertia and angular velocity, and it is produced by a moment acting for a period of time. In Equations 2.16 and 2.17 there are composite terms contributing to a change in angular momentum. Angular momentum will increase if M increases. Since M is Fd, we can increase angular momentum by increasing either force, perpendicular distance between the line of application of force and the axis of rotation, or both. Since it can change with reorientation of body segments, for any given value of $\int M dt$, angular velocity will increase if the body is tucked as tightly as possible to reduce the value I.

Work–Energy Relationship

So far we have produced the important impulse–momentum relationship by multiplying Newton's fundamental equation, $F = ma$, by time. Since movement involves displacement, it would seem to be reasonable to investigate the meaning of multiplication of force by displacement *(s)*. Using an abbreviated form of the kinematic Equation 2.3,

$$v^2 = 2as, \qquad \text{(Equation 2.18)}$$

and the relationship $F = ma$, which can be rewritten as $a = F / m$, we obtain

$$v^2 = 2(F / m)s \text{ or } Fs = mv^2 / 2. \qquad \text{(Equation 2.19)}$$

Equation 2.20 shows the mechanical relationship between work and translational kinetic energy rewritten as

$$\int F ds + mv_i^2 / 2 = mv_f^2 / 2. \qquad \text{(Equation 2.20)}$$

Work is defined as the product of the magnitude of the force and the displacement of the point of application of force in the direction of the force. The direction of displacement in relation to the direction of the force is very specific and important. Figure 2.9 shows three situations for calculation of work.

In figure 2.9*a*, work is Fs since F contributes to the displacement s and since F and s are in the same direction. In figure 2.9*b*, no work is done since F does not contribute to the displacement s because F and s are perpendicular. In figure 2.9*c*, displacement in the direction of the force is $s\cos\theta$ and work is therefore $Fs\cos\theta$. Obviously in figure 2.9, *b* and *c*, there must be other forces involved, but we are simply looking at the work done by the forces shown.

The significance of the work–energy relationship in human movement is that muscles produce force and shorten over a certain distance. They therefore do work to change the KE of body segments. So far we have considered work in a linear or translational sense, which leads to the term translational kinetic energy (TKE)—which is one-half of the product of mass *(m)* and velocity squared (v^2).

Work in Rotation

Since rotation involves motion of body mass, it is not surprising that we use the term rotational kinetic energy (RKE). The work–energy relationship in rotation

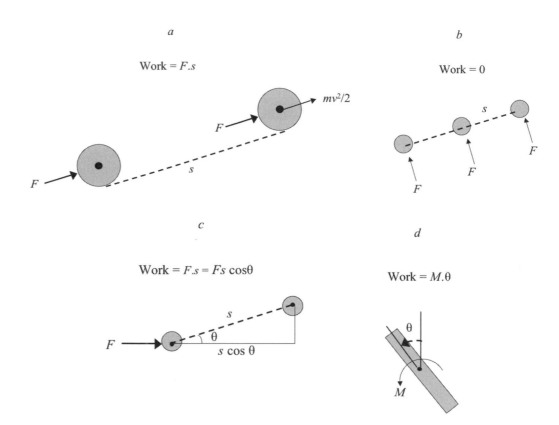

FIGURE 2.9 Work is the product of force and displacement in the direction of the force.

can be obtained by substituting linear terms in Equation 2.12 by rotational terms as follows:

Linear
$$Fs = mv^2 / 2$$ (Equation 2.21)

Angular
$$M\theta = I\omega^2 / 2$$ (Equation 2.22)

Rotational work is therefore the product of moment (M) and the angle (θ) through which the moment is applied. Should M change as angular rotation occurs, Equation 2.22 becomes

$$\int Md\theta + I\omega_i^2 / 2 = I\omega_f^2 / 2.$$ (Equation 2.23)

■ **Key Point**

Work is the product of force and displacement or moment and angular displacement; the dimensions of work are the same in each case.

This work is the cause of RKE, which is one-half of the product of moment of inertia (I) and angular velocity squared (ω^2). Careful inspection of the linear and angular dimensional tables will show that translational and rotational kinetic energy have the same dimensions. This would not be the case if we used degrees as our angular measure rather than radians.

Potential Energy

A final form of energy is potential energy (PE). This is the potential that a body has to do work by virtue of its position. For example, the Shorter Oxford English Dictionary on the edge of a table h meters high has potential energy of mgh. Should the book be displaced slightly so that it falls, it will gain TKE as it falls. As its height decreases, the potential energy will decrease and the KE will increase at the same rate. The converse is true for upward motion of a projected object. As the book lands on your toe, the force applied reduces its KE to zero and work is done on your toe, causing deformation of the bone or soft tissues. Another form of potential energy that is important to human movement is strain energy. Muscle force stretches (or strains) the tendon, which transmits the force to the bone. This represents a temporary storage of energy in the tendon during stretch. Subsequently the tendon will recoil and give up this potential strain energy as KE of movement of the body segment. This exchange of energy has important benefits for human movement as will be discussed later.

Relationship Between Potential Energy and Kinetic Energy

As a body falls, it speeds up. Since gravity is the only force accelerating the body downward (if we ignore air resistance), we are dealing with what is termed a conservative system. In this case, total energy is constant because what the body loses in PE it gains in KE. This relationship can be obtained from Equation 2.3:

$$v^2 = 2as,$$

which, when we replace a with g, replace s with h, and multiply by mass throughout, becomes

$$\text{KE} = \text{PE}$$
$$mv^2 = 2mgh \text{ or } (mv^2) / 2 = mgh. \qquad \text{(Equation 2.24)}$$

It is of significance that doubling the velocity at takeoff leads to a fourfold increase in height reached. So increasing the velocity-producing abilities of body structures and techniques is of paramount importance in such activities as high jumping.

An example of a nonconservative system is a bouncing ball. The KE of the ball immediately prior to contact with the ground is stored as PE or strain energy in the elastic medium of the ball as it reaches zero velocity on the ground. The PE is subsequently returned as KE as the ball leaves the ground. The KE of rebound is always less than the initial KE due to energy loss in viscous components of the ball's material during compression and relaxation. Therefore the ball never bounces as high as it falls initially. The ratio of velocities of rebound and drop (v_R / v_D), which is known as the coefficient of restitution (CR), can never be greater than unity. The CR is related to energy loss through the PE–KE relationship, $mv^2 = 2mgh$ or $v = (2gh)^{0.5}$. Therefore,

$$\text{CR} = v_R / v_D = (h_R / h_D)^{0.5}. \qquad \text{(Equation 2.25)}$$

We can measure the CR of a ball by calculating the square root of the ratio of heights of rebound and drop. The value of CR is not fixed but is dependent upon any factor that alters the viscoelastic characteristics of the ball's material. Candidates are temperature, material fatigue, and amount and rate of strain. For example, a squash ball must be warmed up by striking before play, and it changes its CR with repeated use. Variations of the CR of balls have numerous

The CR of a viscoelastic squash ball increases with temperature but decreases with input velocity.

implications for play and for ball manufacture that have not been investigated sufficiently.

Power

Although the word "power" is frequently misused and misapplied to the effect of force acting on a body, it is defined simply as the rate at which work is being done. In Equation 2.21 the term for work is

W = Fs.

Differentiating with respect to time gives

W / t = Power = Fs / t = Fv.   (Equation 2.26a)

Therefore power is the product of force and velocity, and it is a vector quantity since work, which is scalar, is multiplied by velocity, which is a vector. To be more precise, power is the product of force and velocity of the point of application of the force in the direction of the force. We obtain the rotational equivalent of power simply by replacing the translational terms with rotational terms as follows:

Rotational power = M ω   (Equation 2.26b)

■ **Key Point**

Power is the rate at which energy is changed over time.

The calculation of power can be appreciated if the values for translation s and rotation θ are replaced by velocity v and angular velocity ω, respectively, in figure 2.9. The result is power, and its calculation in translation and rotation is shown in figure 2.10.

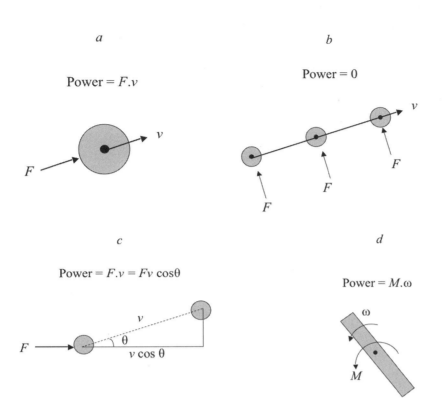

FIGURE 2.10 Power is the product of force (or moment) and velocity (or angular velocity) in the direction of the force.

Summary

At this stage the reader should be armed with sufficient mechanical and mathematical information to understand human movement biomechanically. Although few numbers have been used in this discussion, the reader should understand the meaning of mathematical manipulations. The process of integration of force with respect to time in order to calculate mechanical impulse and the resulting change in momentum, as seen in Equation 2.13, might prove to be the most difficult concept for some readers. All $\int Fdt$ means is summation of successive areas under a force–time curve over discrete, but small, intervals of time dt. Similarly, $\int Fds$ (Equation 2.20) is summation of force over discrete intervals of displacement ds. It is essential that the reader understand this integration process, as we shall be dealing with the manner in which changes in force–time, torque–time, force–linear displacement, and moment–angular displacement integrals modify our motion. Much of the material presented here can be found in Robertson and colleagues, 2004, and methods of calculus can be found in any introductory calculus textbook.

RECOMMENDED READINGS

The reader should have no difficulty finding one of the many books that deal with the mathematical concepts discussed within this chapter. First-year university texts should be sufficient for this purpose.

Mathematics

Gondin, W.R., and Sohmer, B. (1968). *Advanced algebra and calculus made simple.* London: Allen.

Mechanics

Beer, F., Johnston, E.R., Eisenberg, E., Clauser, W., Mazurek, D., and Cornwell, P. (2007). *Vector mechanics for engineers.* New York: McGraw-Hill Ryerson.

Fogiel, M. (1986). *The mechanics problem solver.* New York: Research and Education Association.

CHAPTER 3

Foundations of Movement

This chapter covers the various external and internal forces experienced by the body. External forces are gravity, friction, and impact with the floor and other external objects. Internal forces are those produced by muscle and those induced in structures as a result of muscle action and external forces. Here we give special attention to the biomechanical properties of muscle.

Throughout this book, two major relationships are used to explain and discuss solutions to fundamental human movements. The relationships are impulse–momentum and work–energy. They are easy to write down using one mechanical equation for each relationship. However, mechanics in its simplest form deals with point masses or rigid bodies. Extended mechanics deals with bodies interconnected with frictional joints and springs and dampers. To a large extent the human body can be represented by such an arrangement of rigid bodies, but the only exact model of any body is the body itself. Accepting this constraint we find that there are certain principles that can be applied to human motion, and in many cases the skillful use of the body cannot be achieved without use of these principles. Some of the principles are mechanical and result simply from the manner in which segments are arranged. Others have a strong biomechanical component and depend upon the properties of the force generators in concert with the structures that transmit the forces, namely muscles and their connecting tendons. Muscles that contract can have an effect on motion of segments far distant from them. This phenomenon is explained by recourse to the two major principles just identified.

Gravity

Although gravity is an external force, we deal with it first as a means of introducing the concept of force. Gravity keeps us on the earth. Gravity is one of a number of types of force (others are electromagnetic, nuclear, etc.), and it occurs simply due to the fact that bodies with mass attract each other. Gravity has played a large part in the development of all things that once existed but are now extinct and those that exist in the present. Without gravity, any force that we applied to the ground would produce an equal and opposite upward force that would send us careening off into space. As we stand on the earth, gravity tries to pull us and the earth together. Since we don't disappear into the earth, there are equal and opposite forces of the earth pushing upward on our feet and of us pushing downward on the earth. If we happen to step on quicksand, it is not substantial enough to push up on us and we sink.

Gravity gets less as we travel away from the center of the earth. Eventually gravitational attraction drops sufficiently to a value that just keeps an astronaut's capsule falling in a circular motion around the earth provided that it has an appropriate initial velocity parallel to the earth's surface. At farther distances from the earth, you are on your own until some other body's gravitational field captures you. The force of gravity is not constant but is proportional to body mass. In fact, it is the acceleration due to gravitational force that is constant for bodies close to the earth. From this it follows that the equation governing a falling body (if we ignore air resistance) is

$$F = M \times g \hspace{4cm} \text{(Equation 3.1)}$$

where *F* is the force of attraction of the earth on the body, *M* is the mass of the body, and *g* is acceleration due to gravity—9.81 m per second squared or 32.2 ft per second squared.

Figure 3.1 shows two situations, one without gravity and one with. The main point is that acceleration always occurs in the direction of the resultant force. Readers should refer to addition of vectors as shown in figure 2.4.

A special case of gravitational force is body weight. *F* is what your bathroom scale measures as your weight or the tendency to pull you down, also known as ground reaction force. So, in the absence of acceleration, weight is a force that happily is directly related to your body mass, which is really what we want to measure in the privacy of our bathroom. If your bathroom were on the moon you would appear to have a lower weight despite the fact that your body mass is the same as here on earth. This results from the fact that gravitational acceleration on the moon is about one-sixth of that on the earth because of the smaller mass of the moon. A significant point is that when you step on or off your bathroom scales, or move up and down, the scale fluctuates. However you cannot state that your weight fluctuates. So body weight has a very specific meaning and is no guide to your body mass in all circumstances. The pedantic answer to "How much do you weigh?" should be "X kilograms force [kgf] because my body mass is X kilograms [kg]."

Humans evolved and grow in a gravitational field. Thus the composition of bone is designed to resist the downward force of gravity. This is exemplified by changes in the composition of the bones of astronauts after long exposure to zero gravity. Gravity works against the high jumper and with the diver. Gravity requires us to spend energy in standing, but the energy cost of standing upright is much less than that of standing with the knee joints half flexed, although there is zero motion in each case. In the strict mechanical case, no work is being done in standing, so there should be no difference in energy cost in these two standing postures. This phenomenon will be explained later.

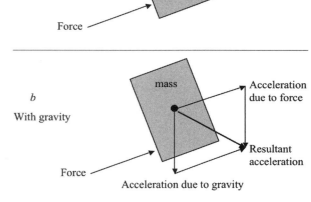

FIGURE 3.1 The instantaneous vector sum of all forces gives the instantaneous acceleration vector of a mass, no matter where the force is applied on the body.

■ Key Point

Gravitational force always acts toward the center of the earth.

Friction

Friction is an external force that comes in a number of varieties but is generally seen as a dissipative force. Friction is generally absent unless we make an attempt to move. Sometimes, but rarely, gravity induces frictional force, for example when we are standing on an icy slope. If the induced frictional force is less than the gravitational force, we slide although we have not attempted to move. We can think of friction as the effect of one thing rubbing on another, be it a shoe on the floor, or air and water vapor molecules rubbing on a golf ball in flight. Frictional force always acts parallel to the direction of motion of objects that are in contact and that slide (or tend to slide) relative to each other. Another name for this type of force is *shearing*, in contrast to body weight, which is *normal* to the surface on which we stand. Friction, like gravity, is useful. It keeps us from sliding around uncontrollably and colliding into each other. When it is unexpectedly absent, as when we encounter unseen ice under our feet, we miss it. Friction is also an annoyance because we are constantly fighting against its resistance to motion. This adds to the energy we have to expend in movement. Friction is dissipative in the sense that it reduces the kinetic energy (KE) in a system and converts it finally into heat. On a large scale it adds to the entropy of the universe. On a small scale it is the reason a squash ball heats up when struck repeatedly as its rubber molecules rub together during deformation. Friction always acts against our attempts to get an external object moving or to keep it moving.

A perfectly (hypothetical) elastic ball will store energy as if it were a spring. This is strain energy acquired as the ball is deformed a certain amount when brought to a stop. This energy will then be released to send the ball up to its starting height. There are no energy losses in this *conservative* system, but perfectly elastic objects are impossible to find. A similar example would be arresting the downward motion of a ball held in the hand and bringing it up with no effort as if the elbow flexors were a perfect spring. Of course our experience tells us that no ball will bounce back up to its original height, and also that we will have to do muscular work to bring a mass up in elbow flexion after the elbow has been extended. As friction is always with us, there is no system that is truly conservative; there are always energy losses. In automobile collisions, for example, energy is lost and appears as both heat and sound energy and permanent deformation of the colliding cars.

A Simplified Representation of Muscle Action

Muscle is the primary source of internal force that moves us. Before we deal with the intriguing biomechanical properties of muscle, it is pertinent to examine what happens when a simplified representation of muscle acts on two masses. This is analogous to an active muscle pulling on two segments of the body that are articulated by a joint.

Figure 3.2 shows two masses (magnitude m and $2m$) that are lying on a frictionless surface and are connected by a spring, which is shortened to less than its resting length in figures 3.2*a* and *d*. This is an ideal spring, which is massless and exhibits a linear relationship between force and length.

The spring resists with a force equal to that required to shorten it, and energy is stored in the form of strain energy. Upon removal of the constraining force, the masses are pushed apart by the spring force (F) as shown in figure 3.2*a*. Here, each mass is unrestrained, while in figure 3.2*d*, the mass on the left is unable to move to the left due to a physical obstruction. When the spring is released, it extends, and the masses in figure 3.2*b* gain both momentum and KE in opposite directions. In figure 3.2*e*, only the smaller mass gains energy and momentum. We obtain the velocities of the masses by rearranging the work–energy equation $\int Fds = mv^2 / 2$. As the masses in figures 3.2*a-c*, are free to move, their momenta are equal and opposite, and the position of the center of mass (CM) of the system does not change. In figure 3.2*e*, the smaller mass has increased velocity, the position of the CM of the system moves to the right, and it is not until the resting length of the spring is reached when $F = 0$ that the mass on the left begins to move (figure 3.2*f*).

■ **Key Point**

Friction dissipates energy that appears largely as a rise in temperature of a system.

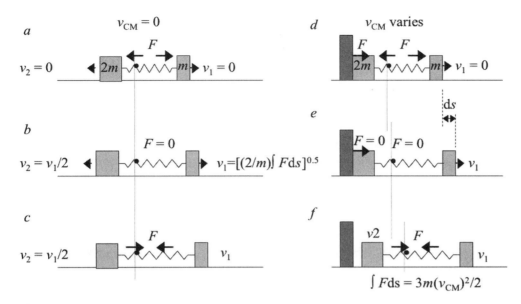

FIGURE 3.2 Internal muscle action leaves the center of mass stationary while the presence of an external reaction allows it to move.

In both cases, momentum and KE are present in the system when the resting length of the spring is reached, and these variables stretch the spring further. It therefore resists with an opposing force that arrests further outward motion of the masses in figures 3.2c and f and begins to return them to their original values. Directions of velocity in figures 3.2c and f are absent since the masses will decelerate and then accelerate in opposite directions over time in an oscillatory manner. The difference between the motion in figures 3.2a-c and figures 3.2d-f is that the former motion shows no change in the position of the total center of mass, while the latter shows displacement of the total center of mass which will continue unless another force stops it. The reason for the difference is that in figures 3.2d-f there is an external force, induced against the physical obstruction on the left, by extension of the previously shortened spring. In the case of this ideal spring, the masses will continue to oscillate. If the spring were to be replaced by muscle, it could be deactivated and there would be no oscillation.

What this simplification illustrates is that the body can gain momentum and KE with respect to the external environment only if part of it is connected to that environment at some stage in muscular contraction. In simple terms this means that to move, we must have something to push against. If we do not, our body parts can change their orientation, but our CM will go nowhere.

Conservation and Transfer of Momentum

Skeletal muscles almost exclusively provide force generation between bones, or segments of the body. Since the muscles are internal to a system of segments there will be an action–reaction phenomenon according to Newton's third law. The simple explanation is that an airborne human can move segments relative to each other by muscular contraction, but this will have no effect on the path of the CM of the body. This is analogous to the simple model shown in figures 3.2a-c. Try the example of standing erect with your hands close to your chest, preferably holding some reasonable mass. Move the mass back and forth, and you will find that this can continue forever (if you have enough persistence). Of course it is easy to lose one's balance in this activity, but this toppling has nothing to do with the back-and-forth movement of the mass. What happens is that as the mass is pushed forward, the remainder of the body moves backward. The force at one end of this "pushing" or "pulling" muscle is the same as that at the other end. So at any instant, the integral of force with respect to time (impulse) is equal in magnitude but opposite in direction. So at any instant, the change in momentum of the mass in one direction is equal and opposite to that of the remainder of the body. Therefore momentum is conserved and is equal to zero at all times. Such is the case in airborne movement.

External forces are required to produce motion in a particular direction. The external force comes from the earth or some other body with which we are in contact. Put simply, action and reaction are equal and opposite in terms of momentum change unless one part of the body is not allowed to react because of the presence of some other object. This is analogous to the simple model shown in figures 3.2*d-f*. Try the same example with your back against a wall, and the effect will be a gain in velocity of the whole body away from the wall. This what happens in jumping with use of the arms as the free, smaller mass. In this case the momentum from the moving mass is transferred to the whole body when the upper limbs become straight.

These effects can be seen in rotational motion. Jumping off the ground and then sweeping the arms in a horizontal plane about vertical axes through the shoulders results in opposite rotation of the remainder of the body about a vertical axis. Again the analogy is in figures 3.2*a-c*, where momentum is conserved. Alternatively, sweeping the arms in the same manner before taking off gives the system angular momentum (all in the arms) that is shared by the whole body when the arms stop in relation to the remainder. Again we see transfer of momentum from moving segments of the body to the whole body. Transfer of momentum is a universal technique used in jumping to enhance both vertical and angular velocity at takeoff.

Conservation and Transfer of Energy

At any instant when a muscle is producing force and shortening simultaneously, the integral of force with respect to shortening displacement is work. This phenomenon has been discussed with reference to figure 3.2. In figures 3.2*a-c*, the individual masses have oppositely directed velocities and momenta that sum to zero because of their vectorial nature, and the CM of the system does not change position. Despite no change in position of the system CM, total KE is conserved as the sum of the KE of the individual masses, since KE is a scalar quantity. In figures 3.2*d-f*, the KE in the small mass gets transferred to the total mass following stretch of the spring beyond its resting length. This example of transfer of energy can also occur in a rotational sense when the work done is the product of torque and angular displacement. The arms are frequently used in this manner to add rotational energy to a person attempting a somersault or an ice skater attempting multiple twists about the longitudinal axis after leaving the ice surface. In the latter case the free lower limb is also rotated to add to the total rotational energy.

Transfer of energy is specifically used in throwing and striking skills. The advantage of this phenomenon is that work done by the large muscles of the trunk can be transferred distally across intersegmental joints to enhance the KE of an object located in the most distal part of this segmental chain. Sport activities in which this mechanism is evident include all throws (e.g., baseball pitch and javelin throw) and all strikes (e.g., golf stroke and kicking a ball).

Biomechanical Properties of Muscle

Muscles are our engines. They act in much the same manner as man-made engines inasmuch as they consume fuel, produce force, and generate heat in the process. However, muscle exhibits some interesting properties not shown by other engines. Muscle is constructed from many small fibers in parallel that generally run the whole length of the muscle from tendon at one end to tendon at the other end (there are many exceptions to this general construction). Muscles in other species also have some properties that differ from those in human muscle and that are adapted to their structure and function (see Vogel, 2001; Alexander, 1968). The force produced by muscle is controlled by "playing a tune" on the number of fibers activated. The controlling process is the central nervous system, which either voluntarily or involuntarily decides on the number of nerves down which messages are sent. The equivalent in your automobile would be an engine with millions of cylinders, any number of which could have fuel sent to them to allow variation in force output. This would be cumbersome for the automobile; muscle is neater than that.

Much is known about the complex electrochemical events involved in muscular contraction. Within the muscle fibers is a complex architecture of microfibers that produce force by a mechanical process not universally agreed upon by muscle biomechanists. But a great deal is known about the mechanical output in terms of how force varies with neural input, muscle length, and muscle shortening and lengthening speed (Chapman, 1985). These properties are germane to an understanding of human movement. Figure 3.3 shows these relationships.

Figure 3.3a is a very simplified indication that force increases when the number of active nerve fibers increases (the force–activation relationship). In 3.3b, active force is seen to be maximal at some intermediate length; it decreases as the muscle is set at shorter and longer lengths (the force–length relationship). While the general shape of this relationship is fixed, the length for optimal force occurs at different joint angles depending on where in the body the muscle is situated and thus the function it has to perform. The broken line in 3.3b represents passive elasticity in the muscle due to connective tissue. It is represented as a parallel elastic component

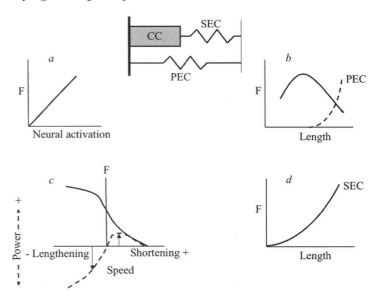

FIGURE 3.3 A Hill-type model of muscle in which force produced by the contractile component (CC) is a function of activation, length, and speed of shortening, along with the force–length relationships for the series elastic component (SEC) and parallel elastic component (PEC) (Hill, 1970).

(PEC) with non-linear properties. The fact is that as a muscle is stretched it begins to resist at some initial length, and further stretching results in increasing resistance or force. Again the joint angle at which this resistance begins depends on where in the body the muscle is situated. Individuals who undergo correctly performed stretching exercises will be familiar with this resistance and also will know that it can be modified somewhat by exercise.

Figure 3.3c shows a very important relationship that is implicated in the dynamic functioning of muscle. As the muscle shortens faster and faster, the maximal force of which it is capable decreases. Alternatively, the faster it lengthens, the more force it can produce. The broken line in 3.3c indicates that the negative power produced is greater than the positive power at any given speed of lengthening or shortening, respectively. If the lengthening speed becomes too great, the muscle will rupture. Part c shows what is generally known as the force–velocity relationship of muscle. This is implicated in energy generation and dissipation by the human body, as we will see later when dealing with the questions "Why can you land safely from more than three times the height to which you can jump?" and "Why can a sprinter not accelerate forever?" Figure 3.3d represents the behavior of the series elastic component (SEC), which becomes stiffer as the force produced by the contractile component (CC) of the muscle becomes greater. The force tending to stretch the series elastic component comes from the contractile component. The model at the top of the figure is our muscle model, which shows the components and their properties. It must be stressed that the term *contraction* indicates the presence of the force-generating process and is not used in its popular sense of getting shorter or smaller.

Muscle is extremely strong. Various values have been calculated for the force produced by a given amount of muscle. The amount of muscle we are talking about is not its mass but its cross-sectional area, perpendicular to the line of action of the muscle force. A value of about 25 N per square centimeter of cross section has been published (Nigg and Herzog, 1994, p. 175). This means that a piece of muscle the size of a pencil can hold up a weight of about 800 g or 1.8 lb (think of the weight of a 2 lb bag of flour). Your calf muscle (actually a number of muscles), with a cross-sectional area of about 80 cm^2 (12 in.^{2}), can sustain a force of around 4000 N or 400 kgf or 880 lbf (depending upon the individual). Therefore standing with our whole weight on one foot with the heel off the ground is no challenge whatever to our calf muscles, even though the muscles are at a mechanical disadvantage (see figure 3.4 which follows). Nor is this a challenge to our Achilles tendon, which is thick and tough. The fact that this tendon sometimes ruptures attests to the great force of which the calf muscles are capable in strenuous movements.

Muscles also seem to have memory, but not in the popular sense of the word. It has been known for many years that electrically stimulated frog muscle produces a greater force following stretch than if it has not been stretched (Edman et al., 1978). Unfortunately, this effect decays rapidly with time, so any benefit to be derived from active stretch is seen only immediately following stretch. This effect is also seen in human muscle, but the enhancement is less and the decay of force much faster (Thomson, 1983). This "force enhancement" effect is obviously useful and is a minor contributor to the beneficial stretch–shortening cycle (SSC) of muscle discussed later. It is part of the reason we use a "backswing" or "countermovement" in some activities. There is no use in stretching an active muscle and expecting it to perform better half an hour (or minute or second) later.

One final quality of muscle as a force producer is the type of muscle fiber. Among many classifications of muscle fiber, the predominant division is between red and white fibers. Red (or slow) fibers develop force slowly and can sustain repetitive activity by using oxygen that is delivered to them by the blood vascular system. White (or fast) fibers act rapidly but cannot do so for many repetitions because of the type of chemical energy they use. We use red fibers to develop force for sustained activities. White fibers develop force for high acceleration.

It would be remiss not to recognize the role of A.V. Hill in bringing the "bio-" into "biomechanics." He began with an interest in sprinting, and in attempting to understand the factors determining sprinting speed he progressed deeper and deeper into the mechanical and physiological properties of muscle. Today most investigators in biomechanics use Hill's work as a basis for understanding how to perform human activities and how to drive their simulations of these activities. Hill's book, *First and Last Experiments in Muscle Mechanics* (1970), is essential reading not only for its wealth of information, but also for a historical perspective on how information was gathered with the limited techniques available at that time.

Use of Muscular Force

Muscular force is a continual presence in most of the equations governing human movement, whether it is viewed from the impulse–momentum or the work–energy perspective. In some cases insufficient muscular force does not allow a maneuver to be performed. The gymnast cannot sustain a handstand unless the wrist and finger flexors can produce sufficient isometric force. Insufficiency of force can lead to injurious situations when a manual worker has to lift a load that is near maximal force capacity. Inability to generate sufficient force will limit the rate at which body segments can be reoriented. In this case the number of rotations about principal axes of the body will limit the aesthetics of a dive. The possession of great muscular strength and endurance is therefore of benefit to all individuals who move in our gravitational and frictional environment. The biomechanical analyses in previous chapters have shown how muscular force fits into the performance of a variety of activities. These should form a basis on which practitioners can determine what type of muscular strength is required of those under their care. How to train to enhance muscular capabilities is a subfield of the study of human biomechanics but does not fall within the aim of this book. In addition to the instantaneous force a muscle can produce, muscle is also a performer of work and a source of power.

Muscular Work

In a popular sense of the term we need energy to do work. This energy comes from the food we eat, after it has been converted into chemical energy. Chemical energy is used by our muscles to develop force, which when applied to our bones makes the body segments move. The amount of work done by the muscles

■ Key Point

An activated muscle produces force as a function of activation, length, speed of shortening, and events preceding the time at which force is measured.

In some cases, insufficient muscular force does not allow a maneuver to be performed, as can be seen in the sport of weightlifting.

is the force multiplied by the distance through which the muscle shortens. The subsequent movement of the segment represents mechanical energy, often called kinetic energy or energy by virtue of motion (or speed).

Three specific names are given to types of muscular contraction. The first is "concentric" (toward the center) contraction, in which the muscle force pulls the two ends of the muscle together. Force and motion occur in the same direction; the result is positive work being done to increase mechanical energy. The power in this case is positive, indicating an increase in energy. This is what is happening in the "shortening" part of the graph in figure 3.3c.

In the "lengthening" part of the same figure, muscular force is tending to pull the muscle ends together while the muscle is lengthening due to some external force. This is known as "eccentric" (away from the center) muscular contraction. Here force and motion occur in opposite directions; the result is work being done to decrease mechanical energy. We stand up or lift an object using concentric muscular contraction; we sit down or let an object down slowly using eccentric muscular contraction. Of course, if we use no muscular contraction, we flop down or drop the load and the energy cost to us is zero. It so happens that eccentric muscular contraction is more energy efficient than concentric contraction (Van Ingen Schenau and Cavanagh, 1990). This explains why it is less tiring to walk downhill than uphill. The third type of contraction, "isometric" (same length) contraction, occurs without any overall change in length of the muscle.

Here force times movement is zero, and no mechanical work is being done. Yet micromovements of structures inside the muscle are occurring, and this metabolic work requires chemical energy and liberates heat. Unfortunately, there are losses of energy as heat during energy conversion. This is what makes us sweat when we exercise, and why simply to survive we must take in more food energy than we expend in doing work.

Muscle Power

Muscle power is the product of muscle force and shortening velocity and represents the rate at which work is being done (see figure 3.3). Power output from muscle is maximal at some intermediate shortening velocity. In most activities in which the objective is maximization of velocity of the CM or of an object held in the hand, it would be inappropriate to reduce shortening velocity because any force, no matter how small, is useful to maintain acceleration. However, when muscular contraction can be mediated through machinery, it is possible to keep the shortening velocity in the middle of its whole range. Such is the case in cycling; the presence of a gear system allows intermediate magnitudes of shortening velocity irrespective of the speed of the bicycle over the ground. This effect is no different in principle from the use of gearing in automobiles.

Undoubtedly this phenomenon at least partially explains why distance running can be sustained for a long period of time while sprinting is a short-lived activity. This phenomenon is also probably related to the choice of combination of stride length and stride frequency used in distance running. The author is unaware of any research that has measured and used properties of muscle of a specific individual to determine the maximal efficiency generated by an optimal combination of these variables.

Stretch–Shortening Cycle of Muscle

In many activities, maximization of muscular work is of primary importance. As muscular work is the integral of muscular force with respect to muscle shortening, and as shortening is largely limited by the anatomical range, it is of prime importance to produce the greatest average force possible throughout the range. The problem the muscle faces is that as the force produces acceleration, the velocity of shortening increases and the force decreases according to the force–velocity relationship of muscle. In addition, part of this shortening phase is occurring when muscle activation is rising but not yet fully developed. However, there is a strategy that can offset the effect of this apparent self-defeating mechanism, and it uses events prior to the shortening phase of the activity. It is known biomechanically as the "stretch–shortening cycle of muscle" and also popularly as a "backswing" in throwing and striking activities (Chapman, 1980).

The mechanism is quite simple. In arresting movement in a direction opposite to the desired final motion, the muscle is producing force against a backward-moving load. This represents eccentric contraction, in which the force-producing capabilities of muscle are greater than in the concentric mode. Consequently the final motion of muscle shortening begins with a high force. In addition, the stretching induces a relatively short-lived force enhancement in the muscle and also results in strain energy being stored in the series elastic muscular structures.

■ **Key Point**
"Isometric" means constant length; "concentric" means muscle shortening; and "eccentric" means muscle lengthening.

Also the muscle has time to be activated fully in the eccentric phase. The final result of these factors is a greater than average force during shortening and an enhancement of the muscular work.

The stretch–shortening cycle is not confined to throwing and striking activities, but is seen in all activities in which maximization of muscular work is the aim, even in weightlifting.

How to get the most out of our muscles is very much dependent on the aim of particular human activities. Therefore this question is addressed in the chapters on fundamental movements.

Rotational Effect of Force

Human movement is by and large rotational. This means that our aim of traveling in a straight line or on any other path is achieved by rotation of joints. Although we have dealt with force as a straight-line concept, it is the rotational effect of force that is of importance. The arrangement of structures of the ankle joint is shown schematically in figure 3.4.

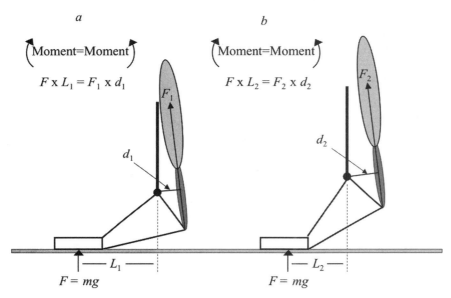

FIGURE 3.4 The moment arms d_1 and d_2 depend upon joint position during standing on one foot with different degrees of ankle flexion.

The turning effect of a force is known as "moment of force" or simply "moment." Moment represents the turning effect of a force that can also have an accelerating effect on an object. The term "torque" is used frequently but should be strictly confined to the situation in which the net translational force is zero and only a turning effect exists. Moment is defined as the force multiplied by the perpendicular distance between the line of action of the force and the axis of rotation. This perpendicular distance is frequently referred to as "moment arm" or "lever arm." In figure 3.4a, the clockwise moment due to the upward force from the floor (F) multiplied by perpendicular distance L_1 is balanced by counterclockwise moment due to the muscle force (F_1) multiplied by d_1. Each equation can be rearranged as follows:

$$F_1 = F \times (L_1 / d_1) \tag{Equation 3.2}$$

Since L_1 is much bigger than d_1, L_1/d_1 is greater than 1. Therefore F_1 is much greater than F. Inspection of the foot gives L_1/d_1 a value of about 3, so the muscle force F_1 will equal about three times body weight.

Clearly muscles and tendons are strong, and the connection where the tendon is inserted into the bone has to be equally strong. If we are jumping, the muscles will produce much more force than three times body weight. Figure 3.4b shows different distances L_2 and d_2, so the muscle and tendon force is likely to be dif-

■ **Key Point**

Human muscular contraction generally occurs at a mechanical disadvantage.

ferent between figure 3.4*a* and figure 3.4*b*. This is a characteristic feature of joint motion: The geometry of joint position combined with the force–length relationship of muscle gives us the feeling that certain postures are easier to sustain than others. As we move from fully crouched to fully erect postures, the stressfulness of the task varies throughout knee extension. It is particularly stressful in the midrange of extension but becomes easier as we approach the vertical. This is a feature weightlifters are familiar with. The success of the lift is governed not by the ability to hold up the load while standing but by the ability to pass this "sticking point" of knee extension. This also explains why stationary standing with a knees-bent posture is more energy consuming than standing with straight knees, despite the fact that no external work is done in either case. The internal work done is converted into heat, more so with knees bent than straight. You can easily verify this personally.

The reason for the bony prominences near the ends of bones where muscles insert was alluded to in chapter 1 in connection with the tibial tuberosity of the knee joint. The schematic elbow joint shown in figure 3.5 indicates what would happen if we did not possess tuberosities and if the ends of the long bones were not enlarged.

In figure 3.5*a* the muscle force *F* acts parallel to and along the surface of the humerus. Since there is no perpendicular distance (lever arm) between *F* and the axis of rotation, no moment is produced. The moment due to the load on the hand will not be sustainable, and the elbow joint will dislocate. In figure 3.5*b* the tuberosity and the enlarged end of the humerus move *F* away from the bone, giving a lever arm of *d*. The moment then is *F* times *d*, and it will enable the load on the hand to be sustained. The term "mechanical advantage" describes the ratio of the load moved to the force applied in such a leverage system as joints. In general, human joints are at a mechanical disadvantage because the moment arm for the muscle is less than the moment arm for the external load. However, the radial tuberosity and the pulley-like capitulum of the humerus increase the mechanical advantage.

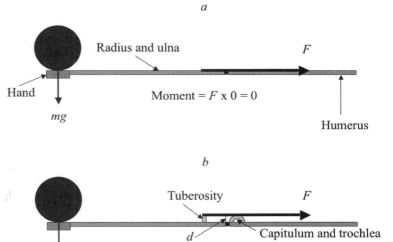

FIGURE 3.5 Tuberosities and condyles near joint axes increase lever arms of muscle.

Force Applied Externally

So far we have dealt with the static case (no movement), but muscles are for movement also, which is the dynamic case. In the dynamic case we must specify direction of movement. Consider the situation in which a person is stationary in a crouched position and then moves to stand erect. If we consider upward as positive, we see a period of initial positive acceleration followed by a period of negative acceleration (or downward acceleration). The "movement" we are

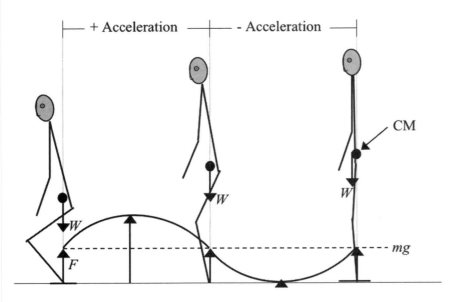

FIGURE 3.6 What your bathroom scale sees as vertical force as you rise from a crouched posture.

considering is the motion of the CM of the body. A stylized graph of upward (or positive) motion of the CM with respect to time is shown in figure 3.6.

The forces acting are the upward force at the feet *(F)* and the downward force or weight *(W=mg)* due to gravity. *W* stays constant since we have no control over gravity. Prior to movement upward, *F* and *W* are equal in magnitude but opposite in direction. No acceleration occurs, and we have no change in either velocity or displacement of the CM. When the muscles begin to contract, the force at the feet increases over that of gravity, and upward acceleration occurs. When *F* equals *W* again, acceleration is again zero. Following this, *F* drops below *W* and deceleration occurs, and velocity drops to zero. If you perform this action on your bathroom scale you will see the described fluctuations in force at your feet. For those not familiar with mechanics, it is often hard to understand that displacement can increase when velocity is decreasing and acceleration is negative. The fact is that while velocity may be decreasing (deceleration), there will still be a positive velocity and a resulting increase in displacement of the body. Careful inspection of figure 3.6 is necessary to lay the groundwork for understanding the remainder of this text.

It may come as a surprise that we have just undertaken the process of calculus. Calculus is difficult only when taught with unfamiliar mathematical symbols. Standing up is a simple act, so why complicate it with this calculus material? The reason is that calculus is necessary to analyze all mechanical actions, both simple and complicated. Note that we have used the terms "upward" and "downward" and "acceleration" and "deceleration." Force not only has a number of newtons but also has a direction. So an upward acceleration is equivalent to a downward deceleration. This concept is often difficult to grasp for those new to mechanics. One convenient way to describe motion is to pick a direction that we define as positive and use the terms positive and negative to apply to the concept of changing velocity (acceleration or deceleration). So if we choose upward as positive, speeding up in that direction, as in jumping, means positive acceleration, and slowing down in that direction due to gravity means negative acceleration.

Forces Acting on a Body Segment

The foot shown in figure 3.7 illustrates the varied forces that can affect the motion of a body segment. These are due to muscle-tendon, ligament, cartilage, joint friction, gravity, and other external forces such as ground reaction force. Ligamentous force affords joint stability and correct relative motion between segments. The many muscle-tendon forces afford complex control of motion. Cartilage affords a large area of contact to reduce pressure; it determines the direction of the net force vector that acts between adjacent joint surfaces. This force is known as joint contact force (often erroneously termed bone-on-bone force); and it reduces sliding friction. Gravitational force acts through the segmental CM. The various external forces can occur anywhere on the segment, as in toe kicking compared with instep kicking. While such qualities afford humans great complexity of motion, they make the problem of biomechanical analysis somewhat insurmountable. For example, it is difficult to identify individual ligaments when one is dissecting a joint. The force vector between cartilage surfaces is unknown, as is the frictional force. Although identification of the direction and point of application of a muscle-tendon force is reasonably accurate, the relative activation levels among a group of muscles are unknown. As a result it is difficult to specify the motion of a segment from knowledge of the forces being applied in a forward dynamic analysis. Similarly, it is difficult to apportion the forces in individual structures when one is calculating net force from motion in an inverse dynamic analysis. A common solution is to consider the forces of unknown magnitude and position on a segment as a single joint reaction force. Likewise, the individual moments created by muscles are considered as a single moment. The next section shows how this is done.

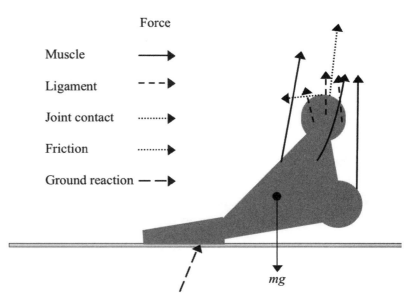

FIGURE 3.7 The many sources of force affecting acceleration of the foot segment.

Inverse Dynamic Analysis

As just stated, inverse dynamic analysis requires the assumption of a single joint reaction force at each end of a body segment and a net moment about each joint center. For the sake of simplicity, we will use a simple rodlike segment as shown in figure 3.8.

As the major aim here is to describe the process of the analysis, the multiplicity of equations is omitted. Those who wish to delve into the mathematical details of inverse dynamic analysis are referred to Robertson and colleagues, 2004.

The reference point in figure 3.8 is point O, at which there is a ground reaction force F_{GR} acting at an angle θ_{FGR} to the horizontal. The segment (assume a de-footed

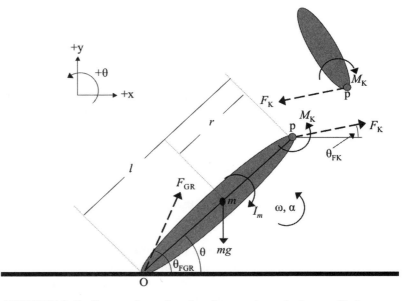

FIGURE 3.8 Conventions for direction and symbols used in inverse dynamic analysis.

leg) is defined by its end points at O and p of distance l, and the distance between p and the CM is r. The actual segment length can be measured on the body, and the values of r and m can be obtained from published tables (see Robertson et al., 2004). There is an unknown force F_K acting at p at an unknown angle θ_{FK}, and an unknown moment M_K acting in an unknown direction. Both F_{GR} and θ_{FGR} can be measured using a force plate, and the positional data of O and p and their derivatives can be obtained from film or some other optoelectrical device and split into their x- and y-coordinates. The x- and y-coordinates of the position of m and derivatives can then be calculated. From these data, the values of both ω and α can be obtained. The true inverse dynamic analysis can then be performed with the aim of obtaining F_K and θ_{FK}. The analysis begins by summing forces acting on the segment as follows:

$$F_{GR} \cos \theta_{FGR} + F_K \cos \theta_{FK} = m\ddot{x} \quad F_{GR} \sin \theta_{FGR} + F_K \sin \theta_{FK} - mg = m\ddot{y}$$

$F_K \cos \theta_{FK}$ and $F_K \cos \theta_{FK}$ can then be obtained; and since $\theta_{FK} = \tan^{-1}(F_K \sin \theta_{FK}) / (F_K \cos \theta_{FK})$, θ_{FK} can be obtained, finally yielding F_K. The values of F_K and θ_{FK} then become known inputs in the subsequent process of obtaining the force at the proximal end of the next segment.

The net moment applied to the segment is due to the sum of the moments created by the forces at each end about an axis through m plus the free moment due to muscles crossing the knee. This sum can be abbreviated as follows:

$$M_{FGR} + M_{FK} + M_K = I_m \alpha$$

where M_{FGR} and M_{FK} are the moments due to the forces at each end of the segment. This allows M_K to be obtained, which becomes an input in the opposite direction at the distal end of the adjacent segment.

This process can be carried from segment to segment through the body to get all joint forces and moments. One problem with this approach is that any error made in the calculation of segmental end-point forces and moment will be carried into the calculations for the next segment and so on from segment to segment. The information obtained in this manner is more or less useful depending upon the aims of the analysis. It is valuable for appreciating the manner in which energy can be transmitted between segments and for identifying mechanical energy savings that may occur as energy is transferred from kinetic to potential forms within a segment. It is obviously of no value in determining individual muscular contributions to a net force or moment. This type of analysis can also be applied in three dimensions with greater complexity of the mathematics, but with more complete analysis of the many nonplanar activities which we perform.

Intersegmental Motion

It is obvious that muscles act upon body segments to produce motion. What is less obvious is that muscles not attached to a segment can influence its motion. The explanation for this phenomenon lies in the fact that segments are connected at joints. For example, the hip flexors, as their name indicates, are located to produce a hip flexor moment about an axis through the hip joint. But if the lower limb is unconstrained, the leg and foot will also move. This occurs through a mutual

■ Key Point

The grouping of all unknown forces acting on a segment into one single force at the segment end is a matter of mathematical convenience.

equal and opposite force at the knee joint axis. Figure 3.9 shows a hip flexor force resulting from hip flexor muscle activity.

This generates a hip flexor moment tending to produce motion of the thigh in a counterclockwise direction around a horizontal axis through the hip joint. The resulting upward and forward motion of the knee joint leads to a force in the same direction pulling on the leg. As there is no mass at the knee joint axis, there is an equal and opposite reaction force from the leg on the thigh. The equal and opposite forces are internal to the body and do not contribute to any change in position of the CM of the body as a whole. The effect of F acting at a perpendicular distance d from the CM of the leg is to produce a clockwise moment. Such moment increases the angular velocity (ω) of the leg in a clockwise direction, with a resulting lifting of the leg and foot. This description would be fundamentally correct in principle if the hip flexors were the only active muscles. In practice, there are many more muscles that could be active than are depicted. In that case the description would need refinement.

There are two major lessons to be learned from this analysis. The first is that because of segmental interaction, the observed motion of a body segment may be at odds with that expected from the name of an active muscle. The second is that muscles far from a given body segment of interest can contribute to the motion of that segment. The latter is more the rule than the exception, particularly in athletic activities in which maximization of muscular work is essential to success.

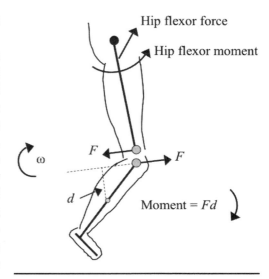

FIGURE 3.9 Moments created by muscles and joint forces can move a body segment.

■ **Key Point**

Muscles far from a body segment have a significant effect on the motion of that segment; this is usually advantageous.

Calculation of Internal Events From External Measurement

Compare the measurement we can make of a moving body segment with the sources of movement described in the previous sections. The major point here is that there are fewer external measurements to insert into equations of motion than there are sources of movement. This leads to the problem of mathematical indeterminacy, which does not allow us to calculate the magnitudes of the forces in all force-producing tissues. The culmination of the effect is that we can determine only a single net force and a single net moment at each end of the segment, as discussed previously. Such a limitation affects certainty about the sources of intersegmental motion and their specific locations.

Two techniques have been used to improve the calculation of the contribution of individual tissues to motion. One is the internal application of a force transducer to the Achilles tendon to register tendon force during locomotion. Such a method is particularly invasive and yields information only on the plantarflexor force (and moment) due to the triceps surae group of muscles. Unfortunately, other muscles that can contribute to the plantarflexor moment go unrecorded with this method. This technique has been discussed extensively by Gregor and Abelew (1994). Another method uses a physiological model of the system in which the musculoskeletal architecture of the system provides a physical model (Pierrynowski, 1982). The muscle model is then used to generate individual muscle forces according to their anatomical ability to contribute to a net muscle moment. Electromyographic activity of these muscles is used to verify their contributions.

Still another approach is to use the model to decompose a net moment into individual muscle forces based on the size and location of the muscles (Caldwell and Chapman, 1991). The application of these techniques is endless. One example is the determination of force in extensor carpi radialis brevis to investigate the implications for injury in a backhand tennis stroke (Reik et al., 1999).

Role of Biarticular Muscles

Most, if not all, mammals have muscles that cross one, two, or many more joints. The numerous muscles of the spine include examples of mono- through multiarticular muscles. In general, the joints that allow intersegmental movement all have monoarticular and biarticular muscles. The reason for this architecture has intrigued biomechanists for many years, and a number of theories for its existence have been developed and tested. Clearly, biarticular muscles are of benefit; otherwise they would represent an evolutionary mistake. Yet the question of their biological significance remains. Some obvious areas in which biarticular architecture can be useful are enhancement of control of our motor system, for reasons of safety; distribution of loads among different muscles; and facilitation of the distribution of mechanical energy among segments.

The change of length and velocity of a monoarticular muscle is directly related to the angle and angular velocity of the joint that it crosses. Its ability to produce force under the influence of the force–length and force–velocity relationships is therefore related to the kinematics of the joint. As the biarticular muscle crosses two joints, the kinematics of both joints is implicated in the muscle's force-producing capacity. The gastrocnemius muscle crosses the ankle joint as a plantarflexor and the knee joint as a flexor. During simultaneous plantarflexion and knee extension, the muscle can be acting isometrically. This result arises from the fact that the muscle shortens during plantarflexion but lengthens during knee extension. If the joint angular velocities are appropriate, the muscle will experience no change in length. As it will be acting isometrically, its force-producing capacity will be high according to the force–velocity relationship. It is also conceivable that the muscle could be acting eccentrically with even greater force available. Returning to the isometric condition brings into question the amount of work that the muscle can perform. With no shortening, no work will be done; and any muscular work required to raise the body in this type of knee-ankle motion will have to be generated by monoarticular muscles.

It has been suggested that this mechanism is an economic means whereby work done by the large proximal monoarticular muscles such as the gluteus maximus is transported distally (Van Soest et al., 1993). Other investigators have disputed this mechanism (Pandy and Zajac, 1991). Arguments of this nature will eventually be solved, since it seems that

Stretch of active muscle, in this case the right pectoralis major, always enhances physical performance.

the universal appearance of biarticular muscles among species is unlikely to be an accident with no purpose. As is seen in many branches of science, the solution may arise from an unexpected source. The fact that the greatest proportion of muscular tears (pulled muscles) in the lower limb occur in biarticular muscles may hold the clue to their role.

Summary

A great deal of biomechanical information concerning the internal and external forces acting on the body has been presented in this chapter. Gravity is an external force over which we have no control. External friction is of consequence only when we aim to move under contraction of our muscles. In fact, our attempts at most horizontal movement would be fruitless without friction. Muscle receives by far the most attention in this chapter as it is the primary internal organ of our movement. Muscle is a complex mechanical organ inasmuch as the instantaneous force that it produces is governed by voluntary and reflex control and by its current and past kinematics. Any given level of muscular force can be achieved by many combinations of the governing variables.

As we shall see in later chapters, muscles are used in specific temporal and spatial ways that are governed by our goals. When our aim is to move as fast as possible, we produce the greatest force we can for as long as possible in order to generate the greatest momentum in body parts. However, the greater the force, the greater the acceleration and the sooner we reach anatomical limits of muscle shortening. The consequent decrease in time can be offset by use of the stretch–shortening cycle of muscle, which enhances the force. Following the generation of momentum in body parts we use muscles connecting segments to maximize whole-body momentum through the principles of conservation and transfer of momentum. Similarly, we use the principles of conservation and transfer of energy to redistribute muscular work produced by the greatest force applied over the greatest displacement possible. This redistribution generally occurs distally, from the strong proximal muscles to the smaller distal segments.

Part of this chapter is devoted to the rotational effect of muscular force since we are by construction largely rotational machines. Muscles act upon bones, which form the framework of our body segments. Muscles connect adjacent segments, or in some cases segments at some anatomical distance from each other. Therefore the discussion included ways of expressing, visualizing, and calculating forces and their effect on motion between segments. The question of why we have biarticular muscles was raised but not answered.

RECOMMENDED READINGS

Alexander, R. McN. (1992). *The human machine.* New York: Columbia University Press.

Hay, J.G., and Reid, J.G. (1988). *Anatomy, mechanics and human motion.* Englewood Cliffs, NJ: Prentice Hall.

Herzog, W. (2000). *Skeletal muscle mechanics.* Chichester, New York: Wiley.

Ivancevic, V.C., and Ivancevic, T.T. (2006). *Human-like biomechanics: A unified mathematical approach to human biomechanics and humanoid robotics.* Dordrecht (Great Britain): Springer.

Nigg, B.M., and Herzog, W. (Eds.) (1994). *Biomechanics of the musculo-skeletal system.* New York: Wiley.

Nigg, B.M., MacIntosh, B.R., and Mester, J. (2000). *Biomechanics and biology of movement.* Champaign, IL: Human Kinetics.

Nordin, M., and Frankel, V.H. (2001). *Basic biomechanics of the musculoskeletal system.* Philadelphia: Lippincott, Williams & Wilkins.

Rose, J., and Gamble, J.G. (2006). *Human walking.* Philadelphia: Lippincott, Williams & Wilkins.

Vogel, S. (2001). *Prime mover.* New York: Norton.

Yamaguchi, G.T. (2001). *Dynamic modeling of musculoskeletal motion.* Boston: Kluwer Academic.

PART II

Fundamental Human Movements

This part of the book deals solely with the biomechanics of the majority of fundamental human movements. The movements are considered fundamental if they aid in survival, are part of everyday life in areas such as recreation and occupation, and do not involve interaction with machines that have external sources of power. They are analyzed by means of a problem-solving approach in which the major aim of the task is identified, the major mechanics underlying this aim are presented, and the biomechanical structures that can solve the mechanical aim are examined.

The chapters in this part explore underlying principles of human motion that are mechanical in nature and that result simply from the manner in which segments are arranged. Other principles have a strong biomechanical component and depend upon the properties of the force generators in concert with the structures that transmit the forces, namely muscles and their connecting tendons. In the absence of published research, many solutions are examined by means of simulation using a "standard" human. It is important to stress that the data from the simulations do not apply to any individual and that the "standard" human is only an approximation of the real thing. Therefore the major focus is on the principles and patterns of motion. While a particular fundamental movement is the basis for each chapter, examples are drawn from sport-specific, occupational, and medical areas for the purpose of enhancing understanding of the fundamental movement. In addition to the mechanical description of the fundamental movement, sections in each chapter address how the movement may be enhanced by technique or training and how safety and avoidance of injury can be largely ensured. Successful understanding of part II will arm readers with the tools and concepts necessary to analyze biomechanically activities of their choice, as well as to read in a scientifically educated manner the vast amount of literature concerning human movement.

CHAPTER 4

Balance

tanding is the most common form of active balancing. Development of the ability to stand erect on two feet has been considered a mechanical factor that led to our relatively rapid advancement over our cousins, the apes. Standing is used to enable us to see farther than if we stood on four legs, to perform a number of daily tasks, and also as a precursor to bipedal walking. While the apparatus that we possess for standing appears to be unnecessarily complex, it's the only thing we have, and it is necessarily complex to enable us to perform other actions. Standing would appear to be a precarious activity were it not for the fact that it is one we are familiar with. It is precarious because we have to maintain balance on two jointed posts (lower limbs) that are enlarged in a fore-and-aft direction at their extremities (feet), and we perform other activities while doing so. But many other forms of balancing are more challenging than standing.

Balance can be defined as a state of static equilibrium that is disturbed only by external force. However, as we shall see, there are other forms of balancing in which stability can be maintained only with the use of our muscles. In fact, the only true form of static equilibrium is lying on the ground without muscular involvement.

Toppling simply refers to a loss of balance. In some activities, toppling is a requirement in preparation for a further activity, as in moving from standing to walking. In other activities, toppling is something to be avoided. The discussion of toppling in this chapter emphasizes either avoidance of toppling or regaining balance after toppling begins.

Aim of Standing

The problem to solve in standing is to keep the position of the center of mass (CM) of the body vertically above and within a certain prescribed area on the ground known as the base of support, as shown in figure 4.1.

The symbols AP and LA refer, respectively, to anterior-to-posterior and lateral dimensions of the base of support. In other forms of balancing such as the

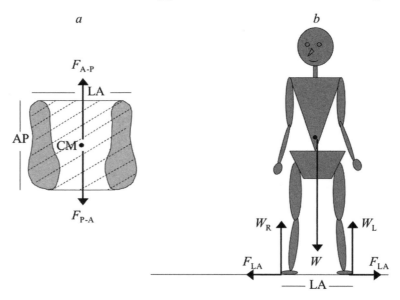

FIGURE 4.1 Forces maintaining zero acceleration of the CM in standing.

handstand, the area of the base of support is severely reduced, which reduces the allowable motion of the CM in the transverse plane.

Mechanics of Standing

Static equilibrium requires the CM of the system to be located vertically above the base of support with zero velocity and with all forces in all directions summing to zero.

Horizontally   *Vertically*

$$\sum F_{FEET} = 0 \qquad \sum F_{FEET} - mg = 0$$

(Equation 4.1)

This requires equality of forces $F_{A\text{-}P}$ and $F_{P\text{-}A}$ (figure 4.1a) and equality of F_{LA} in each direction (figure 4.1b). Similarly, the forces in the vertical direction must add up to zero and $W_L + W_R = W$ (figure 4.1b).

Biomechanics of Standing

The requirement for static equilibrium is hardly applicable in human balancing because of the multisegmented construction of the body. In the human case, there are numerous intersegmental forces that must sum to zero. Certain flexibility is allowed in human standing such that the position of the CM can move both horizontally and vertically within the base of support without the person toppling over. Moving the CM from a vertical position above the center of the base of support, and then to the right and left limits and back to the center, requires the integral of force with respect to time to be zero:

Horizontally   *Vertically*

$$\int F_{FEET} \cdot dt = 0 \qquad \int F_{FEET} \cdot dt - mgt = 0$$

(Equation 4.2)

Equations 4.1 and 4.2 are simple, as is the control of the body motion they describe. People can perform many tasks with their hands and arms by moving the arms laterally without having the position of the CM move outside the limits of the base of support. The whole of the upper body can be moved similarly if the knees are flexed. The golf drive incorporates both of these actions as seen in figure 4.2.

Force, velocity, and displacement of the CM begin at zero. A negative force then imparts a negative velocity to the CM, followed by a positive force, which reduces the negative velocity to zero. Here the CM is at its greatest negative displacement; note how the positive force is almost at its greatest value. Then begins a decline in the positive force accompanied by an increase in positive velocity that peaks when the force is zero and the displacement is back to zero. This is when the ball will be struck. The subsequent negative force reduces the positive velocity to zero, leaving the displacement of the CM at its greatest forward position. Throughout the stroke the CM has moved considerably laterally without exceeding the limits of the base of support. To achieve this maneuver, the integral of force with respect to time is zero so that the velocity which begins at zero must end at zero. During this time there has been a net lateral displacement of the CM.

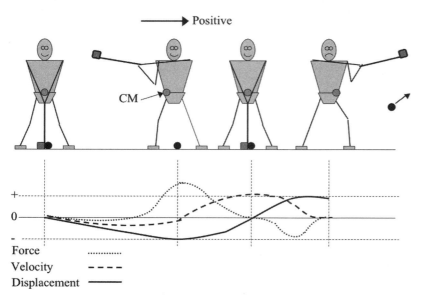

FIGURE 4.2 Kinematics of the CM when it moves laterally within the base of support.

In golf this lateral displacement is known as a "weight shift." It should be termed "lateral mass shift," as the concept of weight is the vertical force of attraction due to gravity. The original term does have meaning in that lateral transfer involves locating the CM over one foot and then the other. Consequently each foot in turn will experience increases and decreases of vertical force. Yet weight shift is not the reason for the maneuver but simply a consequence of applying horizontal force in order to change the position of the CM within the base of support. Figure 4.2 is a very stylized representation of a golf drive, intended to make the salient points about balancing. Do not copy this action if you intend to improve your golf.

From a purely mechanical point of view, the energy cost of standing in different postures is zero. The value of zero arises from the fact that the height of the CM is constant so there is no change in potential energy, and the absence of motion indicates no change in kinetic energy. We know that solid objects such as statues can stand for hundreds of years. But as humans are not single solid objects, the energy cost becomes a biomechanical problem. For example, there are small areas of contact between articular surfaces in the knee, ankle, hip, and other joints involved in standing. Therefore it is probable that a vertical line through the CM of the body does not pass through the articular contact areas. In this case a moment about the joint is created that has to be counteracted by muscular contraction. As the muscles are neither shortening nor lengthening, they do no external work. The work comes from movement of small structures inside the muscle, and the evidence for this is that heat is liberated. Even in erect standing there are small amounts of body sway that periodically require small amounts of muscular contraction. Therefore an energy cost is present even in the quietest standing. Figure 4.3 depicts the feet, legs, and thighs supporting a single block representing the mass of the upper body (the weight of which is mg) above the hip joints. Figure 4.3*a* shows an erect posture, and figures 4.3*b* and *c* show different degrees of stooping.

In figure 4.3*a,* the CM of the upper body (mass = m) lies vertically above the lower limb joints. While there will be upward reaction forces at the hip joints totaling mg, there is no lever arm d to produce a moment. Such is not the case in figure 4.3*b,* where the lever arm d results in a clockwise hip joint moment of mgd. The hip extensors must be recruited to counteract the clockwise moment in order to avoid motion of the upper body. Therefore the downward force pressing on the femoral articular processes will equal mg plus some force contributed by the hip extensors. The latter will lead to a counterclockwise moment approximately equal to mgr created about the knee joint axis, and knee joint extensor activity will be required to oppose it. Clearly each moment depends on the degree of stooping, which affects the magnitudes of d and r. As the trunk approaches horizontal, the moments will increase in magnitude because d and r will increase (figure 4.3c).

Of course the muscles produce only force, but as described earlier, these forces contribute to the moment at the joint, which is why it is convenient to use the term muscle moments. The muscle moments at the hip and knee are counterclockwise and clockwise, respectively, or more properly hip extensor moment and knee extensor moment. Obviously the body would collapse in the absence of the moments, with both the hip and knee joints flexing. This analysis could be extended to the ankle joints.

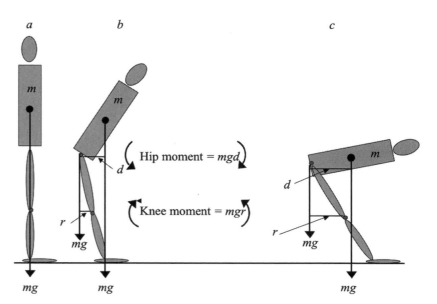

FIGURE 4.3 Different postures require different moments about joint centers to maintain stability.

If we were to look inside the model of the human in figure 4.3, it would be necessary to take into account all of the possible joints, including the intervertebral joints. As we progressed up the spine we would find that the moments became smaller because of the decreasing mass of the body above the joint. The moment at the joint between the first cervical vertebrum (in the neck) and the articular surface on the occipital bone at the base of the skull would be influenced only by the mass of the head.

Different moments are required to maintain different postures. This means that different muscle forces are necessary. The relationship between the posture and the required muscle force is not simple, since muscle lever arms vary with posture (see d_1 and d_2, figure 3.4 in chapter 3). Different postures also require different muscle lengths. As muscle force also varies with muscle length, certain postures will tax the muscles greatly, particularly if the person is holding a heavy weight. The same argument can be made for energy cost. In requiring greater muscle forces, certain postures require motion of many more muscular structures in the contractile apparatus, and the energy cost increases as a consequence. A simple experiment of prolonged standing with knees half flexed will soon bring on perspiration and knee extensor pain that would not be experienced in erect standing. Perspiration is evidence of heat production and therefore of greater internal (or metabolic) work. Standing is easy to perform and easy when viewed from the strict mechanical point of view of zero external work done. Biomechanically, the internal work depends in a complex way on posture.

■ Key Point
The energy cost of maintaining a balanced posture varies with joint angles, which determine the joint moments required.

Variations of Standing

Particularly dramatic forms of balancing can be seen on many pieces of gymnastics apparatus and also in the circus. Balancing on the hands is probably the simplest to understand but is very difficult to perform. A comparison between a handstand and simple standing on one's feet is shown in figure 4.4. The total cross-sectional area of the ankle extensor muscles varies greatly among individuals but is on the order of 130 cm² (20 in.²), giving an available force of about 3250 N. So we can apply 3250 N of force to the calcaneus (or heel) bone through the Achilles tendon. The lever arm for the Achilles tendon is about 5 cm (2 in.). So the moment the ankle extensors can produce is about 3250 × 5 or 16,250 N.cm. A moment of this magnitude makes it easy for us to avoid toppling forward. The equivalent in the handstand is the wrist flexor moment. The cross-sectional area of the wrist flexors (including finger flexors) is about 20 cm² (3 in.²), producing a force of about 500 N. The lever arms for the various tendons of the wrist and finger flexors are very small but average out at about 2 cm (0.08 in.). The available moment is therefore 1000 N.cm, which is about 6% of the ankle flexor moment. A further complicating factor is that the foot is longer than the hands. Since the area of the base of support is smaller in the handstand, the CM of the body cannot move forward or backward as much in the handstand as in normal standing. The handstand is clearly a precarious activity from a muscular perspective alone, so it is not surprising that gymnasts have muscular forearms.

■ **Key Point**

Difficulty in balancing is determined by the size of the base of support and the moment-producing capacity of the muscles.

Female ballet dancers appear to be able to balance "en pointe," or in the more common phrase, on their toes. In fact they are not balanced statically because the area of the base is a very small block of material in the front of their extended shoes. In addition, the action is performed with the ankle joint fully plantarflexed, which locates the center of the ankle joint almost vertically above the point of floor contact. Careful observation of ballet dancers performing this maneuver reveals that slight movements are used to avoid toppling, and this is also the case when they are standing on both "toes."

Handstands in gymnastics are done from a variety of initial postures, but most end up in a vertical position with the toes pointed. Consequently the movements up to the handstand are done deliberately and slowly. This strategy is even more important when the gymnast balances on one hand only. In this case the musculature controlling the handstand is halved, making the performance doubly difficult. While forearm muscular strength is the prime source of balance in the handstand, strategies are available to resist toppling when the performance is not constrained by the requirements of gymnastics judges. Tumblers in the circus, on stage, or in movies often use flexion and extension of the knee joints to play a tune on angular momentum. The tightrope walker uses a long pole for a similar reason. Such strategies are the subject of the section on toppling later in this chapter.

A particularly difficult case of balance is exhibited by weightlifters. Consider the "snatch" lift, in which the load is lifted from the

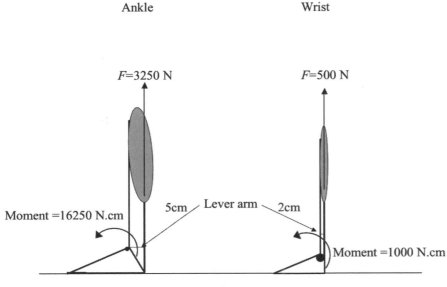

FIGURE 4.4 Ankle plantarflexor moment and wrist flexor moment required to maintain balance.

floor to arm's length above the head in one continuous movement. The difficulty lies not in the size of the base of support but in the moments required to maintain the load above the head. Since there are many joints in sequence between the floor and the load, there are many possibilities for the gravitational force through the load not to coincide with the joint center. Such is the case as the load is lifted, but the difficulty is to attain the correct position of the load when it stops moving upward. Should the arms not be straight when the load becomes stationary, the relatively small extensor muscles of the elbow joint will be unable to supply sufficient moment to counteract that produced by gravity acting on the load. The skill is to "lock" the arms straight at the instant when the load stops moving, or just before.

Enhancement of Standing

As the size of the base of support is critical to maintaining balance, is there anything we can do to increase it while standing on two feet? Of course the feet can be spread farther apart both laterally and also in the direction of the sagittal axis. The direction in which the feet may be spread apart depends on the nature of the balancing task and the direction from which external disturbing force is expected. If a person attempts to disturb your balance by pushing

Balance is more difficult as the area of the base of support decreases.

on your chest, greater stability will result from adopting a fore–aft spread of the feet. In this way the disturbing force has to act over a greater distance in order to disrupt balance. More work is therefore required of the disturbing force because work is the product of force and displacement.

It is a popular misconception that a person who is small, and who therefore has a low position of the CM, will have inherently increased stability. Stability will be lost when the position of the CM migrates outside of the base of support, irrespective of whether the CM is in a high or low position. We consider this question further in the section on toppling.

■ **Key Point**

Static balance depends on the magnitude and position of any supported load.

Safety of Standing

Even in apparent static balancing, there is some sway of the body that is difficult to see. This sway requires the generation of some horizontal force at the feet as shown in Equation 4.2. It is therefore essential that the footwear the person is using provide sufficient friction to facilitate such horizontal forces. This may appear obvious under normal standing conditions, but even small amounts of sway can lead to slippage during standing on ice. Anyone who has been in a lift lineup when skiing has seen someone slip over for no apparent reason. The reason, of course, is that our normal sway pattern in standing cannot be accommodated by the low friction available when we are standing on skis. We see a similar effect in dynamic balancing such as walking when we encounter a patch of oil or ice under the supporting foot. This is why the presence of oil patches is unacceptable in the workplace or anywhere else when people may be carrying loads. The load may survive, but the carrier may not.

Handling loads requires special attention in this discussion because a load in the hands changes the position of the CM of the load-carrier system. First, it is dangerous to pick up a load from a stationary posture when the CM of the body is vertically above the outer limits of the base of support. Adding a load that is located initially outside of the outer limits of the base ensures that the CM of the system will shift outside these limits also, with the obvious consequence of toppling. An obvious strategy is to place one foot in the direction of the load in order to locate the combined CM of the system vertically within the base of support. If the feet cannot be located under the combined CM, the load should not be lifted.

The following example indicates how care and attention must be paid to the magnitude of the load lifted and also to the technique used to lift.

Practical Example 4.1

Part 1

Figure 4.5 shows an individual supporting a load in the hands. The segmental masses and horizontal displacements between the vertical through each mass and the ankle

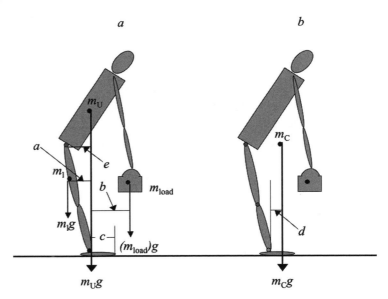

FIGURE 4.5 Relating moments created by individual segment masses to the moment created by the combined CM.

joint are upper body, head, and arms (m_U, zero); lower limbs (m_l, a); and the load (m_{load}, b). The horizontal displacement between the ankle joint and the toes is c. The combined CM of the system is (m_c). Calculate the maximal load that can be held before toppling forward begins.

The following constants are given:

$$m_u = 54 \text{ kg}$$
$$m_l = 26 \text{ kg}$$
$$m_{load} = ?$$
$$a = 25 \text{ cm}$$
$$b = 35 \text{ cm}$$
$$c = 20 \text{ cm}$$

Solution

Toppling will begin when a vertical line passing through the combined CM (m_c) is located outside of the base of support, that is, at a greater horizontal displacement than c from the ankle joint.

Since the positions of the separate masses determine the position of the combined CM, the sum of the moments created by each mass equals the moment created by the combined CM. Assigning clockwise moments as positive, the equation representing this relationship is as follows:

$$(m_u g \times zero) - (m_l g \times a) + (m_{load} g \times b) = (m_c g \times d) \text{ kg.cm}$$

Note the negative sign in the equation indicating a counterclockwise moment due to the mass of the lower limbs. Canceling g throughout and rearranging the equation to

$$d = [(m_u \times zero) - (m_l \times a) + (m_{load} \times b)] / m_c \text{ cm,}$$

and recognizing that m_c is the sum of all the separate masses, we obtain

$$d = [(m_u \times zero) - (m_l \times a) + (m_{load} \times b)] / (m_u + m_l + m_{load}) \text{ cm.}$$

To reduce the number of symbols we insert their values as follows:

$$d = [- (26 \times 25) + (m_{load} \times 35)] / (54 + 26 + m_{load}) \text{ cm;}$$

$$d = [- (650) + (m_{load} \times 35)] / (80 + m_{load}) \text{ cm}$$

This cannot be solved with two unknowns and one equation, but we know that to avoid toppling, $d < c$, or $d < 20$ cm (8 in.). Therefore,

$$[- (650) + (m_{load} \times 35)] / (80 + m_{load}) < 20 \text{ cm;}$$

$$- (650) + (m_{load} \times 35) < 1600 + 20 \times m_{load};$$

$$15 \times m_{load} < 1600 + 650 = 2250;$$

$$m_{load} < 2250 / 15 = 150 \text{ kg}$$

Therefore it would be possible to hold a load of 150 kg or approximately 330 lb, but any greater load would induce toppling. The great majority of individuals could not hold 330 lb off the ground in this or any other way without severe consequences. What this analysis does reveal is that much lesser loads than this can be held in this posture without toppling. It also reveals how a weightlifter is able to lift such large loads without losing balance. Furthermore, judgment is required when one is attempting to lift a load off a table. In this case the force applied to the load increases gradually such that the opposing moment applied by the load increases gradually. This effectively moves the position of the combined CM toward the toes. Therefore loads of unknown magnitudes should be lifted with a gradual increase in applied force so that the force can be reduced should the combined CM fall outside of the base of support. The obvious danger is when the load is lifted with a rapid increase in force and has to be dropped as the lifter loses balance. This is only one consideration in a number of possible strategies for lifting.

Part 2

What hip extensor muscle moment would be required to sustain the given posture during holding of the calculated load? If the average lever arm of the hip extensors is 10 cm (4 in.), what will be the total hip extensor muscle force?

Solution

Clockwise moments are created by the upper body mass m_u and the load m_{load}. The hip extensor moment (HEM) is

$$\text{HEM} = m_u g e + m_{load} g (e + b).$$

Assigning a value of $e = 30$ cm, so $(e + b) = 30 + 35 = 65$, we obtain

$$\text{HEM} = (54 \times 9.81 \times 30) + (150 \times 9.81 \times 65);$$

$$\text{HEM} = 111538 \text{ N.cm or } 1115.38 \text{ N.m.}$$

This moment is supplied by the product of the hip extensor force (HEF) and the lever arm as follows:

$$\text{HEM} = \text{HEF} \times \text{Lever arm};$$

$$\text{HEF} = \text{HEM} / \text{Lever arm} = 111538 / 10 = 11153.8 \text{ N} = 1134 \text{ kg Force}$$

Such a huge value indicates the great strength of weightlifters' muscles as well as the strength of tendons and the supporting bones.

Practical Example 4.2

Through what angle does each of the persons shown in figure 4.6 have to be displaced for balance to be lost in the frontal plane, assuming that they take no evasive action?

Solution

Loss of balance occurs when the CM reaches a vertical line through the lateral edge of either foot. For figure 4.6a, θ_A is obtained from $\sin\theta_A = 0.25 / (0.8^2 + 0.25^2)^{1/2}$.

$$\theta_A = \sin^{-1} (0.2983);$$

$$\theta_A = 17.35°$$

For figure 4.6b we need to calculate the vertical position of the combined CM as follows:

$$80 \times 1 + 150 \times 2.4 = (80 + 150)d, d = (440 / 230) = 1.913 \text{ m};$$

$$\text{Sin } \theta_B = 0.15 / (1.913^2 + 0.15^2)^{1/2};$$

$$\theta_B = 4.483°$$

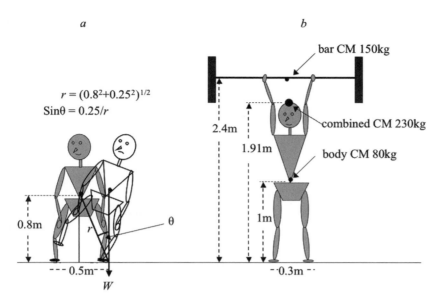

a

b

$$r = (0.8^2+0.25^2)^{1/2}$$
$$\mathrm{Sin}\theta = 0.25/r$$

bar CM 150kg

2.4m

1.91m

combined CM 230kg

body CM 80kg

0.8m

r

θ

1m

0.5m

W

0.3m

FIGURE 4.6 *(a)* The angular displacement θ required to shift the combined CM to the lateral margin of the base of support with a low CM compared with *(b)* the high combined CM in weightlifting.

The lower the combined CM, the greater the angular displacement before loss of balance occurs. Person A has more flexibility in determining the height of the CM and the width of the base of support than does the weightlifter. The latter can take little evasive action, as the requirement of the event is to finish in an erect posture. Both individuals are most likely to lose balance in a sagittal plane since the anterior–posterior distance of the base of support is much smaller than that in the frontal plane considered here. Clearly people should be very careful when bearing loads at great heights above their CM.

Aim of Toppling Avoidance

Once toppling is begun, the aim becomes how to stop it. Figures 4.7*a* and *b* illustrate the initiation of toppling in the frontal plane and the sagittal plane, respectively. In both cases the moment *Wd* is responsible for increasing the angular momentum and kinetic energy about the point of rotation at the feet. The problem is to oppose the effect of the moment *Wd*, which if left unattended will increase as *d* increases and result in rapid increase of the angular velocity of toppling.

Mechanics of Toppling

Equation 4.3 encapsulates the mechanics of toppling by utilizing the relationship between angular impulse and angular momentum.

Angular impulse = Change in angular momentum

$$\int Wd \times dt = I\omega_f - I\omega_i$$

(Equation 4.3)

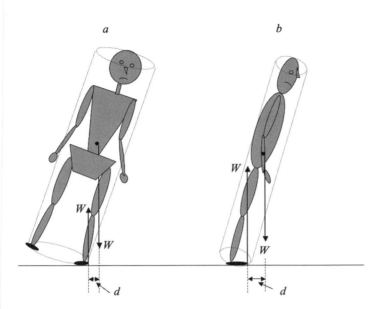

FIGURE 4.7 Toppling is initiated when the vertical line through the CM lies outside the base of support.

where I is the moment of inertia of the body about an axis through the feet and ω_f and ω_i are final and initial angular velocities, respectively. The integral sign $\int$ indicates that we are calculating the area under the moment–time curve. A complicating feature of Equation 4.3 is that as toppling proceeds, the value of d increases. So even though W is constant, the integral, and therefore the angular momentum, increases at an ever-increasing rate. Such an effect makes early intervention in avoidance of toppling most appropriate. If intervention is too late, no amount of avoidance will arrest the toppling. The problem is to counteract the gain in angular momentum and to do so only with forces internal to the body. In preparation for the forthcoming discussion, remember that the value I, calculated about an axis through the feet, can be broken down according to the parallel axis theorem into $I_G + mr^2$ (see Equation 2.14 in chapter 2, "Essential Mechanics and Mathematics").

An alternative analysis of toppling is through the work–energy relationship. While we generally think of work as the integral of force with respect to displacement, it is equally well considered the integral of moment with respect to angular displacement. The units and dimensions of translational and rotational work are exactly the same. Equation 4.4 shows how the work done due to the moment created by gravitational force about an axis through the feet leads to an increase in the kinetic energy of the body in its rigid state.

Work = Change in energy

$$\int W \mathrm{d} \times \mathrm{d}\theta = (I_G + Mr^2)(\omega_f^2 - \omega_i^2)/2$$

(Equation 4.4)

Note that as θ increases from the vertical, the value of d increases, leading to an ever-increasing rate of doing work. Note also that a change in body configuration will change I_G, mr^2, and d; the consequence is that ω will change.

Biomechanics of Toppling

The gain in angular momentum cannot be counteracted if the body maintains a fixed configuration resembling the cylindrical mass shown by the broken lines in figure 4.7. The only manner in which the body can be stopped from toppling will be to transfer the gaining angular momentum to some moving part of the body. Since the body is segmented it can approximate a cylinder plus a wheel, the wheel representing the rotating arms with its axle located at the shoulder (figure 4.8).

This rather busy figure is quite simple in concept. In figures 4.8*a* and *b*, the body has angular velocity ω_B. In figure 4.8*a*, the arms are rotated with angular velocity ω_A in the frontal plane in a clockwise direction (i.e., increasingly darker shades of gray), including motion in front of the body and overhead. In figure

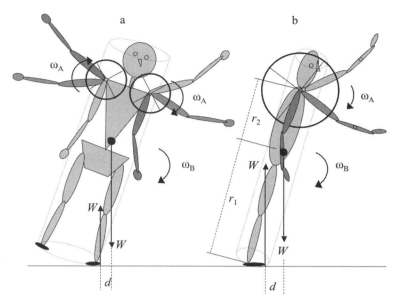

FIGURE 4.8 Motion of the wheel represents motion of the arms where the angular momentum of the whole body can be located.

4.8*b,* the same direction of arm motion occurs, although it cannot be quite circular due to restrictions of shoulder joint motion. The total instantaneous angular momentum (AM) of the wheel-body system (figure 4.8*b*) about a horizontal axis through the feet is shown in Equation 4.5.

$$AM = \left[I_B + m_B r_1^2 + m_A (r_1 + r_2)^2 \right] \omega_B + I_A \omega_A \qquad \text{(Equation 4.5)}$$

If the wheel (arms, A) is rotated faster and faster until its angular momentum equals the total system angular momentum, the angular momentum of the remainder of the body B will equal zero. If all of the body's angular momentum is due to arm motion, the remainder of the body will have no angular momentum and it will remain stationary instantaneously. A simplified depiction of this case is shown in figure 4.9, where the arms are used to move us from situation *a* to situation *b*.

With use of the word "instantaneously," a very important point is raised. Although the body may be at zero angular velocity, it will be leaning. Therefore there will be a value *d* as shown in figure 4.3, which means that there will exist a moment due to gravity and a subsequent gain in angular momentum as *d* increases with time. The consequence is that the arms must be rotated faster and faster so that the resulting rate of increase in their angular momentum will either equal or exceed the rate of increase in the body's angular momentum.

We should also appreciate that the muscles that rotate the arms are applying force through a displacement in a complex manner in order to increase the rotational kinetic energy of the arms. If the body is to remain stationary, the work done by the shoulder muscles will have to increase over time at the same rate as gravity is working in order to increase the rotational energy of the body. If the work done by the shoulder muscles exceeds that done by gravity, the body will rotate backward in the direction of erect standing. While the arms appear to be the best candidates for avoidance of toppling, any other suitable part of the body

■ **Key Point**

Muscular contraction can move angular momentum between body segments.

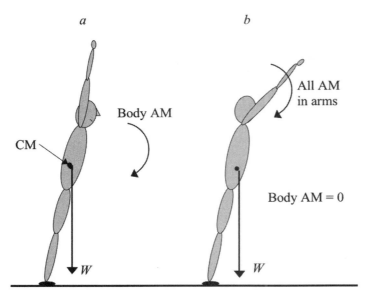

FIGURE 4.9 Transfer of the rigid–whole-body angular momentum to the arms only.

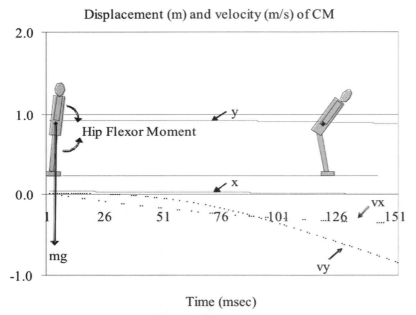

FIGURE 4.10 Computer simulation of the use of hip flexor moment to avoid falling forward.

can be used to replicate their effect. Such an effect can be seen in figure 4.10, which shows the results of a computer simulation using a two-segment body.

The upper segment is analogous to a rigid configuration of all of the body above the hip joints. The lower segment represents the lower limbs kept rigid. So the two segments are joined at the hips, and the hip flexor muscles are responsible for the relative orientation of the two segments in this simulation. At time zero the body has a forward lean of 88° from the vertical and it is stationary. Therefore the vertical through the combined CM passes in front of the ankle joint, as shown by the curve x, and the individual will simply fall unless some muscular effort is made. At this time a hip flexor moment of 250 N.m is produced, and it decays with time as the angular velocity of hip flexion increases. The work created by the hip flexor moment produces rotational kinetic energy and segment translational kinetic energy. This results in a backward or negative horizontal velocity of the combined CM (vx). The CM eventually is positioned behind the ankle joint, and forward toppling is avoided. Although not shown in the current simulation, unless further action is taken the individual will topple backward. This can be avoided by use of the hip extensor muscles to produce the opposite effect, leaving the individual standing erect (not shown). Because of the rotational nature of segmental motion, the vertical displacement (y) and velocity (vy) of the CM are also shown to be significant in this movement.

Some readers will remember performing such a maneuver as children when the class bully pushed them into the swimming pool. You will see gymnasts attempting to disguise this action after a rather poor, unstable landing. The righting effect would be greater if the arms had also been rotated in a clockwise

■ Key Point

When leaning, the body can be righted only if angular acceleration of segments increases their angular momentum at a rate greater than the rate at which gravity increases the toppling angular momentum.

direction in figure 4.10. This action is yet another example of one that appears to be hardwired into our system, or at least learned without instruction at an early age. Try a gentle push on one of your nonlitigious friends and see what happens. But choose your friends carefully and make sure that they know which way to rotate their arms.

Variations of Toppling

People voluntarily put themselves in many situations, usually of an acrobatic nature, in which toppling avoidance is of paramount importance. If performing a handstand on the floor is difficult, it might be assumed that performing a handstand on gymnastics rings would be more difficult. It certainly is difficult to maintain perfect stability with no movement of the body or ring cables. However, a little movement of the hands makes it surprisingly easy to avoid toppling, for several reasons. Should the gymnast tend to topple in a specific direction, the muscles of the shoulder joint can be used to move the hands in the same direction. In this way the base of support is easily placed under the vertical through the CM. The inertia of the mass of the body above the shoulders is little affected by the small horizontal force causing hand movement. Unfortunately, incorrect judgment of the required force can lead to toppling in the other direction and subsequently to large swinging movements. This situation is similar to balancing a pencil on one finger and moving the finger when toppling begins. Toppling avoidance obviously cannot be performed in this manner when the hands are firmly fixed to the floor as in a normal handstand.

So far we have considered avoidance of toppling due to gravity. Toppling can also occur as a result of other external forces that can vary both in magnitude and in direction. The avoidance of toppling caused by such forces can be summarized simply. Orient the base of support and the position of the CM in such a way that the greatest amount of work is required of the toppling force. This simply means elongation of the base of support in the direction of the oncoming force, which therefore must be applied over as great a distance as possible. Also lower the CM so that the toppling force has to do the greatest amount of work in increasing the potential energy of the body. It is often purported that smaller individuals are more stable than their taller fellows. The explanation given is that those who are smaller have a lower center of gravity. This explanation is only partly true. In fact the taller person has longer lower limbs and can therefore achieve a greater area of base of support than the smaller person. Given the degree to which the human body can change the height of its CM and the size of its base of support, there is no evidence that a squat person is more stable than a tall person. Toppling avoidance depends more on the movements made than on the height of the human body.

Using angular momentum of the arms aids in balance.

Toppling from the end of a high board is a frequent occurrence in diving. It is included here as a means of both clearing up a frequently mistaken observation and introducing the concepts of local and remote angular momentum. In rare cases, a toppling diver begins in a handstand position and maintains a straight body position throughout the remainder of the dive. Immediately prior to leaving the board, the diver has angular momentum that is maintained subsequently, since leaving the board does not add any further moment to the system. Yet many observers claim that the diver's angular velocity increases because the position of the axis of rotation changes from one through the board to one through the CM. While it is true that the axis of rotation changes its position, the angular velocity does not increase, for the following reasons. When calculating angular momentum, we do so with reference to some fixed or nonaccelerating axis of rotation. Using the end of the diving board as our fixed axis, we calculate an angular momentum as $(I_G + mr^2)\omega$. Remember that the moment of inertia term derives from the parallel axis theorem. This can be split into two parts:

$$I_G\,\omega = \text{Local (to the CM) angular momentum}$$

$$mr^2\,\omega = \text{Remote (from the CM) angular momentum}$$

■ **Key Point**

Local angular momentum of a body segment is that calculated about an axis through the segmental CM. Remote angular momentum is that due to the CM's having a velocity that is not directed through some remote point.

Since there is no moment while one is airborne, the local term remains constant; and if the body configuration is maintained, the moment of inertia remains unchanged, so the angular velocity must be constant. What does change is the remote term because there is a moment created by gravity with respect to the axis at the end of the board. You can verify the constancy of angular velocity by drawing lines along the length of the body at equal time intervals on film or video images of a diver who maintains a straight body position before and after leaving the board. As you will see identical changes in the angle of the line at each time interval, you will conclude that the angular velocity is constant.

The rider of a unicycle is supported essentially on one point where the tire contacts the ground. Therefore toppling can occur in any direction. Toppling in the frontal plane can be avoided by lateral movements of the arms as described previously. The dangerous motion is toppling in the sagittal plane. Any person who has attempted to ride a unicycle knows how quickly such toppling occurs. Fortunately there is a direct mechanical connection between the muscles of the lower limb and the ground through the pedals. The analysis of this motion is given in the example at the end of this chapter.

Enhancement of Toppling Avoidance

From the preceding analysis it should be clear that strong shoulder and hip flexor muscles enhance our ability to avoid toppling. An activity that is inherently more dangerous than simply toppling onto our hands is shown in figure 4.11.

The advantage of using a pole is due to its large moment of inertia. Not only does the balance pole add mass to the system; it also distributes the mass away from the axis of rotation between the hands. Such a large moment of inertia, when multiplied by a relatively small adjusting angular velocity, yields a large angular momentum. Therefore the angular momentum possessed by the toppling tight-

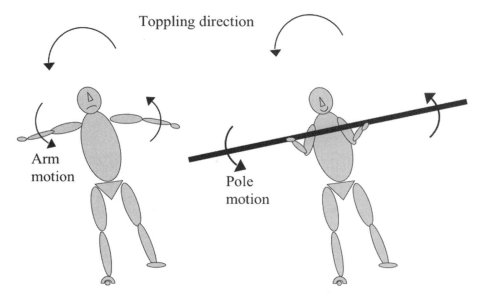

FIGURE 4.11 The increased moment of inertia due to a pole allows the creation of an increased angular momentum to counteract toppling.

rope walker can be transferred more easily to the pole than to the performer's arms alone, with much less angular velocity required of the pole (see West and Griffith, 2004). Some tightrope walkers in the circus use an umbrella instead of a long pole. The mass and moment of inertia of the umbrella are smaller than those of the pole, but the umbrella helps to maintain balance in a different manner. The umbrella has considerable resistance to movement through the air. When it is held overhead and moved sideways, the transverse force produced has an equal and opposite effect at the shoulder, as if another person were attempting to keep someone upright by pushing the individual. With this technique angular momentum considerations are minimized, but force due to air resistance is maximized.

Toppling avoidance can be enhanced by movement. If a person doing a handstand begins to topple, one hand can be moved in the toppling direction so the base of support is elongated in that direction, and further toppling can be arrested. Unfortunately this represents failure if the requirement of the activity is to perform a stable handstand. Therefore the technique used to avoid toppling depends on the nature and aim of the activity.

Toppling Safety

Safety in avoiding toppling, apart from staying away from tall buildings, depends on a number of factors. It is necessary to be aware constantly of the magnitude and direction of potential external forces. The base of support should be oriented with its greatest dimension in the direction of expected external force. Footwear and the floor surface should provide sufficient frictional resistance. Ensure that a working floor is clear of obstacles, since toppling due to tripping leaves the contact foot behind the CM of the body. In this case extra time is required to restore that foot to a position in front of the CM. When

■ Key Point

Avoidance of toppling is made easier if a large mass can be moved, whether the movement is translational or rotational.

one is pushing objects, one should test their resistance to motion so that they do not move away rapidly. Free segments such as the arms should be rotated in the direction of toppling in order to transfer the whole-body angular momentum in the arms. Restoration of balance can be achieved in this manner only if the arms are accelerated.

Practical Example 4.3

Figure 4.12 shows a stationary unicyclist instantaneously leaning forward at an angle υ to the vertical.

Part 1

Describe how the angle θ can be maintained at a constant value.

Solution

Gravity creates a clockwise moment Wd about a horizontal axis through the axle of the wheel. Therefore an equal and opposite moment must be produced by the force of the feet on the pedals. In this way there will be no net moment to change the angle θ.

Part 2

Calculate the horizontal velocity of the unicyclist after 2 s given the following constants:

m	=	80 kg
r	=	0.3 m
l	=	1.0 m
θ	=	5.0°
υ	=	?

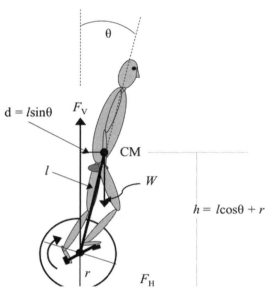

FIGURE 4.12 Free body diagram of forces acting on a unicyclist.

Solution

Let us approach the problem by asking a question. Why should the unicyclist gain horizontal velocity? This can occur only through acceleration. Therefore there must be a horizontal force (F_H). Such a frictional force is a reaction to the tire tending to push backward on the ground due to pedaling in a clockwise direction. While this explanation facilitates an approach to biomechanical analysis, it does not give us the answer. We need some equations. The sum of the moments must equal zero if the angle θ is to be maintained constant. We can resolve the moments about any convenient axis. As W creates no moment about the CM, an axis through the CM is preferred.

> *Clockwise moment = Counterclockwise moment*
> $$F_V \times d = F_H \times h = F_H(l\cos\theta + r)$$

(Equation 4.6)

Rearranging gives F_H, which is required for calculating horizontal acceleration.

$$F_H = F_V\, l\sin\theta\, / \,(l\cos\theta + r)$$

Since there is no vertical acceleration, $F_V = W$ and we obtain

$F_H = W \, l\sin \theta \, / \, (l\cos \theta + r)$;

$F_H = 80 \times 9.81 \times 1.0\sin 5° \, / \, (1.0\cos 5° + 0.3)$;

$F_H = 52.77$ N.

And since $a = F / m$, a becomes $52.77 / 80 = 0.66$ m/s². As the horizontal acceleration is constant, we can simply multiply it by time to obtain the change in velocity (v) from its initial value of zero. Therefore,

$v = 0.66 \times 2 = 1.32$ m/s.

This value is similar to a fast walking pace, but if the current conditions are maintained, the horizontal velocity will become very large. This increase will be accompanied by an increase in angular velocity of the wheel as the unicyclist attempts to maintain a constant moment through the pedals. Such conditions cannot be maintained because the faster the muscles shorten, the smaller the force they can produce. This biomechanical limitation requires the unicyclist to produce a large moment as early as possible in order to produce a counterclockwise rotation with a consequent vertical orientation. The advantage the unicyclist has is that backward toppling can also be corrected as the pedals are fixed directly to the axle, with no freewheeling capacity. If you watch motorcycle racing, observe how long the winner can celebrate by raising the front wheel off the ground. You will appreciate that the motorcycle engine gives a mechanical limitation in much the same manner as our muscle engines do.

Part 3

What would be the type of arm motion required to maintain the current angle of lean of the unicyclist shown in figure 4.13 without horizontal motion of the CM occurring?

The wheel of mass m_W and radius of gyration k replace the arms.

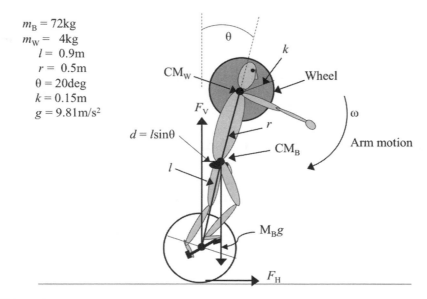

$m_B = 72$kg
$m_W = 4$kg
$l = 0.9$m
$r = 0.5$m
$\theta = 20$deg
$k = 0.15$m
$g = 9.81$m/s²

FIGURE 4.13 Arm motion required to maintain the posture shown without horizontal motion of the CM.

Solution

For the body to remain stationary, there must be no net gain in angular momentum of the system. Therefore the potential gain in angular momentum due to gravitation moment must equal that created at the same rate by the arms (wheel) in the same direction, as follows:

$$m_B gl\sin\theta + m_W g(l + r)\sin\theta = m_W k^2(d\omega / dt);$$

$$(d\omega / dt) = (m_B gl\sin\theta + m_W g(l + r)\sin\theta) / m_W k^2$$

And since $d\omega / dt = \alpha$, the following angular acceleration of the arms is required:

$$\alpha = [72 \times 9.81 \times 0.9 \times \sin20 + 4 \times 9.81 \times (0.9 + 0.5)\sin20] / 4 \times 0.15^2;$$

$$\alpha = 2625 \text{ rad/s}^2$$

This is clearly an impossible objective for a human being, because after 1 s the arms would have to be rotating at 2625 rad/s or 417 revolutions per second. The preceding equations indicate that success could be obtained only for very small angles of θ, very large arm mass and moment of inertia, or very tiny people with huge arms and lots of shoulder muscle. Even if the angle θ was only 2°, an angular acceleration of 268 rad/s², giving 268 rad/s or 42 revolutions per second after 1 s, would be required. The solution to avoidance of falling forward is best obtained as shown in Parts 1 and 2 of this question rather than in this last part.

Summary

Balancing on two feet is not a particularly precarious activity for intact humans, and it is largely dependent on the dimensions of the base of support. Balance is enhanced by increasing the size of the base of support in the direction of the disturbing force. Balancing becomes more precarious when the area of the base is reduced, as in a handstand. Balance is also enhanced by reducing the vertical height of the CM. However, maintaining a crouched position is costly in terms of muscular energy requirements due to the force–length relationship of muscle and the increased joint moments when the limbs are in a flexed posture. Balancing is also precarious when a load of unknown magnitude is lifted rapidly, and more so when the nature of the supporting surface provides small frictional resistance. Neuromuscular dysfunction also interrupts balance because of the difficulty in producing appropriate patterns of force that keep the CM vertically above the base of support.

Toppling is simple; avoid restoring forces, and gravity will do the rest. Avoidance of toppling depends on the nature of the disturbing force. One can offset angular velocity of toppling by transferring the toppling angular momentum to movable segments such as the arms. An angle of lean can be maintained constant only if the angular momentum in other segments increases continually, as gravity is constantly increasing the toppling moment. Restoration of the body to balance above the base of support requires generation of angular momentum in body segments at a rate greater than that generated by gravity. One can also explain this phenomenon by saying that the muscular work is being done at a greater rate than the work done by gravity. Toppling can also be avoided by appropriate movement that changes the size of the base of support at the foot–ground interface. Toppling and its avoidance can occur only in the presence of sufficient friction at the foot–ground interface. If friction is insufficient we experience slipping, which is more dangerous and is the subject of a separate chapter.

RECOMMENDED READINGS

The universal nature of standing, balance, and toppling generally confines literature on this subject to specific areas of study. Readers are encouraged to search the literature in their particular area for information on activities that are important within that area.

West, B.J., and Griffith, I. (2004). *Biodynamics: Why the wirewalker doesn't fall.* New York: Wiley-Liss.

CHAPTER 5

Slipping, Falling, and Landing

S lipping, falling, and landing represent a biomechanical continuum inasmuch as slipping causes falling, which requires landing. While slipping is largely an accidental occurrence, many falls and landings are also intentional, particularly in gymnastics.

Slipping is usually accidental, and it accounts for a large proportion of injuries to humans. Slipping occurs when resisting force is insufficient to keep two surfaces from sliding relative to each other. Although we generally think of slipping as a lack of resistance to motion of the foot on the floor, slipping can equally well mean loss of handgrip and general relative motion between any body surface and an external surface. As the conditions for slipping are mechanically straightforward, the following section deals primarily with slipping avoidance.

Aim of Slipping Avoidance

The aim in slipping avoidance is to ensure that there is sufficient frictional resistance between two surfaces. Figure 5.1 shows the situation of foot–ground contact in which the aim is to move the center of mass (CM) of the body to the right from a stationary position. Clearly a force is required to accelerate the body to the right, and this force arises from the tendency of the foot to slip to the left. If slipping does not occur, the CM will be accelerated to the right. The force N is known as the normal force, being perpendicular to the surface, and it is equal to mg in the current case. Should there be upward acceleration of the CM, N will be greater than mg.

Mechanics of Slipping

In the static case, there is a coefficient of limiting friction (μ). The greater this value, the greater will be the frictional force prior to slipping. If the force F rises sufficiently for the ratio F/N to exceed the coefficient of friction, slipping will occur when

$F/N = \mu$ = coefficient of limiting friction

$$F / N > \mu. \qquad \text{(Equation 5.1)}$$

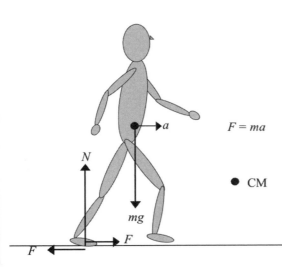

FIGURE 5.1 Slipping occurs when the ratio $F/N > \mu$, the coefficient of limiting friction.

For horizontal motion, N is equal to mg, and it is known as the normal force since it is normal or perpendicular to the surfaces in contact. The coefficient of friction depends only on the nature of the surfaces in contact and not, for example, on the area of contact. If some material is located between the surfaces in question, then the ratio F/N depends on the separate coefficients between each surface and the intervening material.

Sliding friction or kinetic friction is somewhat more complicated than static friction. In general, the kinetic coefficient of friction is less than its static counterpart. For surfaces to slide over each other, work must be done, and a consequence is modification of the molecular interaction between the materials composing the surfaces. The speed of sliding also affects the frictional resistance during movement. Intervening materials, whether solid or liquid, also affect the coefficient of kinetic friction. This is the basis of lubrication whereby synovial fluid separates the two hyaline cartilage surfaces of a movable human joint.

Biomechanics of Slipping

Friction costs us metabolic energy, but it is necessary in order for us to move in any other than a vertical direction. In other words, friction represents a means by which horizontal forces can be produced to enable us to move. Slipping occurs when there is too great a horizontal force F so that F/N tends to become greater than μ. Therefore we are more likely to slip when applying a large horizontal force F because the ratio F/N is large and may exceed μ. The task of lifting a box vertically is less prone to slippage than is pushing a heavy cart horizontally. The gradient of a slope on which one is standing is also pertinent to slipping, since the ratio F/N is considered relative to the surface, not to the vertical. Figure 5.2 shows how the ratio F/N is calculated for the situation of standing on a sloping surface.

There are only two external forces acting on the person: mg due to gravity and a vertical reaction force equal to mg at the foot–ground interface, both acting through the CM. Therefore there is neither translational nor angular acceleration of the CM. The force at the foot–ground interface can be resolved into components perpendicular and parallel to the surface. The perpendicular component is the normal force N, and F represents the frictional component parallel to the surface. The ratio F/N at the instant of slippage therefore becomes

$$F / N = mg\sin\theta / mg\cos\theta = \sin\theta / \cos\theta = \tan\theta = \mu.$$

So no slipping will occur when one is standing on a slope if the ratio F/N is less than the coefficient of limiting friction. It is no surprise that increasing the slope increases the chance that F/N is equal to or greater than μ. However, there is a question of the sensitivity of slipping to change in slope angle, since $\tan\theta$ is nonlinearly related to μ. For example, a change in θ from $0°$ to $5°$ requires the value of μ to increase by 0.09; from $20°$ to $25°$ μ must increase by 0.1; from $60°$ to $65°$ μ must increase by 0.41. When μ is small, as in the case of standing on ice or snow, a very small value of θ is all that is necessary to induce slipping. A rather unexpected occurrence is when prolonged standing on ice or snow produces pressure that raises the freezing point of water; the snow therefore melts, μ decreases, and our feet slip from beneath us, causing confusion in the ski lift lineup.

Even during walking, the chances of slipping are ever present, since the ratio F/N

■ Key Point

Force parallel to a surface divided by force perpendicular to the surface when slipping occurs is known as the coefficient of limiting friction μ.

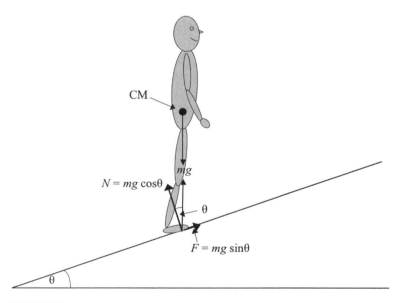

FIGURE 5.2 Parallel F and perpendicular N components of gravitational force.

Falling is made safe by landing on a compliant surface.

■ Key Point

The value of μ varies with the molecular properties of surfaces in contact and with any intervening lubricant, but does not vary with area of contact between surfaces.

varies throughout the stance phase (see figure 6.3 in chapter 6). The ratio F/N is particularly variable at heel strike and at toe-off, which common experience indicates are danger points in the stance phase.

Variations of Slipping

Slipping can also occur between the hand and the surface of something that is being held. A pole-vaulter holds the pole overhead with no resistance to its horizontal motion at one instant, and at the next instant the motion of the pole is arrested severely when the pole is placed in the box in the floor. A very large force results between the hands and the pole in a longitudinal direction. Vaulters use a strategy of wrapping the pole in tape and covering the wrapping with a fluid that increases the coefficient of friction between the hand and the pole. Gymnasts who perform on various pieces of apparatus adopt a similar strategy, using a form of chalk on the hands plus a sheet of leather affixed to the palms. In almost all athletic activities that use gripping, some form of material intervenes between the hands and the apparatus. An unfortunate consequence of increasing friction in this way is that the skin of the palm is subject to shearing stress. The calluses on gymnasts' hands represent the body's way of building up tissue to withstand such stress.

A problem with sports such as tennis, squash, badminton, and golf is perspiration in the palms of the hands. The implement cannot be affixed rigidly to the hand, as adjustments are required for different types of strokes. In the past, leather grips were the traditional form of covering for a racket, but leather has largely been replaced with synthetic material that can help with perspiration in racket sports and with rain in golf.

Enhancement of Slipping Avoidance

In this section enhancement refers to enhancing our chances of not slipping. One solution is to ensure that the ratio F/N does not exceed μ. This can be achieved in two ways: by reducing the force F parallel to the slipping surface or by increasing the force N perpendicular to this surface. Reducing F may result in poor human performance in many activities. In sports, maximizing F is usually the major aim. In industry, limiting F may reduce productivity by reducing the overall speed of movement. Alternatively, for people in older age groups, who may suffer extreme injury as a result of falling, reducing F is essential. They achieve this by reducing walking speed through reduced step length. The biomechanical reason for short step lengths is that the angle of inclination of the lower limb at heel strike is close to perpendicular to the slipping surface. This posture combined with a small horizontal velocity requires a small value of F in comparison with N.

Another way of reducing slipping is to use footwear that has an appropriate value of μ when interacting with the surface upon which one is moving. The changes in sport shoes that have taken place over the last 40 years are partly a testament to this need. This strategy is appropriate during moving on a plane, unyielding surface. If sports are performed on a yielding surface such as grass, the slipping conditions change. One reason is that grass contains water, which is squeezed out of the leaf when pressure is applied. The water provides a lubricating medium that reduces μ. The solution here is not to rely on poor friction, but to apply cleats (in North America) or studs (in Britain) to the plantar surface of the sport shoe. The sliding motion of the shoe on the grass is therefore arrested by a force perpendicular to the studs and parallel to the surface. Slipping is easily stopped with this strategy except when very soft ground cannot resist the force perpendicular to the studs.

Slipping Safety

One feature of the use of friction is that for a given coefficient of friction, the greater the body weight mg, the greater can be the frictional force prior to slipping. Such an observation argues for paying greater attention to the frictional characteristics of footwear worn by less massive individuals.

The nature of the surface upon which motion is taking place is paramount to safety. Measurement of the coefficient of limiting friction on a wooden squash floor demonstrated that almost any type of running shoe was adequate in providing the frictional requirements for the game (Chapman et al., 1991). Safety was not compromised when the floor was wetted to simulate drops of perspiration, provided that the floor was not sealed with an impermeable paint. However, a serious reduction in μ was observed when dust was scattered on the surface. This information further emphasizes the requirement that the surface must be well maintained and cleaned of any substance not intended to be present.

■ **Key Point**

Different activities require different frictional characteristics, which can be varied with the use of appropriate covering of the surfaces such as glove material and paints.

■ **Key Point**

Many sporting activities benefit from enhancement of friction, particularly during throwing, catching, and swinging on gymnastics apparatus.

Unfortunately, too high a coefficient of friction may also be injurious. Experimental evidence in support of this statement will always be difficult if not unethical to obtain. We must resort to biomechanics for our suggestions and assume that sliding is sometimes better than maintaining rigidity of the foot on the surface. The problem is simply that a high decelerating force may lead to high shear forces at the ankle joint unless the foot slips. In this case, disruption of ligaments and even articular cartilage and bone may occur.

Where manual handling is the nature of the task, people should use gloves. Gloves serve the purpose of giving a predictable coefficient of friction and also protection from skin burns should slipping occur.

Practical Example 5.1

A hospital worker of mass $m = 80$ kg (215 lb) is pushing a cart of mass 60 kg (160 lb) at a constant speed up a ramp of slope $\theta = 25°$ as shown in figure 5.3. The rolling resistance of the cart at this speed is $F_C = 20$ N.

Part 1

If both feet are in contact with the ramp and sharing body weight equally, what is the minimal value of the coefficient of friction μ required to avoid either foot slipping?

Solution

To calculate μ, we require the forces parallel to the ramp (F_1 and F_2) and their respective normal forces (N_1 and N_2). Since the force parallel to the slope due to gravity on the worker is shared equally, F_1 and F_2 are equal, as are N_1 and N_2. $F_1 + F_2$ equals the total opposing force, as there is no acceleration, as follows:

$$F_1 + F_2 = mg\sin\theta + W\sin\theta + F_C;$$
$$F_1 = F_2 = (mg\sin\theta + W\sin\theta + 20N) / 2$$

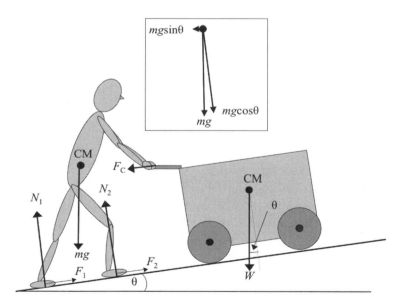

FIGURE 5.3 Forces applied in pushing a cart uphill.

Since the normal force on the worker due to gravity is shared equally,

$N_1 = N_2 = (mg\cos\theta) / 2.$

For either foot not to slip,

$F_1 / N_1 = F_2 / N_2 < \mu.$

The minimal coefficient must then be

$\mu = (mg\sin\theta + W\sin\theta + 20N) / (mg\cos\theta);$

$\mu = [(80 \times 9.81 \times 0.423) + (60 \times 9.81 \times 0.423) + 20] / [80 \times 9.81 \times 0.906];$

$\mu = 0.817$

Part 2

If the person lifts one foot while progressing up the slope, will slipping occur with the present coefficient of friction?

Solution

Since the entire load is borne by one foot, the ratio F/N on that foot will remain the same as before and the person will not slip.

Part 3

If the worker presses down or pulls up on the handle of the cart, will slipping be either enhanced or diminished?

Solution

In the previous case, the resisting force was assumed to be parallel to the ramp surface. Pressing down on the handle creates an upward force that partly supports the gravitational force on the worker. Consequently N will be reduced and the ratio F/N may exceed μ provided that F is unaltered. Pulling up on the cart will have the opposite effect. The present effect need not be conscious, as a changed angle of lean on the part of the worker will affect the direction of the pushing force.

This problem shows that there are numerous ways of pushing a cart and that a high value of μ is no guarantee of slipping avoidance. However, some simple measurements of pushing in different ways should provide a range of values for μ that are acceptable for most situations. It is also apparent that the gradient of the slope plays a large part in the likelihood of slipping, with steeper slopes requiring greater values of the coefficient of static and dynamic friction.

Aim of Falling and Landing

Falling is the easiest of our maneuvers. All that is required is to remove support fully or partially, and gravity does the rest. Falling is usually accidental in normal daily maneuvers, and it is the result of a mistake that has mechanical consequences. Our mistakes include not seeing either sudden drops or rises in the level of the floor and mistaking a lower frictional surface for one that is normal in friction. Other agencies inducing falling are forces from very high winds or from nasty people and automobiles (very nasty). Loss of consciousness is another factor. Fortunately, some falling maneuvers are intentional in many sports. The reasons for falling are numerous and varied.

The aim in falling and landing is to position the body so that landing can occur safely. In routine daily activity, this means landing on the feet. In some sports it means landing on some part of the body other than the feet. In all cases, the aim is to dissipate the body's kinetic energy so that body tissues are not subject to energy failure levels.

Mechanics of Falling and Landing

When a body falls from a stationary position at a height h above some convenient level (usually the ground), the potential energy mgh that it had initially is lost and is converted to kinetic energy $mv^2/2$ immediately prior to ground contact. This kinetic energy is then dissipated by an upward force from the ground representing work done $\int Fds$ to bring the body to rest as follows:

$$PE \text{ prior to fall} = KE \text{ prior to impact} = \text{Work done on body}$$
$$mgh = mv^2/2 = \int Fds$$

(Equation 5.2)

So the work required is directly related to the height of the body above the ground because m and g are constants. The work–energy approach is used in preference to that of impulse–momentum because the height through which to fall is more easily appreciated than the time to fall. When the body contacts the ground, generally there is no displacement of the point of application of the force at the ground. This would imply that the force does no work to reduce the kinetic energy, since work is the product of force and displacement. If the body is a perfectly rigid mass landing on rigid floor, microamounts of displacement will occur, probably leading to sound energy and permanent damage to the body and floor. Heat is also a by-product of such contact. In a segmented body there will be intersegmental forces applied through their individual displacements. It is the work done by these forces $\int Fds$ that dissipates the energy in human landing.

Falling can also include a horizontal component of velocity of the CM, which is unchanged while one is airborne unless the velocity is so great that air resistance becomes a decelerating force. In this case the events during landing depend upon the direction of the path of the CM at impact, the frictional characteristics of the landing surface, and the position of the CM relative to the point of impact on the extended body. We consider these factors, along with the effect of rotation at impact, in the following section on biomechanics of the fall and the landing of a segmented human body.

Biomechanics of Falling and Landing

The segmented nature of the human body allows some flexibility in landing posture as shown in figure 5.4. Here two configurations of the body are shown immediately prior to landing.

The configuration in figure 5.4*a* clearly has not allowed the CM to fall as far as in figure 5.4*b*. The kinetic energy of the body is therefore less in figure 5.4*a* than in figure 5.4*b* at touchdown. The strategy adopted in figure 5.4*a* also allows a greater

■ Key Point

Arresting a fall requires work to be done by a force acting in a direction opposite to the direction of the fall.

displacement s of the CM downward before the fully flexed posture of the body is reached. As work is ∫Fds, a greater allowable value of s results in a lower average force for a given amount of kinetic energy to dissipate as is seen in the left lower panel. Here the gray-shaded areas represent ∫Fds or the work done. The displacement of the CM s is achieved by motion of the segments of the body. Since there are many segments, there are many combinations of intersegmental motion that will have the same effect on landing. However, downward displacement of the CM of the trunk segment by means of knee and hip flexion will be more effective than an equal downward displacement of the CM of the arms. The reason is simply that the greater mass of the trunk contributes to a greater change in position of the whole-body CM than does the smaller mass of the arms. It should be stressed that landing is an eccentric muscular activity that results in motion of the CM in the direction opposite to that of the landing force.

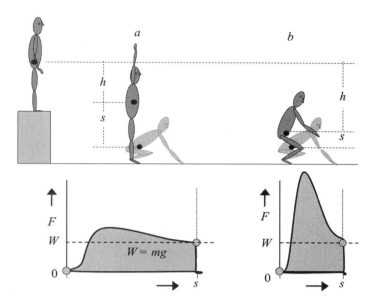

FIGURE 5.4 Increasing the displacement of the CM on landing reduces the peak force.

While landing can be analyzed biomechanically and studied from the point of view of energy dissipation, the ability to adopt a landing strategy requires knowledge of the presence of falling. Reaction in the form of developing muscular force takes time. Time is required for the body to sense instability by means of the sensory organs of balance. Further time is needed to determine the nature of the fall and therefore which muscles to activate. Muscular activation also takes time, as does the rise of force to a protective level. Sometimes it is better to fall a longer than a shorter distance. While this seems counterintuitive, the statement is justified because the longer distance allows us time to react and call upon muscular preactivation. Falling a shorter distance onto a sharp object can produce injury before we are able to avoid it. Anyone who has stepped into an unseen depression in the ground or who has miscounted the number of stairs while descending will have experienced the effect of being unprepared in a muscular sense. The hardwired nervous system is expecting mechanical conditions that require a well-known pattern of muscular activity. The small drop gives us insufficient time to modify this pattern, and the shock can be felt traveling up the body. Alternatively, a very long falling distance allows us plenty of time to react but may result in more kinetic energy than we are able to dissipate by our muscles.

In some purposeful activities, falling is intentional. Gymnastics and judo are typical examples. In gymnastics, falling (or dismount) from a height ends routines such as those seen on the horizontal bar, asymmetrical bars, vault, rings, balance beam, pommel horse, and aerial maneuvers in tumbling. The aim is to land in a stationary posture without tripping. The problems that the human machine faces are twofold. The first is to dissipate safely the kinetic energy just prior to landing (dissipation is frequently known as absorption). The second is to arrest both linear and rotational motion.

■ **Key Point**

As all body segments can be displaced, many muscles can share the workload of landing.

■ **Key Point**

Both translational and rotational work need to be done to arrest the fall of a rotating body.

Variations of Falling and Landing

Landing from a fall that has vertical and horizontal components of velocity presents the individual with a more complex problem than has been considered so far. Figure 5.5 shows vectors (*v*) that are parallel and of equal magnitude and that represent the velocities of the CM at touchdown with three orientations of the body.

At the instant of touchdown, each of the bodies has the same translational momentum (*mv*) but different angular momentum (*mvr*) with reference to the point of contact with the landing surface. The change in analytical approach from energy to momentum is made necessary because we need to consider rotational direction. Energy analysis precludes this aim since energy is a scalar variable without direction. Angular momentum is a vector variable that has direction. From right to left, the figures have clockwise, zero, and counterclockwise angular momentum. A frictional force (*F*$_F$) and a normal force (*N*) will be induced by the kinetic conditions present at the instant of contact and the intersegmental motion during contact. The exact outcome of each landing is unknown, as it will depend on the magnitudes of *N* and *F*$_F$, which will change with time depending upon intersegmental motion. Such motion is in turn dependent on the sequence of intersegmental muscular contraction.

Since energy dissipation is the aim, the probable motion following contact will be toward a crouched position as discussed in relation to figure 5.4. The forces *N* and *F*$_F$ combine to produce a force vector that, if it acts directly through the CM, will have no effect on the angular motion of the body. In this case, the light-gray figure will fall backward in clockwise rotation, the medium-gray figure will not rotate at all, and the dark-gray figure will fall flat on its face. If the frictional force dominates over the normal force, the possible counterclockwise angular impulse created may counteract the initial clockwise angular momentum of the light-gray figure, thus giving a safe landing. The same force effect for the medium-gray figure will create counterclockwise rotation, and the dark-gray figure will increase its original angular momentum. Again it must be stressed that the actual outcome of the landing will depend upon how the muscles are used to produce various intersegmental motions. Of equal importance is the possibility of a subsequent takeoff and further airborne motion following the landing. For example, the dark-gray figure would be in a good position to perform a forward somersault with its current counterclockwise rotation unaffected if only a brief normal force acted during landing. This would require little or no frictional force during the initial landing.

■ **Key Point**

Friction plays a large part in successful landing when horizontal motion is involved.

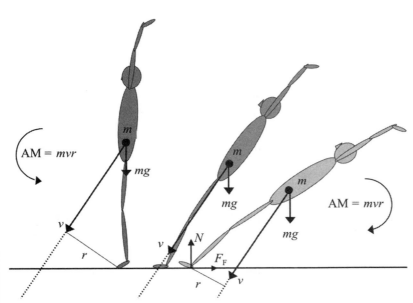

FIGURE 5.5 The direction of the velocity vector in relation to the axis of rotation determines the initial angular momentum.

Enhancement and Safety of Falling and Landing

Enhancement and safety are combined in this chapter since we are attempting to enhance the likelihood of landing safely. Safety refers to obviating stress in the body tissues that is sufficient to produce either temporary or permanent damage.

As discussed previously, damage occurs as a result of exceeding the ultimate stress in the tissues, which we can translate as too high a landing force. Damage can also occur even when the ultimate stress is not exceeded if the present stress is sustained for too long a period of time. How can these two possibilities be obviated? Eccentric muscular work is required in the form of application of a force in a direction opposite to the movement. Our capacity for landing is greatly enhanced by virtue of increasing muscular force as muscle lengthening speed increases (see figure 3.3c in chapter 3). Therefore one can achieve dissipation of a fixed amount of energy by varying both muscle lengthening and muscle force. For example, it appears to be preferable to use a small muscle force multiplied by a large amount of lengthening to produce the effect shown in the lower left of figure 5.4. Yet the use of a large force over a small displacement cannot be discounted provided that the duration is very small. The choice between these strategies cannot be guessed at; it can be deduced only from knowledge of the properties of the tissues involved. The preference for the strategy utilizing low force and large displacement is further supported by the fact that the low muscle force reduces compressive stress in adjacent tissues such as bone and articular cartilage.

Events following the landing will also determine the nature of the landing. If a gymnast is required to follow the landing with another upward jump, the short-duration landing will be preferable for reasons having to do with the stretch–shortening cycle of muscular contraction. We know that high muscular force is produced during the stretch-shortening cycle, which is why it is used in rebound types of activities. As the resulting ultimate stress level of tissues may be induced, it is not surprising that most injuries occur in this situation. In the situation in which safety and not elegance of performance is the requirement, figure 5.4 shows how the muscles of the arms may be used if the hands are allowed to touch the floor, thus contributing to the arresting force. If elegance of landing is primary, the landing using high force and small displacement will win, as the landing will finish in a more erect posture.

Biomechanically there is no ideal solution to safety in landing in gymnastics, as the pursuit of elegance pushes us further toward tissue damage. One possible strategy for gymnasts is to opt for landing with deep knee bends (large muscle lengthening) during training so that they can repeat other aspects of the performance many times. Whether such "practice landing" conflicts with the "performance landing" is a question for gymnastics coaches. Landing with the upper limbs above the head followed by their purposeful downward motion allows greater vertical displacement of the CM. In this way the shoulder adductor muscles are contributing to $\int Fds$ and thus reducing the peak force at the feet. The alternative might be a very erect landing followed by a lengthy and painful period of lack of practice due to healing of a broken leg. Another strategy is to land on a very compliant surface such as foam rubber during practice. The foam rubber

■ Key Point

It is safest to allow a large displacement of the CM when arresting a fall.

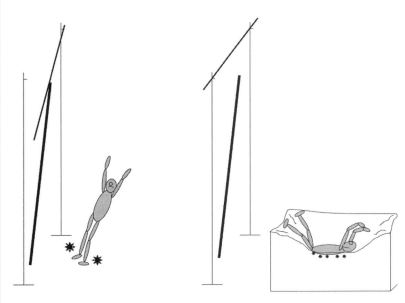

FIGURE 5.6 Compliance and area of the landing surface determine the likelihood of injury.

■ **Key Point**

Compliant energy-dissipating materials located between the person and the ground obviate the need for large amounts of muscular work to be performed on landing.

■ **Key Point**

Ground irregularities produce small landing areas, which lead to high tissue pressures.

increases the d*s* part of ∫*F*d*s*. Mechanically, mats are compliant objects that deform in response to force. The force multiplied by mat deformation represents dissipation of energy. Placing the mats between the feet and the hard ground therefore results in an extra energy-dissipating mechanism that reduces the muscular contribution to energy dissipation. The biggest landing mat is found in the sport of pole-vaulting, as shown in figure 5.6.

Not only do the best pole-vaulters fall from about 6 m (near 20 ft), but the nature of the event sometimes makes the body position on landing uncertain (particularly if the pole breaks). Usually body rotation is needed to get over the bar, and the subsequent landing is on the back. Therefore the landing mat is a huge block of compliant material, usually some type of foam rubber, that can deform over a great depth. As this occurs, the peak force of landing is small and can be easily handled by back structures. The back has a large surface area. As pressure is force divided by area, a large surface area means that any given part of the back is subject to small pressure. In the absence of pole-vault landing mats, vaulters must land on their feet. In this case, ankles are particularly subject to injury as can be attested to by your author, who vaulted in the time when landing mats were actually sandpits. Bedded-down sand is very incompliant.

So far we have considered landing on a plane surface. In the event that the surface is not a plane, the question of pressure becomes important. The following equation relates force *F*, pressure *p*, and area *A* of the contact surface.

$$p = F \: / \: A$$ (Equation 5.3)

All tissues can sustain a certain amount of pressure without damage. When the surface area of contact is small, the pressure can rise to too high a level, as occurs when we step on an upturned nail. It is clear that any attempt to increase the surface area of contact can be beneficial to safety. Under normal conditions we can walk on a plane surface in our bare feet. When we traverse pebbles, the total area of contact is greatly reduced, leading to local areas of high pressure. A similar effect occurs with even a very small piece of grit in your shoe.

When we land, there is usually rotation of the body in addition to downward motion. This would be the case when one is being pushed over a safety rail. Unfortunately, practice is not available prior to accidental falls. In this case we must depend upon luck or the following two strategies, which good school physical education classes should teach. The first is to make the area of contact with the ground as large as possible, as do participants in judo. These athletes dissipate energy by making ground contact with the whole length of the arm as confirmed by the accompanying loud slapping sound. They perform this action vigorously, which further reduces load on other parts of the body. Presuming that most of us are not experts in judo, the other strategy available to us takes the form of conscious rolling. By this means we are constantly changing the area of the body that is applying force to the ground. Force is therefore applied over a body displacement leading to reduction of both translational and rotational kinetic energy. The message is to roll, if you can, when you fall over. Do not attempt to dissipate energy by using the small area of one hand; the resulting high force may break any number of the bones in the wrist region. Figure 5.7 shows an individual with initial translational kinetic energy (TKE = $mv^2 / 2$) just prior to landing.

Due to friction at the feet, the body rotates about a horizontal axis through the feet, thus converting some of the TKE into rotational kinetic energy (RKE = $I_G\omega^2$). The RKE can be used to somersault and roll over the ground following an airborne phase. The successive displacement of the point of application of force will further reduce TKE. Finally eccentric action of the hip and shoulder flexors can further reduce the RKE.

Dissipation of energy in landing is best achieved when the maximal number of body parts makes contact sequentially in rolling.

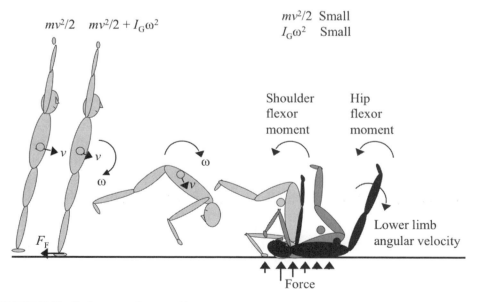

FIGURE 5.7 Safety is enhanced by small ground reaction forces produced sequentially when a person rolls while landing.

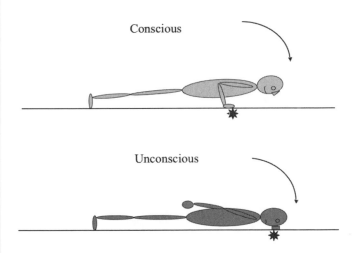

FIGURE 5.8 Possible points of contact during falling in conscious and unconscious states.

■ **Key Point**

Conversion of translational to rotational energy enhances safety by increasing the displacement over which force acts and by decreasing pressure on any single body part.

There is one situation in which none of this advice is useful, and that is when a person falls due to fainting as seen in figure 5.8.

In this situation you cannot see, you are aware of nothing, and you cannot turn on your muscles. The statement "The reason she avoided injury was that she was unconscious and therefore relaxed" is often heard, particularly with reference to sporting falls. This is clearly nonsense, because without the conscious use of muscles, the best energy dissipater in the human machine is ineffective. Other tissues such as bones, ligaments, and even skin and fat have to dissipate the energy, and their ability to do that pales in comparison with the ability of muscle. The major strategy for avoiding serious injury in unconscious falls is luck.

Injuries will occur due to tissue failure in falling. Yet it is better to break your arm when using arm muscles consciously than to fall on your head unconsciously as shown in figure 5.8. This figure is somewhat unrealistic, because muscular activity is required to maintain a straight body configuration. Another fall, shown in figure 5.9, is the result of computer simulation with no moments applied by the muscles.

In fact, a small moment is required to stop the simulated heel from going through the simulated floor (the body is more adaptable than a computer simulation). All segments begin at zero velocity with the CM positioned exactly vertically above the ankle joint. The system simply collapses downward in the absence of moments created by the muscles, but there are forces in the vertical and horizontal directions at the feet due to the inertial force created between segments. As no moments are being applied, the segments have considerable velocity in the lowest position. Clearly there is considerable danger of injury at this point. Other simulations were performed with the CM not balanced exactly above the ankle joint, and the change in configuration during falling was disastrous for the joints. In this case the joint ligaments would be subjected to enormous strains.

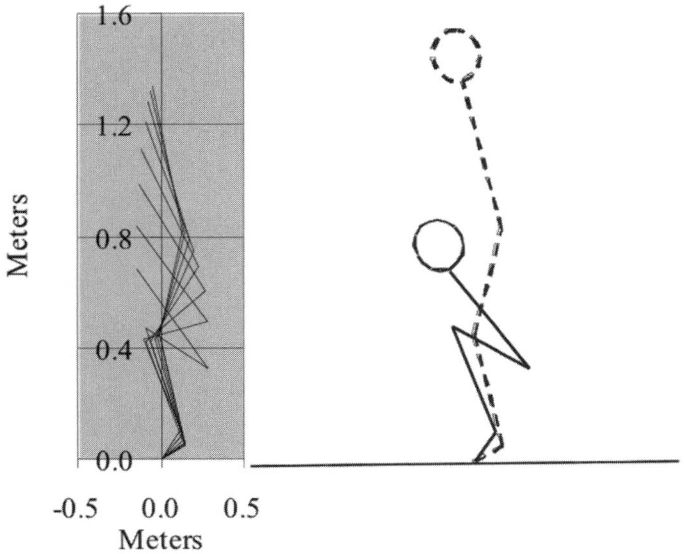

All types of falls are dangerous to persons who are aged, primarily because of low muscular strength and also due to the possibility of osteoporosis. Osteoporosis is a disease that markedly alters the mechanical properties of bones. The consequence is a very low level of energy to failure. What might be an insignificant fall for a small child (children's tissues allow them to bounce) can result in a broken femur in an aged person with osteoporosis. The strategy in this case becomes more one of avoiding the fall than of invoking the body movements required to dissipate energy.

FIGURE 5.9 Simulation of unconscious collapsing when no muscular force is available.

Movement must be made slowly and deliberately, and there should be no surprises awaiting the person. Therefore it is clear that health care facilities for the aged must pay constant attention to environmental conditions such as frictional characteristics of the floor and ensure that stairs and other objects are clearly marked.

■ **Key Point**

Muscle is the best energy dissipater the body has because of its great force and lengthening characteristics.

Practical Example 5.2

A person who is standing erect falls forward to land on the hands as shown in figure 5.10. As the hands contact the floor initially, the arms are straight and the force at the feet is zero.

Part 1

If the body is finally brought to rest after an angular displacement of θ_2, calculate the average force on the hands given the values shown in figure 5.10.

Solution

This is a problem concerning the generation and dissipation of mechanical energy. Gravity produces RKE of the body about a horizontal axis through the feet, and the muscles of the arms work eccentrically to dissipate the RKE. The force vector at the feet is partly gravitational and partly frictional during the fall; but since it is fixed, it acts simply as a constraining force but does no work. The first problem is therefore to calculate the RKE gained about an axis through the feet because this is the amount of energy to be dissipated by the arms.

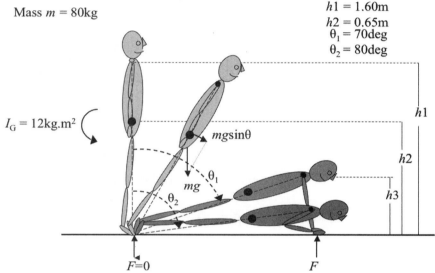

Mass $m = 80$kg

$I_G = 12$kg.m²

$mg\sin\theta$

mg

θ_1

θ_2

$F=0$

F

$h1 = 1.60$m
$h2 = 0.65$m
$\theta_1 = 70$deg
$\theta_2 = 80$deg

$h1$

$h2$

$h3$

FIGURE 5.10 A fall arrested by the arms.

The RKE is the integral of moment due to gravity with respect to angular displacement and is calculated between the final value ($\theta = \theta_1$) and initial value ($\theta = 0$) as follows:

$$\text{RKE} = \int_{\theta=0}^{\theta=\theta_1} mgh_2 \sin\theta d\theta$$

$$\text{RKE} = -mgh_2[\cos\theta]_0^{\theta_1} = -mgh_2[\cos\theta_1 - \cos 0]$$

$$\text{RKE} = -80 \times 9.81 \times 0.65 \, [\cos(70) - \cos(0)] = 335.65 \text{ J}$$

This is RKE possessed after a displacement of θ_1. From θ_1 to θ_2, gravity continues to act to add to the present RKE, but this is opposed by a moment due to the force (*F*) acting at a perpendicular distance of h_1. Assuming F to be the average force and therefore constant, we obtain the following equation:

$$Fh_1 = \text{RKE} + [-mgh_2 (\cos\theta_2 - \cos\theta_1)];$$
$$F = [335.65 - 80 \times 9.81 \times 0.65(-0.168)] / 1.6;$$
$$F = 263.34 \text{ N}$$

It should be noted that the value of F is approximate because it is not acting exactly perpendicular to the longitudinal axis of the body during the arresting motion. Another feature of interest is that the force will begin at zero and end at some value smaller than the calculated average force F. Therefore the peak force will be much greater than the average value calculated and may exceed the ultimate tissue strength while the average value does not. Most reasonably fit people can perform this action without injury, but it could be disastrous for someone who is older, unfit, and unskilled.

Part 2

Calculate the velocity of the shoulder joint marker immediately prior to ground contact.

Solution

The velocity of the shoulder marker is due to angular velocity gained by the body, due in turn to the acting moment produced by gravity. In Part 1 we used energy as our approach to the solution, but we may also be tempted to use angular momentum gained (AM) from angular impulse. Knowing AM allows calculation of angular velocity (ω). The angular momentum would be obtained from

$$\text{AM} = \int_0^{t\,\text{at}\,\theta_1} mgh_2 \sin\theta \; \mathrm{d}t$$

The problem with this approach is that we have no knowledge of the time when ground contact occurs. If we were to film the activity we would know the time, but in that case we could measure the velocity of the shoulder marker directly. So we return to RKE as our approach, noting the following identity:

$$\text{RKE} = (I_G + mh_2^2)\, \omega^2 / 2$$

where $(I_G + mh_2^2)$ is the moment of inertia about an axis through the toes. Therefore, from Part 1 where RKE = 335.65 J:

$$335.65 = [12 + 80 \times (0.65)^2]\, \omega^2 / 2;$$
$$\omega = \{2 \times 335.65 / [12 + 80 \times (0.65)^2]\}^{1/2};$$
$$\omega = 3.83 \text{ rad/s}$$

Using the identity for circular motion:

$$v = h_1\, \omega;$$
$$v = 6.13 \text{ m/s}$$

So we obtain a rather large value for velocity of the shoulder joint marker.
As a further point of interest, we can expand the following equation:

$$\text{RKE} = (I_G + mh_2^2)\, \omega^2 / 2;$$
$$\text{RKE} = I_G\, \omega^2 / 2 + mh_2^2 \omega^2 / 2$$

where RKE represents the total kinetic energy (TKE) from the only source, which is gravity. Using the identity for circular motion again,

$$v_{CM} = h_2\omega, \text{ therefore } h_2^2\omega^2 = v_{CM}^2,$$

we obtain

$$\text{TKE} = I_G\, \omega^2 / 2 + mv_{CM}^2 / 2;$$
$$\text{TKE} = \text{RKE}_{CM} + \text{TKE}_{CM},$$

which is the sum of the RKE about the CM plus the TKE of the CM.

Summary

Slipping occurs when the ratio of force parallel to the sliding surface over force normal to the surface exceeds a certain value known as the coefficient of limiting friction μ. When slipping occurs, the coefficient of dynamic friction not only is less than μ but also varies with the speed of slipping. When we are walking on a dry surface wearing normal footwear, there is little chance of slipping, and avoidance strategies are not needed. If the necessary horizontal force at the feet is large, we are liable to slip. If the surface on which we perform has a low value of μ, because it is either wet or soiled by some low-friction fluid, we are in great danger of slipping. Clearly sporting and working environments are places where we would expect to find the greatest incidence of slipping. Playing and working surfaces are therefore required to be maintained appropriately. An attendant problem is that when we begin to slip, the value of μ decreases and it becomes more difficult to arrest movement in the slip. It is clearly better to avoid the onset of slipping than to attempt avoidance after motion is present. Increasing the coefficient of friction to very high levels or otherwise arresting motion of the foot on the ground can lead to shearing forces at the ankle joint. Shoes have been developed to protect our feet from the ravages of large frictional forces during ambulation. This argues for the development of gloves for specific manual handling occupations, and wearing such gloves should be a requirement, not a suggestion.

Falling happens, and the problem faced becomes one of dissipation of energy. The problem is best overcome by landing in an erect posture with the CM as high as possible, followed by the greatest possible excursion of the CM downward. In this way peak force can be minimized to a level below tissue failure force. Peak force can also be minimized by landing on a compliant surface where much of the landing energy is dissipated in deformation of the surface. Friction also plays a part in safe landing so that the feet can be the contacting structure with the floor. Low friction removes foot contact, and the result may be landing on a less well designed part of the body such as the greater trochanter of the femur. A large area of contact with the ground lowers the pressure on any one part of the body, as does rolling following landing. Shoes with fairly rigid soles also distribute the pressure over the plantar surface of the foot.

Falling accidents are simply the result of unforeseen circumstances. The advice to avoid injury is to keep your eyes open, be aware of potential danger, keep your muscles in good order, and learn rolling strategies. Also, do not faint. The only biomechanical strategy to cope with injury due to fainting is always to stay close to some strong, alert friends. Choose your friends carefully.

RECOMMENDED READINGS

Chapman, A.E., Leyland, A.J., Ross, S.M., and Ryall, M. (1991). Effect of floor conditions upon frictional characteristics of squash court shoes. *Journal of Sports Sciences* 9, 33-41.

Walking and Running

Bipedal walking is the primary method of getting from here to there, apart from the initial stages of shuffling, rolling, crawling, and other methods that are part of our developmental process. The importance of this primary form of human locomotion is exemplified by the enormous amount of research done on the topic. Walking is more precarious than standing because it requires support on one foot at a time during a specific stage in the walking cycle. Running occurs later in the maturation of a human being, and it is complicated by the presence of an airborne phase in which there is no ground contact with either foot. Every person runs at some time in his or her life. The development of alternative means of transport makes running less widespread than when humans had to run to get food or to avoid predators. Yet running is well established in sport, where it is a goal in itself or a means to an end. Trained runners are unlikely to be injured unless they meet some unseen obstacle or overdo training. On the other hand, casual runners are subject to danger due to lack of the necessary physical requirements, lack of skill, and many other factors. Normally running does not require teaching unless it involves some specific, expert activity, but teaching may be necessary to obviate potential injury brought on by poor running style.

Aim of Walking

The general aim of walking is to move the mass of a jointed segmented body horizontally from one place to another, more or less at a constant speed. This aim must be achieved in a gravitational and frictional field. The general aim is to propel oneself with the least amount of expended energy. Change in speed is also one of our aims, as is change of direction to avoid objects in our path. Such changes are frequently necessary when we are dealing with undulations and other variations in the walking surface.

Mechanics of Walking

The general mechanical problem in bipedal locomotion is to produce sufficient vertical impulse to counteract that due to gravity, during both the double-support phase and the stance phase on one foot, while doing sufficient muscular work to bring the nonstance lower limb forward in preparation for the next foot strike. Bipedal locomotion is not a trivial task either to perform or to engineer. For a few pennies, it is possible to purchase a toy figure that can perform the relatively useless function of walking down an incline. Yet attempts to manufacture a robot that can perform like a human have cost many millions of dollars, and such robots are inflexible in terms of the functions they can perform. The problem is not in the construction of the robot but in the mechanisms that sense its movement and control it. The subject here is the mechanics of walking. For the purpose of analyzing the mechanics of walking, we will confine ourselves initially to movement of the center of mass (CM).

We can analyze motion of the CM using either impulse–momentum or work–energy relationships. Motion of a point mass horizontally requires an initial horizontal application of force F for a period of time t to generate momentum mv as follows:

$$\int_0^t F dt = m(v_t - v_0)$$

Once momentum is achieved, it will continue forever unless an opposing impulse is applied. Therefore subsequent impulses must sum to zero. Walking also uses motion of body segments, so changes in angular momentum are a consideration as well. Because of the cyclical nature of walking, the kinematics at any stage in the cycle are almost identical to those at the same point in the following cycle. The angular momenta must therefore sum to zero throughout the cycle. The same could be said for energy in a strict physical sense if we could store the energy required to slow down motion and release it to speed up motion. The fact is that both production of mechanical impulse and production of mechanical work are costly in terms of metabolic energy. The problem is how to minimize this cost.

Friction at the foot–ground interface does not represent energy dissipation unless the foot slides, because point of application of the frictional force does not have a displacement. Yet friction as a force of constraint is necessary to stabilize the foot and allow energy to be exchanged from translational to rotational as shown in figure 6.1.

This figure ignores gravity and shows a body in pure translational motion, traveling with a horizontal velocity and experiencing a force that stops motion of the point where force is applied. This is equivalent to applying a parallel force through the CM and a moment created by the force about an axis through the CM. The subsequent motion is then rotational and translational. Energy loss is due to the force acting in a direction opposite to the displacement ds, while the body gains energy by an equal amount through the moment M acting in the same direction as the angular displacement dθ. This kind of energy exchange can be expected to some extent during the support phase in walking. The example of figure 6.1 is approximate in order to illustrate energy exchange. It applies only when θ is small, and would need some refinement to yield a complete analysis of the energy exchange.

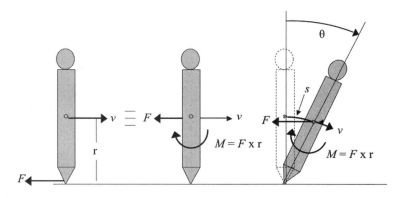

Energy loss = $\int F ds$ Energy gain = $\int M d\theta$

For r = constant:

$\theta = s/r$ so $d\theta = ds/r$; so $\int M d\theta = \int M ds/r = \int Frds/r = \int F ds$

FIGURE 6.1 Energy can be exchanged from kinetic to potential forms without loss.

■ **Key Point**

The cyclical nature of walking means that both rotational and translation mechanical impulses must sum to zero over a complete cycle.

Biomechanics of Walking

Humans are equipped to walk with two jointed lower limbs in order to move the mass of the upper body. Considering angular motion only in the vertical plane of progression, the ankle, knee, and hip joints imply three degrees of freedom controlled by three net joint moments. If we consider that each moment can be either flexor or extensor, we approach six control mechanisms. If we further consider the role of muscles that can produce a moment under concentric and eccentric contraction, there appear to be 12 control mechanisms. Obviously there are many combinations of muscular contraction that will fulfill the role of walking control, but not all will be efficient. The exact objective function in walking is unknown. It would seem that minimization of energy expenditure would be a useful objective function, and the increased efficiency of eccentric over concentric muscular contraction must be considered in this context. But a secondary aim might be to minimize the perturbations of the head, where the organs of balance and orientation within a frame of reference are located. Kinematic and kinetic analysis of walking shows us how we solve the walking problem, but it does not tell us which, if any, objective function is being solved. However, there are certain phases and aspects of walking that can be explained by recourse to experimental work.

We know that walking consists of successive periods of double and single support at the feet and that a recovery phase brings the nonsupporting lower limb from behind to in front of the body in preparation for the next support phase. First we will deal with walking at a constant speed in a straight line. In fact, this does not happen, even if the speed of the CM of the body is identical at identical points in the walking cycle. The CM changes speed in both horizontal and vertical directions throughout the cycle due to the changing forces exerted on the ground in the fore–aft direction. The jointed nature of the segments also leads to considerable changes in angular momentum and rotational and translational kinetic energies. The reason for these changes is that the lower limbs must undergo some pendular-like motion in order for the leading foot to be in a position in front of the body's CM to provide support as the CM travels horizontally. These limbs support an upper body that is far from being a point mass and has joined articulated segments that are necessary to produce the motion we know as walking. Not having a complete upper body can lead to "abnormal" walking.

The problem is to keep walking while supporting the body successively on each lower limb with a vertical force, as well as providing sufficient horizontal propulsive force to counteract energy dissipation due to friction at heel strike. The vertical impulse will be the result of vertical force, which will undulate above and below the body weight value. But its net value will be equal to mgt created continually by gravity. The horizontal impulse will result from the anterior–posterior (AP) force, which will sometimes be applied in the forward direction by use of the ankle, knee, and hip extensor muscles. The AP force will sometimes be applied backward due to friction acting against a forward-moving foot. The net horizontal impulse will sum to zero if a constant average walking speed is to be maintained. While this characterization does not give us the objective function in walking, it does state the biomechanical problem of walking.

Biomechanics of Heel Strike

Figures 6.2 through 6.4 are pertinent to much of the following discussion and should be inspected together and compared frequently. This section could well be titled "How to Hit the Ground Successfully." Figures 6.2 and 6.3 show that at the instant of right heel strike (RHS), there is no force on the right foot. When the heel strikes the ground, the two major components of force that develop are vertical F_V and fore–aft or anterior–posterior F_{AP}, which respectively oppose gravity and arrest the forward motion of the foot. Fortunately at this time the opposing foot is still on the ground until it leaves at left toe-off (LTO). This serves to oppose the retarding force on the right foot. Observation and experience of walking show that when the rear foot slips backward, the forward motion of the body is arrested, as only the heel striking force of the forward foot is acting. Also, when the front foot slips, the force on the rear foot is not opposed and the person falls forward. These opposing forces induce friction at the feet and cost the system muscular energy but do not contribute directly to forward movement. Rather they give the system dynamic stability. The cost to the muscles is well worth it, for without such frictional force we would be sliding uncontrollably.

A second problem of heel strike is the shock of landing on the heel. Were the lower limb not jointed, both components of force would rise rapidly, leading to potential injury as described in chapter 5, "Slipping, Falling, and Landing." At heel strike (HS) the ankle joint begins to experience a plantarflexion moment as the force at the heel is directed behind the ankle joint (figure 6.2). The problem is solved by eccentric contraction of the dorsiflexors of the ankle joint which appears as a negative value on the axis labeled Extensor moment (figure 6.4). In fact HS is anticipated by initiating the dorsiflexor moment prior to HS. Subsequently the dorsiflexor moment rises to its maximum at HS as seen in figure 6.4, in order to let the foot down relatively gently and avoid "foot slap." Fortunately we appear to be preprogrammed to avoid the potential hazards of heel strike, as electromyographic recording shows activity in the dorsiflexor muscle tibialis anterior prior to heel strike. Should HS be delayed by some unseen depression in the floor level, experience tells us that there can be a significant, rapid rise in force that is physically shocking. The reason is that the delay has allowed the dorsiflexor moment to rise too high and the ankle joint has become mechanically "stiff." Various amounts of delay of HS because of various depths of floor depression can have different effects, such as reflex abolishment of tibialis anterior activity. In this case there is no normal cushioning when the heel, or the flat foot, eventually strikes the ground. The figures shown are representative of the many and varied illustrations in the literature that indicate wide interindividual variations in walking biomechanics.

This effect is one not afforded those individuals who have paralysis of the anterior tibialis muscles (flexors or dorsiflexors). Such a foot slapping condition can be seen in some people who suffered from the poliomyelitis disease prevalent in the 1950s and also those who have experienced very severe destruction of nerves due to problems with intervertebral discs in the lower back. This dorsiflexor moment also provides ankle dorsiflexion at toe-off in order to keep the toes from catching the ground. Again this is a problem for those with anterior tibial paralysis.

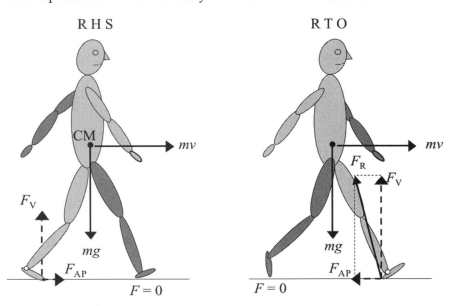

FIGURE 6.2 There are forces acting on the left foot at right heel strike and right toe-off.

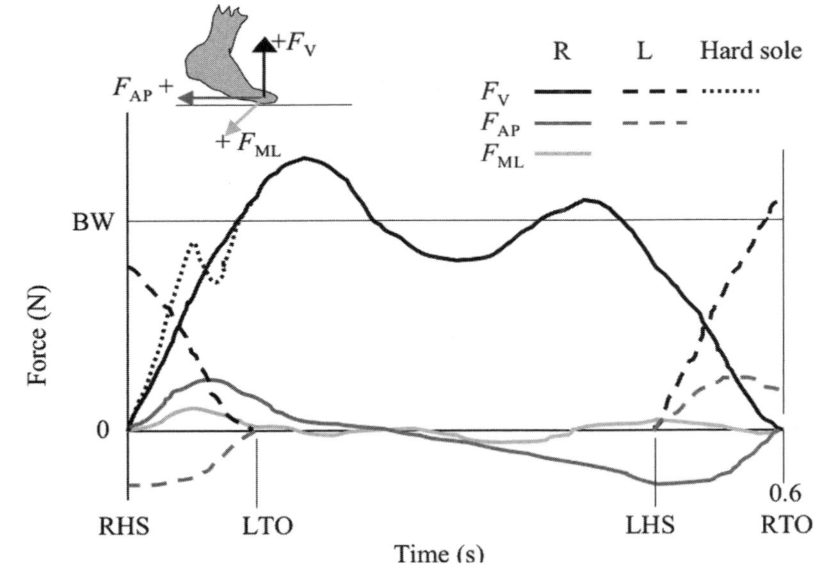

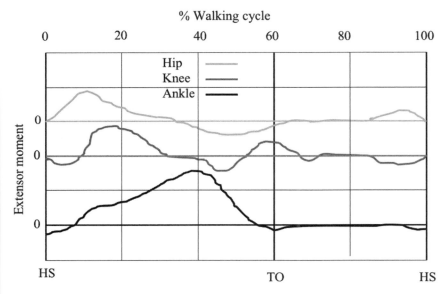

FIGURE 6.3 Note the presence of two periods of double support and change in vertical force profile occurring when hard-soled shoes are worn.

FIGURE 6.4 Note the small eccentric dorsiflexor moment at the ankle immediately prior to and following heel strike and toe-off, and also zero ankle moment when the foot is off the ground.

Biomechanics of Stance

The significant property in the stance phase is the presence of a jointed lower limb. If the lower limb were a single rigid pole, there would be a vaulting effect as the CM passed over the stance foot. Therefore the potential energy attained would be increased over that in conventional walking. A necessity would therefore be considerable kinetic energy at heel strike, probably accompanied by much greater rear-foot force at considerable muscular metabolic cost. This problem is obviated in conventional walking through reduction of the extensor moment at the knee in midstance so the knee joint will flex and thus the CM can take a more horizontal trajectory. There is another period of flexor moment at the knee in the late recovery phase that serves to keep the knee joint from rotating into full extension. Such a knee joint angle could lead to a significant shock to the lower limb, as the knee joint would tend to be locked at heel strike. In other words, the cushioning effect of early knee flexion at heel strike would be absent.

Lower Limb Recovery

Recovery is the process of moving the lower limb that is not in contact with the ground. This must be performed with sufficient ground clearance to avoid stubbing the toes. A further aim is to minimize the height of ground clearance so that minimal muscular work is done in increasing the potential energy of the lower limb as it is lifted. Remember that potential energy is Wh; in this case the letters represent, respectively, weight of the lower limb and the height to which its CM is raised.

The body uses the properties of a double pendulum to deal with this problem in a way that minimizes the muscular work. The process uses the concept of exchange of energy from its potential form (by virtue of height) to its kinetic form (by virtue of motion) and back again. As the toe leaves the ground with the hip joint ahead, the lower limb acts like a double pendulum. It is double in the sense that the thigh and combined leg and foot represent two segments that are joined and free to articulate about the knee joint. A property of such a double pendulum is that when released, the proximal segment initially gains a greater angular velocity than the distal segment. Subsequently the distal segment catches up with a greater angular velocity than its proximal neighbor. If the hip were fixed, the lower limb would be straight as it passed through the vertical. However, the hip joint axis is accelerated forward, which delays straightening of the lower limb until it has passed the vertical. In this way the toe is kept above ground level. Subsequent deceleration of the hip joint axis serves to produce a reverse effect, which is seen as straightening of the lower limb in preparation for heel strike.

Fine interplay of muscular contraction is required for these events to be timed correctly. But the muscular requirements are minimized because the natural conversion between early potential energy to kinetic energy and the later conversion from kinetic to potential energy do not require muscles to lift the lower limb and let it down again. The latter conversion allows heel strike to occur with the angular velocity of the lower limb almost zero. So the small levels of muscular contraction in the recovery phase of walking can be considered more a means of control than a source of energy. The period of free oscillation of a compound pendulum depends on its moment of inertia and length. These values seen for the lower limb provide a period of oscillation that accords well with moderate walking speed and makes walking an energy-efficient means of locomotion.

The ground clearance of the toe of the recovery limb is very small. During walking in grass no taller than that seen in the "rough" on a golf course, the lower limb must be lifted higher than normal. The necessary muscular contraction of the knee flexors therefore requires more energy than normal. The same can be said about walking over uneven surfaces. So any surface other than flat ground can produce problems of both tripping and increased energy cost in the stance and recovery phases of walking.

Use of the Arms

Although we generally consider the arms a reaching mechanism, they are also used advantageously for different functions in a variety of activities.

Try typical walking with your arms held stiffly by your sides or folded across your chest. If you do not attempt compensation, you will notice that the upper body gains rotation around the long axis (nearly vertical) of the trunk. It will require considerable concentration to refrain from compensating, as the body is well used to compensating for all kinds of modifications of segmental motion. A consequence is additional required rotation of the head around the long axis of the neck so that you can keep both eyes looking forward.

When we project ourselves forward during the second half of the stance phase on the left foot, the right leg comes forward. This action gives the body counterclockwise rotation as seen from above. The resulting angular momentum of the trunk and head is compensated by movement of the arms to give angular momentum in the opposite direction (figure 6.5).

The angular momentum created by both arms is needed to compensate for the large angular momentum possessed by the relatively massive trunk and the free-swinging lower limb. In this way the trunk does not experience a gain in rotation, and the head can be kept facing forward with little effort. So use of the arms in

■ Key Point

The jointed nature of the lower limb reduces the potential energy of the body in the midstance phase, which reduces the kinetic energy required at heel strike and attendant high peak forces at the heel that may be problematic.

■ Key Point

There appears to be a considerable saving of energy in the recovery phase because of exchange between kinetic energy and potential energy of the swinging limb.

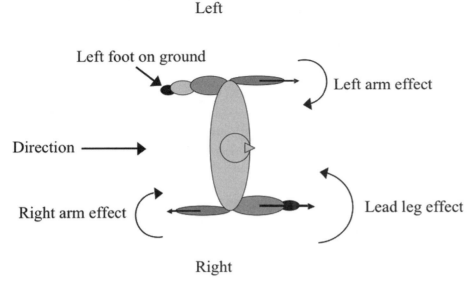

FIGURE 6.5 Competing angular momenta in the limbs to maintain stability of the trunk and head about the longitudinal axis.

■ **Key Point**

Arm motion in walking aids in countering the rotational effect of lower limb angular impulse to maintain the trunk and head facing forward.

walking is essential to avoidance of excessive bodily rotation. Unfortunately, people who have no arms or have paralysis of shoulder muscles must learn to compensate with other muscles or live with an uneconomical gait pattern. To appreciate the use of the arms, try walking with the same-side arm and leg forward for a distance and see what happens. Also try walking fast and appreciate the greater peak angular velocities of both arms and their increased amplitude of motion.

The contralateral motion described is also seen in most animals that walk on four legs, and for the same mechanical reason. While humans retain contralateral motion for running, quadrupeds change from this motion when required to move faster. The reason for this change is the greater contribution of the muscles of the trunk in the running or galloping of quadrupeds.

Center of Pressure

Pressure (F/A) is a mechanical measurement of force F divided by the perpendicular area A over which the force is applied. Pressure is an important concept in human activity, since large applied pressures can do damage to tissues. This is the reason for padding, which distributes a force over a large area so that each underlying piece of tissue sustains only a small force. The center of pressure (COP) in walking is a theoretical point representing the average of the distributed points of pressure between the floor and various parts of the foot. It is the point where we would locate a single force representing distributed forces under the foot. The importance of the concept of center of pressure lies in the fact that it progresses along the foot from the heel, where first contact with the floor is made, to the great toe, the last point of contact in typical walking as seen in figure 6.6.

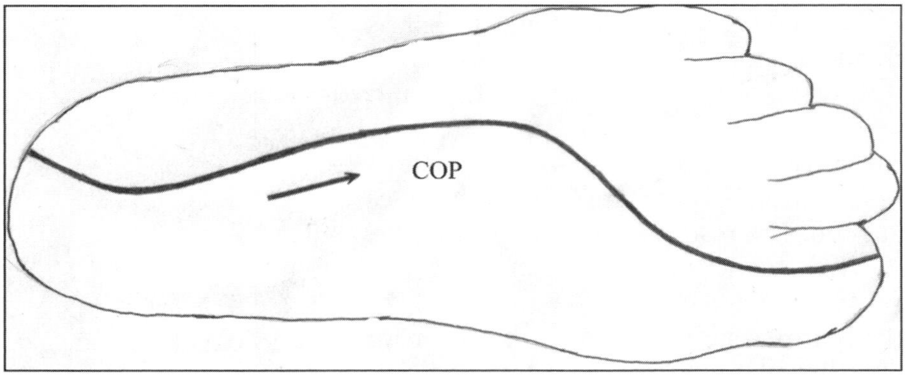

The word "normal" is used specifically to indicate walking with an intact foot as opposed to a foot with abnormalities. The arrow in figure 6.6 indicates the direction of progression of the COP. This pattern varies among individuals with intact feet, and it is well withstood by the

FIGURE 6.6 The average of distributed pressure progresses over the pad of soft tissues under the foot in stance in walking.

architecture of the "normal foot." For example, there is a fat pad under the heel that safely attenuates the initial contact force applied to the heel bone. A sheet of connective tissue under the muscles of the foot, along with muscles, keeps the longitudinal arch of the foot from collapsing during midstance. As the COP passes from the lateral to the medial side of the foot, there is substantial padding under the metatarsal bones. Finally, the great toe also has a pad and has strong flexor muscles that allow it to avoid being forced into a position of excessive extension. Thus the path of the COP is well suited to the complex architecture of the foot.

Unfortunately there are many medical conditions that arise from pressure on the plantar surface of the foot. One condition is the ulceration seen in persons with diabetes. Some conditions are induced by the choice of footwear.

Variations of Walking

Walking on a plane horizontal surface occurred late in the evolution of humans and today is a typical form of walking in developed societies. However, recreational and occupational activities frequently require us to walk on surfaces that are uncommon to us but common in less developed societies.

Walking Up and Down a Slope

You will have noticed that walking up an incline of increasing slope eventually requires a change in walking pattern. The major change is that the leading foot does not land with the almost straight knee joint seen in horizontal walking (see figure 6.7). A consequence of this change is that we cannot take advantage of the energy conservation effect seen in horizontal walking. There is needed extra concentric contraction of the hip flexor muscles, which naturally represents extra work and costs us energy. Following lead-foot landing, work is done concentrically by the hip and knee extensors to raise the CM. All of this concentric muscular work gives a continual increase in potential energy equal to $+Wh$ as seen in figure 6.7. The unfortunate fact is that the force available to us in concentric contraction decreases as the shortening speed increases (see figure 3.3c in chapter 3, "Foundations of Movement"). So concentric work is very costly, and the more so the faster we attempt to do it. Running uphill is an excellent cardiovascular training method as it stresses both the muscles and the oxygen delivery system (the lungs), which converts stored energy from food into muscular energy output. A further requirement of walking uphill is an increased range of ankle joint angular motion. This means that the ankle extensors are lengthening and shortening over a greater length change than is customary on a horizontal surface. Such a pattern of motion can induce pain unless the individual is accommodated to it. Walking uphill requires training if long distances are to be experienced without pain.

Walking downhill might be expected to be the exact opposite of walking uphill from an energy point of view. This would be true if we were a simple conservative mechanical system without energy dissipation. In this case as we lost potential energy continually ($-Wh$), we would gain kinetic energy and move faster and faster. We would soon be out of control and in great danger of injury. To remain in control, we reduce the gain in kinetic energy by eccentric muscular contraction. In other words, we use the same muscles to descend as we do to ascend the slope;

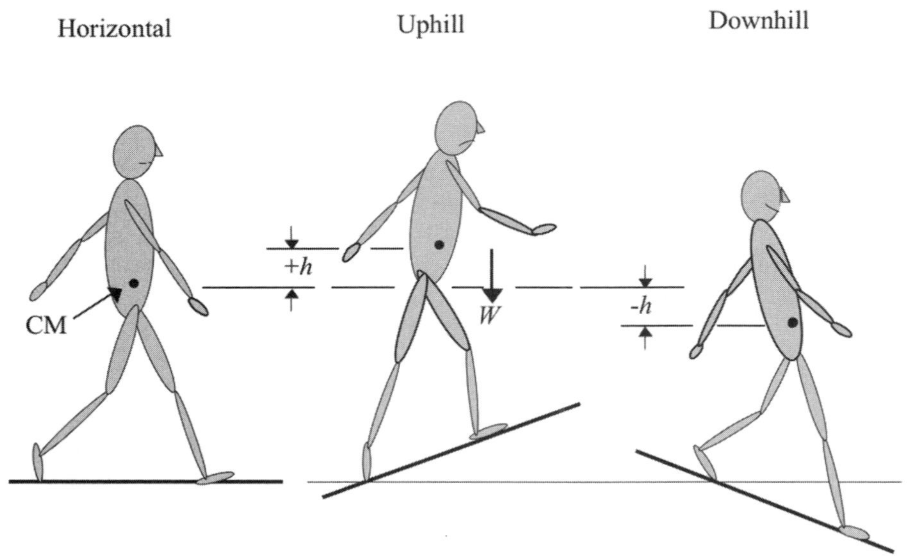

Horizontal   Uphill   Downhill

FIGURE 6.7 Walking uphill and downhill respectively gain PE (+Wh) and lose PE (-Wh).

■ Key Point

In using eccentric muscular contraction, downhill walking is more economical than uphill walking.

■ Key Point

Downhill walking induces higher peak forces than uphill walking; the chance of stress injury is therefore greater in walking downhill.

but in descending, the muscles are being stretched while developing force whereas in ascending they are shortening. During this stretch of muscle, only a small amount of the energy is stored as potential energy in elastic tissues such as muscle and tendon. The remainder is dissipated as heat. Fortunately the force available to us in eccentric contraction is greater than in the concentric mode (again see figure 3.3c), so we are more mechanically efficient in descending slopes and stairs than in ascending them.

In summary, imagine a hypothetical, frictionless ball rolling across a U-shaped valley. It would continue going up and down forever with a perfect exchange between potential and kinetic forms of energy. We are not like that. Walking costs us energy in both the up and down phases—less downhill than uphill per distance traveled.

There is one respect in which walking downhill is worse for our bodies than walking uphill. As previously mentioned, we are able to generate greater muscle forces in eccentric compared with concentric contraction. The danger here lies in the tendency to use the high eccentric force with short muscular lengthening to control our downhill kinetic energy. The higher the force, the greater the stress applied to tissues such as cartilage and ligaments and therefore the greater the chance of injury. Any reader who has run down hills rapidly for an extended period of time will be familiar with the pain that appears in the joints on the subsequent day. Also, the muscles are not immune to stress. The high eccentric force induced leads to postexercise soreness on the following day. This effect has been well documented in research involving weight trainers who let down enormous loads in addition to lifting up smaller loads. It is the eccentric work that does the damage. The advice for hikers and those who train by hill running is to maximize the amount of muscle stretch when descending in order to minimize the peak force and avoid pain, soreness, and potential injury.

Walking on Stairs

A general rule in the construction of stairs is that the sum of the rise (vertical height of step) and run (horizontal depth of step) should be within narrow limits. Departures from this rule make walking up stairs difficult because people have to shorten or lengthen their stride by an uncomfortable amount. Much of what has been said about walking on a slope applies to stairs. A disadvantage of walking up a slope is the large amount of angular motion required of the ankle joint. The advantage of walking on a slope is that the individual can choose his or her own stride length. On stairs this is not possible; stride length is forced upon us by the construction of the stairs. One advantage that stairs afford is a flat horizontal surface upon which to land with the leading foot. In addition to being familiar, the flat surface helps if ice is covering the stairs. Ice on a slope is obviously very dangerous because we may be unable to avoid slipping despite our best efforts. Alternatively, stairs are of no use to those who use a wheelchair. So there is no general rule that supports the choice of stairs over slopes from a biomechanical perspective. Clearly the human body did not evolve to walk up stairs, but the brain evolved sufficiently to construct them.

Racewalking

Racewalking is not so much an exercise in walking fast but an attempt to avoid running. The natural progression when one is increasing speed of locomotion is an abrupt transition from a walking gait with double-foot contact to a running gait with single-foot contact. Such a transition occurs at about 4 m/s but varies among individuals. The task facing the racewalker is to prolong the walking gait well past the speed at which running would be more comfortable. The two variables that contribute to speed of locomotion are stride length and stride frequency. We can increase our speed by increasing either of these variables or both; this applies equally to walking and running. The problem in racewalking is the limit imposed on increasing stride length. It is a simple matter to increase the vertical force at the feet and leave the ground in order to increase the horizontal distance between takeoff and landing on the opposite foot. Unfortunately, this action is running; racewalking requires at least one foot to be in contact with the ground at all times. The racewalker avoids airborne motion by increased rotation of the pelvis about a vertical axis. Not only does this strategy increase the amount of sliding of the articular cartilage of the hip joint, but it also induces increased rotation of the lumbar intervertebral joints. Furthermore, the arm action has to be more vigorous to accommodate the increased angular momentum of the trunk. This description points to the conclusion that racewalking is unnatural. This conclusion is supported further by the observation that the first thing racewalkers do when finishing a race is to break into a trot, which is more comfortable.

Gait in Amputees

Lower limb amputees require prostheses in order to walk. The prosthesis used depends on the extent of the amputation. Below-knee amputation is usually performed in a manner that leaves the muscle crossing the knee joint intact, or at least functional to some extent by means of some clever orthopedic surgery. Therefore the amputee is able to support his or her body mass during the stance phase and also to swing the lower limb during recovery. The major problem is to avoid collapse of the ankle during the stance phase. Force in the calf muscles is required to produce the plantarflexion moment at the ankle that provides forward propulsion at the end of stance. Since there is no calf muscle present, the original solution was a single-segment leg and foot that allowed no articulation of the ankle. Subsequently a stiff ankle was combined with some spring behavior of the foot through insertion of a metal sole similar to a leaf spring. This device is shown very approximately in figure 6.8.

■ **Key Point**

Stairs provide a flat surface perpendicular to the vertical, where frictional requirements are less than when walking on a sloping surface.

Intact   Artificial

Achilles tendon

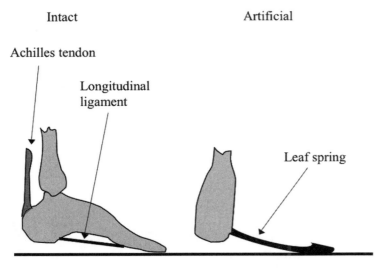

FIGURE 6.8 The combined effect of the plantarflexors and longitudinal ligament is replaced by a leaf spring in below-knee amputees.

■ **Key Point**

Above-knee and below-knee amputees are largely indistinguishable if one is observing only the recovery phase of gait.

This mechanical approximation of a leaf spring to the ligament of the longitudinal arch facilitates storage and release of strain energy as is seen to a small extent in the normal foot. In this manner the stance phase of amputee locomotion approximates the intact pattern of motion.

Above-knee amputation leaves the individual with no knee joint, and an artificial joint must be incorporated into the prosthetic lower limb. Since there is no knee extensor activity, the prosthetic knee must lock or at least must be stable in the stance phase. The characteristic motion is of a vaulting type, with no change in distance between the hip joint and the foot. Fortunately, the recovery phase can appear as it would if the natural limb were intact. The mechanics of this phase involves use of the hip flexors as shown in figure 6.9.

One means by which the energy of the lower limb can be changed is work done by muscles crossing the hip joint, in this case a single hip flexor. Gravity also contributes but for the purpose of simplicity is left out of this analysis. The work done by the hip flexor is the vector product of force F_H and displacement s of the point of application of force:

$$\text{Hip flexor work} = F_H s \cos\theta_H \qquad \text{(Equation 6.1)}$$

Since we require displacement for this calculation, we can use another kinetic variable called power to look at the instantaneous rate of change of energy. As power is the first differential of change in energy with respect to time, we obtain

$$\text{Hip flexor power} = F_H v_H \cos\theta_H, \qquad \text{(Equation 6.2)}$$

which represents the rate at which energy is being added to the lower limb. Acceleration of the inertia of the leg will induce a force at the artificial joint. This force, associated with the velocity of the joint (often termed joint force power), represents power, which is the rate at which energy is being input to the leg:

$$\text{Power at the knee axis} = F_K v_K \cos\theta_K \qquad \text{(Equation 6.3)}$$

As the force at the knee on the thigh is equal and opposite to that on the leg, the power value will be equal in magnitude but opposite in direction. The full picture is that energy is being generated by the hip flexors at a rate shown by Equation 6.2, and this energy is being input to the lower limb. The rate at which the energy of the thigh increases is equal to

$$F_H v_H \cos\theta_H - F_K v_K \cos\theta_K,$$

and the rate at which the energy of the leg increases is equal to

$$F_K v_K \cos\theta_K.$$

This analysis does not allow us to determine in which form the energy appears in terms of partitioning of rotational kinetic energy and translational kinetic energy. Yet it does show how the energy of a distal segment (leg) can be changed by work done by some muscle (hip flexor) far removed from the segment. The actual increase in energy of each segment over a period of time in the recovery phase will be obtained from the integral of each power–time relationship. The opposite effect is seen in the late stage of recovery when the power at the knee joint is negative with respect to the leg. This arises from activity of the hip extensors, which produces a force on the prosthetic knee to achieve angular motion of the leg that appears as knee extension. This effect is so remarkable that it is very difficult to distinguish between intact and artificial gait of the above-knee amputee if one is seeing only the recovery phase.

Problems in Walking

We know that a problem in walking occurs when reduced floor friction is unanticipated. Normally our hardwired system activates the knee extensor and ankle flexor muscles in preparation for heel strike. In this way the knee joint is prevented from being driven into flexion, and our foot does not slap down flat on the floor when foot contact occurs. Without friction there is no resistance to the extensor muscle activity; the lower limb straightens and the foot shoots out in front of the body. A fall is almost inevitable in this case. With prior knowledge of reduced friction, one can consciously override the automatic pattern of muscular activation in walking, thus avoiding the fall. Specifically the knee extensor activity can be reduced, resulting in a deeper knee bend than is normal, and reduced ankle flexor activity can also allow the foot to land flat to improve stability. Furthermore, the step length is reduced in order to minimize the horizontal component of force at heel strike. The result is to reduce the ratio F/N in an attempt to keep it less than the coefficient of friction μ. What is seen usually is a shuffling

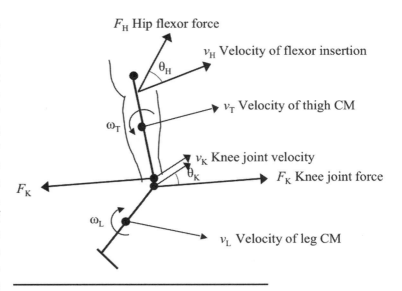

FIGURE 6.9 The leg segment of above-knee amputees can be moved by a force at the artificial knee joint due to a net hip-flexor force and resulting net moment at the hip.

Walking or running up a hill where the slope angle can vary by small amounts with every foot-fall is much more hazardous on a steep, rather than a gentle, incline.

movement of the feet in which no large horizontal forces are created and the low friction encountered can actually aid us in sliding. The best advice when locomoting on ice is not to walk, but to use ice skates (if you can).

The automatic muscular activation in preparation for foot strike also assumes a specific position of the heel contact point with the ground. Should the contact point be higher than expected, heel contact will occur sooner than expected. In this case the extensor muscles of the knee will not have had time to activate sufficiently. The consequence is a tendency for body weight to dominate, and a greater than normal knee flexion occurs. The potential for disaster in this case is directly related to the difference between the expected and the encountered height of the landing surface. A small difference can be accommodated by reflex increase in knee extensor force. A large difference may produce so much knee flexion, possibly into a weak joint position, that recovery is impossible. The alternative problem occurs when we encounter a surface whose height is below that expected. The increased time prior to heel contact will result in greater knee extension than normal, and contact will occur with a relatively straight lower limb. The cushioning provided by the springlike behavior of the knee extensors will be absent. The result will be a high peak force transmitted up the lower limb through the knee cartilages to the articular surfaces of the hip. When walking rapidly, the shock wave can travel up the spine and be felt in the neck. Doubtless every reader will have encountered these situations. The personal and the safest solution is to be aware at all times of the forthcoming floor level. Apart from personal awareness, it is somebody's responsibility either to remove such changes in level or to mark them clearly. Obviously people with impaired vision are particularly susceptible to the problems just described.

■ **Key Point**

Because the complex act of walking is "hardwired" in the nervous system and flexible with respect to changing external conditions, unforeseen conditions create problems in human walking.

Enhancement and Safety of Walking

The rising phase of the vertical force indicates weight acceptance by the foot as shown in figure 6.3. The superimposed broken line shows an early peak that would be the result of using shoes with little cushioning and a hard heel, or walking on an unyielding surface such as concrete. In the latter case, there is greater shock to the foot, which would be transferred up the leg to the knee. So individuals who are given to walking long distances should use a semicompliant shoe or walk on grass in order to avoid the dreaded shinsplints. An extreme case of cushioning is walking on soft sand. While this ensures no rapid changes in force, it does present other problems. Barefoot walking on sand allows irregularities in the surface to move the tarsals, metatarsals, and toes (all bones of the foot) relative to each other. The muscles of the foot and the intrinsic foot ligaments are therefore required to produce stabilization between bones. Muscles become fatigued as a consequence, and ligaments are subjected to increased strain. The solution in long distance walking over uneven terrain is therefore to use shoes with a relatively compliant heel and a firm sole. A great deal of research has been undertaken on the mechanics of the foot, since it is our intimate connection with the external environment in walking and running. Such research includes the problems of rear-foot control and pronation, as well as the specific problems faced by heel strikers versus midsole strikers.

Should your aim be to keep the muscles of your foot in good working order, walking barefoot, particularly on sand, is a good training method. It is the author's opinion that people could avoid many mild foot problems by removing their shoes whenever they are unnecessary, for example at home. Track athletes use very compliant shoes to provide propulsive friction rather than support. They can wear such flimsy shoes because their foot muscles are healthy and strong.

Force in the AP direction rises to an early peak following heel strike, provided that the friction between the heel and the ground is sufficiently high. Evidence for the presence of friction can be seen in the wear pattern of shoes, which occurs primarily at the heel. This force arrests further forward motion of the foot, and the body rotates around the heel. When the foot becomes flat on the ground, rotation is around the ankle joint. As the heel rises, rotation occurs around the joints between the metatarsal (or long) bones and the toe bones. Reduced friction is present during walking on ice or when one hits a patch of oil underneath the heel. In this case the early and late peaks in the AP force pattern cannot occur. This accounts for the altered gait on ice where the stride length is reduced. The reason is that placing the heel well in front of the body can be accomplished only with considerable friction, and the stride length can result only from placing the heel well in front of the body. For much of the middle phase of the stride, when only one foot is in contact with the ground, the AP force is small and the change in fore–aft velocity of the CM is insignificant. It matters little, therefore, whether we are walking on ice in this phase. Near the end of stance, the AP force increases in the opposite direction, which serves to accelerate the CM in preparation for heel strike of the opposite foot. Again this cannot occur on ice.

When the foot properties are modified by such factors as muscle or ligament weakness, disease, injury, or partial amputation, a modified path of the center of pressure will occur and unexpected load will be applied to adjacent parts of the body such as the ankle joint. The use of poorly fitting shoes also modifies the path of the center of pressure so that vertical loads are placed upon areas of the sole that are not well suited to the task. Poor shoes also increase pressure on the sides of the feet, resulting in calluses and other painful and disfiguring conditions. Properly fitting shoes can aid in producing lateral forces that avoid collapse of the longitudinal and transverse arches of the foot and strain in the fascia and ligaments, which are meant to maintain the integrity of these arches.

■ **Key Point**

Walk barefoot when it is safe to do so; avoid shoes that produce local pressure points.

Practical Example 6.1

Table 6.1 shows extensor moment of the knee and leg angular velocity of extension deduced from the stance phase of a person walking. Calculate the work done by the knee extensor muscles during this time period.

Solution

Since we are given only moment and angular velocity, the only kinematic variable we can obtain is power, which is the product of the two variables shown in Equation 2.26b. The integral of power with respect to time will give us work done or change in energy as follows:

$$\int M\omega \, dt = \Delta E = \text{Work done}$$

TABLE 6.1

Numerical Integration of the Product of Knee Extensor Moment and Leg Angular Velocity to Yield Work Done

Time (s)	0	0.05	0.1	0.15	0.2	0.25	0.3	0.35	0.4	0.45	0.5	0.55
M (Nm)	0	30	45	75	100	85	75	90	100	75	50	30
ω (rad/s)	-0.2	-0.18	-0.15	-0.1	-0.06	-0.02	0	0.2	0.25	-0.3	-0.2	-0.1
$M\omega$ (Nms^{-1})	0	-5.4	-6.75	-7.5	-6.0	-1.7	0	18	25	22.5	10	3.0
Work (Nm) = ΔE	0	-0.135	-0.439	-0.795	-1.13	-1.32	-1.36	-0.91	0.165	1.35	2.16	2.49

Total work = 2.49 Joules

The fourth row of the table gives us the product $M\omega$. In the fifth part of the table, the integral is obtained by taking the mean of two adjacent values of power and multiplying by the time interval between them and then adding this result to the previous change in energy. This process is illustrated in chapter 2, figures 2.1 and 2.2. The third part of the table shows negative power during the period 0 s to 0.3 s. In this period energy is being dissipated by eccentric contraction of the knee extensor muscles as the muscle moment and angular velocities are in opposite directions. Following 0.3 s, the contraction becomes concentric and energy is generated.

It should be noted that while a net value of 2.49 J of work has been added to the system, estimations of the metabolic requirements of muscular contraction cannot be based upon this value alone. The fact is that in the period of eccentric contraction in which negative values of work are shown, there is a metabolic cost to the muscles. Only in the case of a perfectly elastic body would we see equality of work done in stretching and work reappearing during recoil. Unfortunately we cannot store energy when walking down a hill and get it back when walking up again. If we could, the period between rising out of bed and returning to it would cost us nothing metabolically.

Practical Example 6.2

Figure 6.10 represents standing (a) with the CM located vertically above the ankle joint and (b) with the heel off the ground.

Integrity of the longitudinal arch is maintained by the total force due to longitudinal muscles, ligaments, and fascia underneath the sole of the foot. If we consider a slow walking speed in which there is insignificant acceleration, we assume that the body is perfectly balanced as in a static case. We shall also assume that friction at the sole of the foot is insignificant.

Part 1

Calculate the force F in the longitudinal structures for the posture in figure 6.10a given the following constants:

$$m = 40 \text{ kg}, l_1 = 16 \text{ cm}, l_2 = 10 \text{ cm}, \theta_1 = 60°, \theta_2 = 36.9°$$

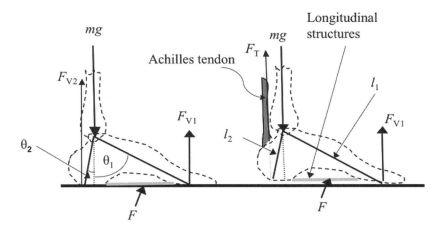

FIGURE 6.10 Schematic representation of the foot which articulates with the tibia at the ankle joint.

Solution

Resolving forces vertically for the static case gives

$$F_{V1} + F_{V2} = mg.$$

Resolving moments about the ankle joint gives, for the forefoot and rear foot separately,

$$F_{V1}l_1\sin\theta_1 = Fl_1\cos\theta_1, \; F_{V2}l_2\sin\theta_2 = Fl_2\cos\theta_2;$$
$$F_{V1} = F(l_1\cos\theta_1 / l_1\sin\theta_1) = F / \tan\theta_1;$$
$$F_{V2} = F(l_2\cos\theta_1 / l_2\sin\theta_1) = F / \tan\theta_2.$$

Inserting these expressions into the original equation gives

$$F / \tan\theta_1 + F / \tan\theta_2 = mg; \; F = mg(\tan\theta_1 \times \tan\theta_2) / (\tan\theta_1 + \tan\theta_2);$$
$$F = 40 \times 9.81 (1.732 \times 0.751) / (1.732 + 0.751) = 205.5 \text{ N}.$$

A significant feature of this result is that if the angles increase incrementally, the force increases dramatically. In other words, having a low arch puts great stress on the longitudinal structures that maintain its integrity. Persons with low arches are therefore susceptible to plantar fasciitis, which is overstress of the plantar connective tissue sheet.

Part 2

Calculate the Achilles tendon force F_T and also F for the posture in figure 6.10b, where the heel is just off the ground.

Solution

As the Achilles tendon force arises from muscles that produce an equal and opposite force on some other part of the body, F_T is internal to the system and serves to increase the compressive force at the ankle joint to $mg + F_T$. Resolving forces vertically,

$$F_T + F_{V1} = F_T + mg; \; F_{V1} = mg.$$

Resolving moments about the ankle joint for segment 1,

$$F_{V1}l_1\sin\theta_1 = Fl_1\cos\theta_1, F = F_{V1}\tan\theta_1; F = mg\tan\theta_1;$$
$$F = 40 \times 9.81 \times \tan(60) = 679.7 \text{ N}.$$

To obtain F_T:

$$F_Tl_2\sin\theta_2 = Fl_2\cos\theta_2; F_T = F / \tan\theta_2; F_T = 679.7 / 0.751 = 905 \text{ N}$$

Raising the heel off the ground significantly increases the force in the structures of the longitudinal arch and requires a very large force in the Achilles tendon.

As the body sways, compensatory changes in the anterior and posterior muscle forces will modify the force in the structures of the longitudinal arch when the heel remains on the ground. Considerable changes in these forces will also be seen in walking, in which there are inertial effects due to acceleration of the foot. Walking can be a particularly difficult and painful exercise for those suffering from plantar fasciitis.

Both the longitudinal and transverse arches can be maintained by footwear that produces lateral forces tending to avoid spreading the bones outward during walking and standing. Unfortunately, our physical evolution did not intend footwear that pinches the foot from the sides. Footwear, particularly if it is ill fitting, can produce some very serious conditions to which most podiatrists will attest.

■ Key Point

Horizontal momenta and angular momenta sum to zero over one walking cycle at a constant walking speed; vertical momentum and momentum due to gravity also sum to zero.

Aim of Running

The purpose of running is to allow us to move faster than we do while walking. The goal is usually to run either as fast or as far as possible. These represent two different aims with the common factor of producing sufficient vertical force to allow us to experience an airborne phase.

Mechanics of Running

Running at a constant speed appears mechanically to be a simple problem. All that is required is a horizontal mechanical impulse that adds up to zero and a vertical impulse equaling that produced by gravity, as follows:

Horizontally   *Vertically*

$$\int F_H dt + \int R dt = 0 \qquad \int F_V dt + \int W dt = 0$$

(Equation 6.4)

where R is the sum of forces of friction and air resistance, and always in the opposite direction to F_H; F_V is upward vertical force; and W is body weight or mg. An exception to these requirements is during acceleration when our net horizontal impulse increases with time and the left side of the horizontal equation is greater than zero. Running at submaximal speed for an extended period of time requires delivery of electrochemical energy to the muscles at the same rate at which the muscles are doing work as follows:

$$dE / dt = \text{Muscle power}$$

(Equation 6.5)

Running as fast as possible requires maximal delivery of electrochemical energy to the muscles so that they can produce maximal work. In this case dE/dt will be maximal. Eventually energy stores within the muscle will be depleted, dE/dt will decrease, and muscle power output will decline. Therefore the effort to run as fast as possible inevitably leads to a reduction in speed. Such an effect can be seen even in a race as short as 100 m in which the finishing speed is less than that achieved earlier in the race. Following this state, rest is required in order to recharge the muscular energy stores. This recharging ability is evidenced by the fact that athletes can perform more than one race per day, the second race often demonstrating a performance superior to that of the first.

Biomechanics of Running

The exact objective function that the body uses to determine running technique is unknown, as also discussed earlier in the chapter in connection with walking. A further complication is that there are more types of running than there are of walking, so we could expect different objective functions depending on the aim of the run. One candidate is minimization of energy cost due to the stress imposed upon the cardiovascular system. Before dealing with the intricacies of the biomechanics of running, it is necessary to describe some general features of the activity.

Running occurs as sustained activity over a long distance and also as brief activity over short distances. Obviously it takes a longer time to run longer distances, as shown in figure 6.11, which was obtained from data of world record performances.

This figure appears to show a straight line, which implies that we can travel at the same speed over all distances. This is misleading because the figure refers to such a large range of distances. The fact is that the faster you run, the less distance you can cover. When the average speed of running is calculated as distance divided by time, we obtain figure 6.12.

Here we see that the greatest average speed is performed in the 200 m race. It seems at first surprising that the greatest average speed is not obtained from the 100 m race.

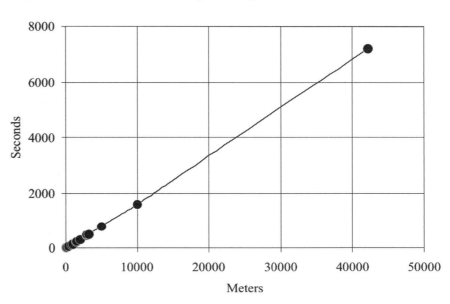

FIGURE 6.11 Running time and distance appear to be directly proportional over this large range of distances.

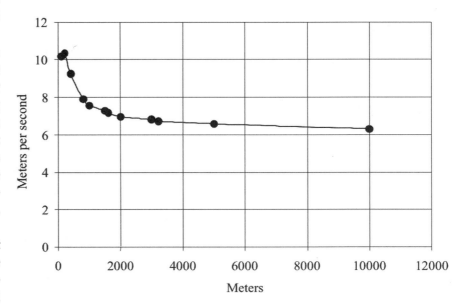

FIGURE 6.12 The 200 m runner has the greatest average speed of all runners.

■ **Key Point**

Maximal average
running speed
is a function of
the duration of
the acceleration
phase and the time
over which readily
available energy
stores last.

There are good reasons for this. The short sprints involve acceleration. Acceleration implies time spent below maximal speed, and this acceleration time represents a greater proportion of the race time in the 100 m compared with the 200 m. The question following logically is why the 400 m is not run at a greater average speed than the 200 m for the same reasons of acceleration. The answer here lies in the energetics of supplying our muscles with fuel and the way in which our muscles use the fuel. The white or fast muscle fibers that we call upon for rapid movements use fuel that is stored in the muscles and that is therefore readily available. This activity is known as anaerobic activity, as it occurs without oxygen. Unfortunately these stores are depleted when we sprint, so that we cannot sustain the same speed. This is another reason the average speed between 200 m and 1000 m races shows a rapid decrease. It also falls, but at a much lower rate, between 1000 and 2000 m. The reason for the lower rate is that we begin to change over from predominant use of white fibers to include use of more red fibers. The fuel required to use white fibers eventually gets restored after a substantial period of rest.

The discussion in "Mechanics of Running" captures the mechanical essentials of running but hardly describes the complex nature of the activity. The bipedal activity of running can be described as one in which progress is made by successive interactions of each foot with the ground, which themselves require that the foot leaving the ground be brought forward for the next ground contact. A further feature is that there is an airborne phase. Running is easier than walking to visualize and to understand biomechanically in the sense that the motion of the body CM is entirely the result of kinematic conditions just prior to landing and the force on the supporting foot during ground contact.

Biomechanics of Support

The support phase in running is that period when ground contact is made with one foot . Figure 6.13 is an approximate representation of forces for rear-foot (RF) and midfoot (MF) strikers, who land on the heel first and land flat-footed, respectively. These patterns are not dissimilar from those seen in walking (see figure 6.3), but the magnitudes of the forces are significantly greater in running.

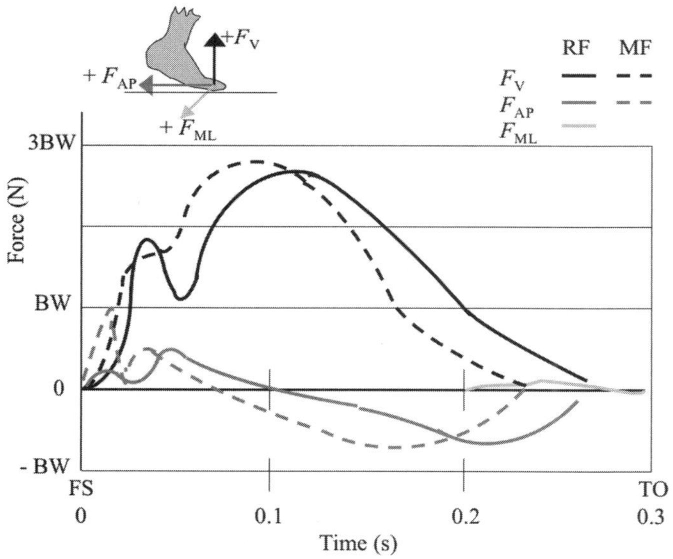

The major difference between the two landing techniques is the sharp initial peak in vertical (V) force exhibited by the rear-foot striker. This difference closely resembles the difference seen during walking with and without a rigid shoe, and its effect is likely to be the same.

Midfoot strikers whose vertical force pattern rises fairly smoothly do not require as substantial an eccentric contraction of dorsiflexors as do rear-foot strikers. The former are likely to transmit a significant load through the ankle and knee joints. On the other hand, the RF strikers load the anterior tibial muscles substantially to lessen the shock to ankle and knee joints. A potential problem for RF strikers is stress placed upon the large insertion area

FIGURE 6.13 The part of the foot that strikes the ground first has a significant effect upon ground reaction force on the foot in running.

of the anterior tibial muscles. Successive loading in this manner is the probable cause of shinsplints, or at least irritation of the periosteum covering the tibia. The biomechanical expression of the shock sustained at foot strike is not so much the sharp peak seen in RF strikers but its differential, namely the "jerk," which is the rate of change of force with respect to time. As the tendinous insertion of the muscle into the tibia is viscoelastic, large jerk values can be expected to put great load onto the viscous component without stressing the elastic component significantly. Unfortunately the viscous element, although it is energy dissipating, does not allow much elongation, and stress is applied to the insertion.

The shape of the vertical force profile will determine the velocity vector of the CM at takeoff. Should the vertical impulse ($\int F_v dt$) be greater than that due to gravity ($\int W dt$), there will be an upward component to the velocity of the CM (see Equation 6.4). This will lead to a downward component at foot strike. While some upward and downward components of velocity are inevitable, large values are uneconomical. A large value represents a large increase in potential energy due to an increase in the work done by the muscles of the lower limbs. Therefore such an undulating running pattern is wasteful of energy. A further effect of this type of running is that it increases the vertical impulse on landing with an associated increase in force and therefore shock at foot strike. The objective function in this case would seem to be to minimize the upward and downward motion while airborne to a level where minimal vertical mechanical impulse and associated muscular work during the subsequent support phase are required.

Lower Limb Recovery

Once the foot leaves the ground, the recovery phase begins. The aim then is to get that foot forward as soon as possible. Muscles get tired in running, not only due to the stance phase in which they are required to support us against gravity and to push off. If this were the case we could sprint almost forever at a high constant speed, as the stance phase would consist only of supporting the body with every step. The fact is that the muscles are used to accelerate the lower limb forward in relation to the CM in the recovery phase. Muscle forces are therefore acting over distances and represent work done (Chapman et al., 1984). We know that there is a cost to this in terms of energy supply. Happily the construction of the limbs enables us to save a certain amount of energy.

Figure 6.14 shows the early (left) and late (right) stages of lower limb recovery, with toe-off (TO) and heel strike (HS) representing the beginning and end of recovery. In the early stage, strong hip flexor moment allows the thigh to pull on the leg at the knee (F_l), where equal and opposite forces occur on the femur (F_t). F_l multiplied by the perpendicular lever arm (d) creates a clockwise moment $F_l d$, which aids knee flexion. So knee flexion is produced as a result of activity of the hip flexor musculature. In sprinting, this effect is so strong that knee extensors need to be activated to slow down knee flexion and avoid the heel bruising the buttock. The reverse is the case in the later stage of recovery, in which the hip extensor musculature induces a force at the knee and provides final knee extension in preparation for touchdown of the foot. Again this effect is large, and knee flexors are required to obviate an extension injury to the joint. Advantage is gained because hip muscles are much larger and stronger than knee flexor and extensor muscles and can do more work. Also the task of lifting the leg and then

■ **Key Point**

The part of the foot that strikes the ground first has a significant effect on the type of repetitive strain injury that may be sustained.

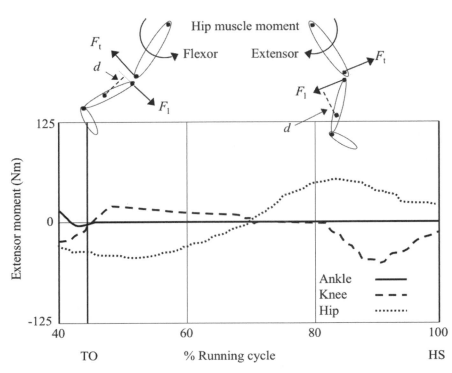

FIGURE 6.14 Hip and knee moments in the recovery phase of running.

extending it is being performed partly by hip musculature. The advantage obtained by pendular motion of the recovery limb as described in the section on walking is largely insignificant in running. The reason is that the period of pendular oscillation of the lower limb is far too long to be effective in the rapid recovery phase of running.

Although the pendular effect is insignificant in the recovery phase, the ability to reorient segments of the lower limb can be used beneficially. The sprinter flexes the lower limb at the knee so the moment of inertia about the transverse hip axis is reduced. Consequently the angular acceleration of the lower limb will be greater for a given hip moment or recovered with the same acceleration by a smaller hip moment. As running speed increases, the rate at which the moment of inertia of the lower limb can be changed will be limited by the rate at which muscular work (power) can be produced both concentrically and eccentrically. Indeed such power considerations appear to be one factor implicated in the limitation of maximal sprinting speed (Chapman and Caldwell, 1983b). In running at a slower speed there is no need to recover the lower limb quickly, so recovery can be made with the greater moment of inertia of a straighter limb. The appropriate rates of knee flexion and subsequent knee extension require fine control of hip and knee musculature.

The effects described are common to all sprinters, yet there are minor differences in sprinting style. Such differences are partly the result of individual differences in size of muscles and length of limb segments.

Sustained Running

A question of considerable importance is, "What determines our running style?" The objective function of minimized muscular impulse and associated minimized muscular work has been alluded to previously. There are other possible biomechanical factors that might determine running style. Running style was investigated in our study of "funny running," in which we filmed running at the same speed with a variety of accentuated styles (Lonergan, 1988). These comprised stiff lower limbs, high knee raises, and others including the preferred style. In all of the funny styles, there was at least one joint moment that was very high compared with that seen in the preferred style. Such a result means that a particular set of muscles would fatigue more rapidly than normal. What is certain is that the joint moments produced in a "normal" running style are all kept within reasonable limits of the individual's capabilities so that no particular set

■ Key Point

In recovery, the hip musculature is responsible for the greater part of the energy required, much of which is transmitted to the lower limb segments by eccentric contraction of the knee musculature.

■ Key Point

Flexion of the knee joint in recovery reduces the moment of inertia of the lower limb and allows greater angular acceleration.

of muscles is stressed over the others. When fatigue begins to set in, it would be ideal if all of the muscles fatigued simultaneously. As a practical consequence of this study, any group of muscles that shows more fatigue than others should be a specific target for training. Should specific training not solve the problem entirely, one should consider changing the running style.

Different types of muscle fibers require fuel in different forms. The red muscle fibers described previously can use oxygen combined with sources of chemical energy in the bloodstream to provide a continual supply of chemical energy to fulfill our mechanical energy requirements. This process provides energy aerobically. As we increase our demand for energy we require increased blood flow. This is why we we breathe deeper and faster when we increase running speed. Provided that the chemical energy is supplied at an appropriate rate, we can generate mechanical energy at a rate that allows us to run. For example, the average speed of running between 3000 m and the marathon varies little within the grand scale of average running speeds. In this type of sustained running we pay as we go. As we run faster, our mechanical energy requirements outstrip the rate of delivery of chemical energy, and we experience what is known as oxygen debt. We cannot deliver oxygen fast enough, and we have to either stop running or slow down. In either case, we continue to breathe as if we were running at the original speed until our oxygen debt is paid and we return to the normal metabolic state.

The human machine is no different from others in which the motors (our muscles) produce waste products during the conversion of chemical to mechanical energy. Unfortunately, we do not blow the waste products out of the end of an exhaust pipe as does our automobile engine. The only way the waste products can get out is through the bloodstream, and this can be a slow process. This is why your muscles feel stiff the day after an unaccustomed long run; pain receptors are in fact irritated by these waste products if they stay within the muscle. The best way to avoid this unfortunate pain or stiffness is to engage in running at a much reduced pace after the prior higher-intensity activity. The pressure changes in the muscle during contraction help to "milk" out the waste products so that they do not dwell within the muscle. This postrace exercise is used by competitive athletes and is known as warming down. It is named as such only because warming up is done before the race. Warming down is a misnomer and has little to do with warming per se.

A certain amount of the energy cost of distance running is reduced by the storage and release of elastic energy. Each time the foot hits the ground, the knee flexes to some extent and the knee extensor muscles and the patella tendon are stretched. Stretch in the tendon indicates storage of strain energy. When knee extension occurs, some of this strain energy is returned to the whole system, which reduces the amount of energy required of the knee extensor muscles. As the knee extensors move from stretch to shortening, the force-producing capabilities are increased by the prior stretch as described previously in chapter 3 in the section titled "Muscular Force." Consequently a smaller percentage of contractile activity is required to perform the task.

Acceleration

The most important direction for acceleration in sprinting is horizontal. So force in the AP direction is necessarily greater than that seen in figure 6.13 for

■ **Key Point**

Muscular fatigue is unlikely to be equally distributed among the lower limb muscles involved in running.

■ **Key Point**

Removal of waste products of running can be aided by gentle exercise following an intensive run.

■ **Key Point**

The stretch–shortening cycle of muscle is implicated in reducing the energy cost of running.

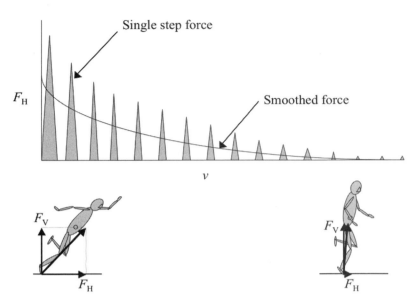

FIGURE 6.15 Acceleration in sprinting occurs in steps rather than following a smooth curve.

running at a constant speed. Horizontal components of force are so large at the beginning of a sprint race that starting blocks are employed. This strategy allows the force to be applied almost normal to the surface of the block so that the runner does not have to rely on limiting friction to supply the necessary ground reaction force. The combination of a vertical force supporting against gravity and the high horizontal force gives a force vector that is much more inclined to the horizontal than during running at a constant speed. Therefore the line joining the foot to the CM must be similarly inclined if a rotating moment on the body is to be avoided. Figure 6.15 shows, among other information, both force and position vectors at the beginning of the sprint.

At this instant the shortening velocity of hip, knee, and ankle extensors is zero, which gives them high force-producing capacity. As horizontal velocity increases, these muscles will be shortening ever more rapidly, so their force-producing capacity will decrease. The effect of this is to produce a more vertically inclined force vector. For the sprinter to avoid falling forward, the accumulated angular impulse must be counterclockwise in this figure. This requires the resultant force vector to pass in front of the CM. The result is a gradually decreasing forward lean of the body. Additionally the time of foot contact will decrease. The overall effect is to reduce the horizontal impulse with each successive step so that acceleration decreases with each successive step. Figure 6.15 also shows in stylized form the horizontal force (and therefore acceleration) with each ground contact plotted against horizontal velocity. The area under each force profile is power, which represents large rates of change of energy early in the race. It is clear that the smoothed force–velocity relationship is of the same form as that exhibited by muscle. This similarity establishes muscular properties as the major determinant of our acceleration profile.

As we know, acceleration represents changing speed, and we need force produced by our muscles in order to accelerate. The greater the force, the greater is the acceleration. So if our activity requires a very rapid change in speed, it is in our interest to invoke the greatest muscle force at the greatest rate at which it can rise. This is the role of the white (or fast) muscle fibers. These fibers do not immediately use oxygen to convert the fuel stored within them into mechanical contraction; the contraction is named anaerobic. As a result, the energy supply to this type of fiber is soon exhausted. Such properties of muscle are characteristic of all animals such as cheetahs and greyhounds that can accelerate rapidly and achieve extraordinary speeds. So we and these other remarkable animals cannot accelerate for long periods of time. This is a good thing since it does not allow us to reach running speeds that could be injurious.

■ **Key Point**

Acceleration is limited partly by the force–velocity relationship of muscle.

Maximal Running Speed

The question of who is the fastest runner in the world is asked frequently. The answer depends upon how you define the "fastest runner." When I ask "How would you find out?" most respondents reply "Look for the world record holder of the 100 m." Unfortunately this answer is not always correct. The objective in winning the 100 m race is to get there first. In fact, it is usually the best accelerator who wins this race because the best accelerators achieve their top speed sooner and spend a greater proportion of the race at top speed. Of course it may happen that they are caught by a runner with a greater top speed near the finishing line. But the prize is for getting there first, not having the greatest speed when you get there. If the criterion of fastest is the greatest average speed, then the 200 m record holder wins the prize as shown in figure 6.12. The most logical definition is greatest sprinting speed, be it for only an instant in time. Unfortunately this value cannot be obtained from data on the time to complete the race. Such a determination would require continual monitoring of the position of the runner with respect to time. So the answer to the question, as I choose to define the fastest runner, remains "We don't know until we define the criterion."

What we do know is that there is a maximal running speed. The limitations to running faster have been covered to some extent with reference to fuel supply, which limits our ability to sustain maximal running speed. The question that remains is whether there are any other factors, either extrinsic or intrinsic, that limit maximal running speed itself rather than limit our ability to sustain it. Clearly our muscular properties must play an intrinsic part. Figure 3.3c shows that there is a maximal speed of muscle shortening at which no force is developed. It is doubtful that this speed of shortening is ever reached in any of our sprinting muscles. But it is likely that we reach stepping frequencies and therefore shortening speeds above which the muscles cannot apply propulsive force to the ground. In other words, the force is used to accelerate and decelerate the limbs in the recovery phase so that the speed of the landing foot is equal to the speed of the CM of the body. Obviously the speed of the foot relative to the ground is therefore zero.

What is it about the muscles that determines this speed? Figure 6.16 shows diagrammatically an analogy between the contractile elements of muscle and the runner's lower limb. Tiny actin and myosin filaments, as shown in figure 6.16a, slide past each other under the influence of cross-bridge motion. In figure 6.16b, the body slides past the ground under the influence of lower limb motion. The analogy is that the lower limbs represent the cross-bridges. There are only certain sites where the cross-bridges can attach and develop force. After rotating they detach and look for another approaching site. There is a complex series of electrochemical events that contribute to the time course of this cycling effect. When a certain speed of shortening is reached, the cross-bridges can only keep up and cannot develop force. It takes little imagination to draw the analogy with the lower limbs and appreciate how maximal running speed is limited. Of course we have only two legs, whereas there are billions of actin and myosin filaments in muscle. The study of theories of muscular contraction is very current, and so far many bits and pieces are known. Yet the whole process still eludes investigators in this field. I have given a very simplistic version of an extremely complex process. I believe that this version captures the essentials for our purposes.

■ Key Point
The best accelerators win short sprint races even though they may not have the greatest top speed.

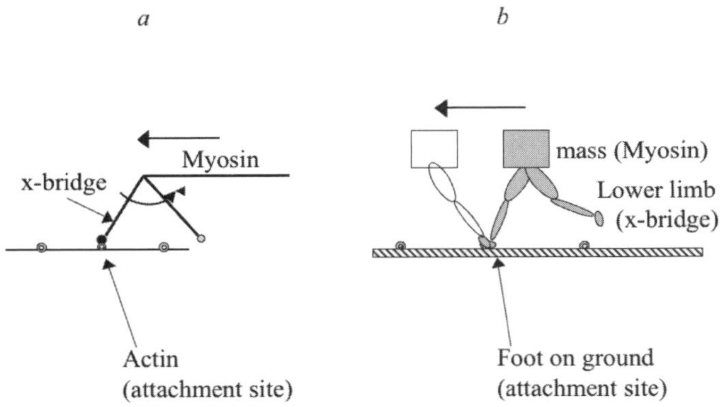

a   b

FIGURE 6.16 Does the muscle shortening mechanism limit maximal sprinting speed?: an analogy.

An extrinsic factor is air resistance, because the faster we travel through the air the greater is the force against us. Some of the force we generate against the ground opposes this decelerating effect, and we lose some force to accelerate our limbs back and forth. Air resistance is not trivial in sprinting. World records are rarely if ever made against a headwind. The sprinter's greatest wish is for a tailwind that is just below the legal limit imposed by the authorities. The authorities have been spoilsports in another area. Readers will have noticed that running tracks lie flat in a horizontal plane. If we sprinted downhill, there would be a small component of gravitational force accelerating us. The opposite is the case uphill. Perhaps it is comforting to know that the ancient Greeks seem to have been responsible for this horizontal stipulation, so we cannot blame any living being.

Energy Cost in Running

Running is costly in comparison with walking. The fact is that we cannot use the pendulum effect in running to the extent we do in walking. The time of one swing of a pendulum is related partly to its length. So each length of pendulum has its own natural frequency of oscillation. This is partly why we have a preferred frequency of steps in walking. Departure from the natural frequency requires work to be done by our muscles. In running the stepping frequency increases to values well above the natural frequency. Therefore we require a great contribution from muscular contraction to accelerate and decelerate our lower limbs over a large range of movement. Added to this cost is the work required to project ourselves into the air as a means of increasing flight time and therefore stride length. As we strive for greater stride length, the angular momentum generated about the long axis of the trunk is increased. To counteract this disturbing motion, we use forward motion of the arm opposite to the leg that is moving forward as in walking. Of course the arm motion must be vigorous, and the work done by the muscles of the upper limb represents a further cost to the human machine. If you have ever wondered why sprinters need those muscular arms and shoulders, this is your answer—they train for this quality.

Fatigue and Running Style

Rapid onset of fatigue is seen particularly in races between 400 and 1500 m, where there is a sprinting component. The shorter races are completely anaerobic, while the longer races use anaerobic sources of energy in the final sprint. It is not surprising that the running style changes during these races due to muscular fatigue. Also style will change differently in different runners because of their individual tolerances to aerobic and anaerobic fatigue. Figure 6.17 shows plots of hip angle versus knee angle at the first 100 m (normal) and at 380 m (fatigued) of a 400 m race (Chapman and Medhurst, 1981).

The diagram is cyclical, beginning and ending at the same point, because time is removed in such a presentation of data. The significant point is that the pattern

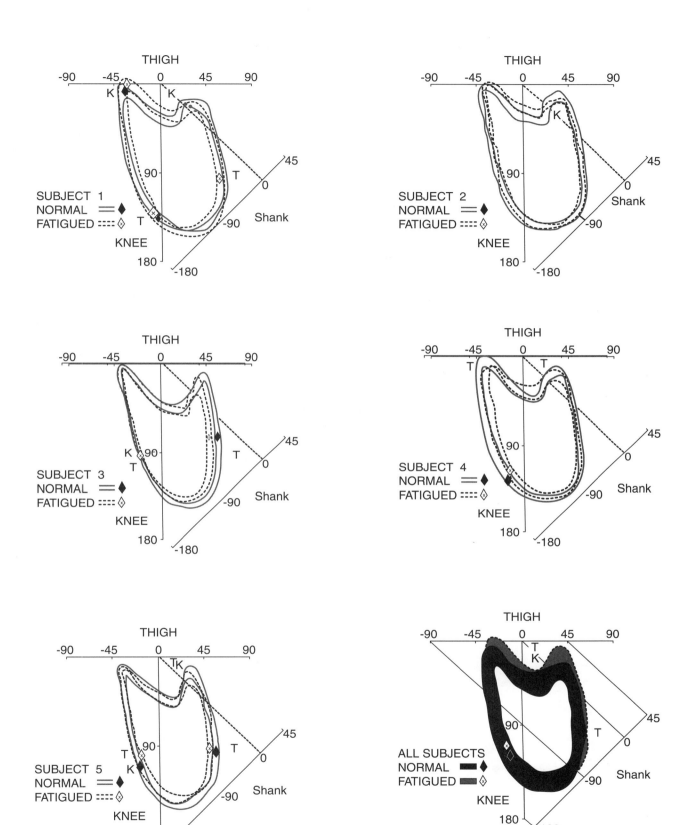

FIGURE 6.17 Fatigue affects different sprinters in different ways according to the hip and knee angular kinematics.

Reprinted from A.E. Chapman and C.W. Medhurst, 1981, "Cyclographic evidence of fatigue in sprinting," *J. Human Movement Studies* 7: 225-272.

of motion changes from the fresh to the fatigued state and does so differently for each athlete. This is clear evidence that muscular capabilities fatigue differently among individuals and that the differences are specific to the individual. Biomechanical analysis has been used to identify a hierarchy of changes that are induced by fatigue (Chapman, 1982). Such analyses should aid in identifying areas of emphasis in training to respond to fatigue. The major conclusion is that the athlete is a machine with capabilities that change during fatigue and should be treated as such if technique is to be changed in response to fatigue.

Coaching wisdom currently stresses that the athlete should attempt to maintain the same style even when fatigued. This makes little sense, since the runner is a different human machine when fatigued than when fresh. Athletes should change their style based on the extent to which specific muscles fatigue. Unfortunately this is a very complex problem, and to my knowledge we do not yet know enough about the machine to be able to investigate and prescribe changes. So the poor athlete is left to trial and error in this context. All is not gloom, though, because athletes and coaches have been using trial and error since competition began and this has proved successful. Sadly, some coaches have never thought of any other approach to training.

■ **Key Point**

The changing pattern of lower limb motion in fatigue suggests specific muscular endurance training to offset this effect or training to modify the motion pattern to accommodate the fatigue.

Variations of Running

In addition to running in a straight line on a horizontal surface, runners are required to change direction and also to negotiate slopes. Bend running is confined mostly to sport and formal exercise. While running speed may not change, running velocity may. Such a statement seems nonsensical unless we are prepared to accept that speed and velocity are not synonymous. Speed is measured in units of distance per second in the direction in which you happen to be traveling at the time. Sprinters may travel at 10 m per second around a bend continuously, so their speed is constant in the instantaneous direction they are going. Velocity is also measured in distance per second, but with the proviso that a particular direction is stipulated at the outset and maintained as the reference direction throughout the run. Since the runner goes around a bend, the direction of motion is changing, so the velocity is changing. We can feel this change in velocity as acceleration toward the center of the bend, and we know that we need a

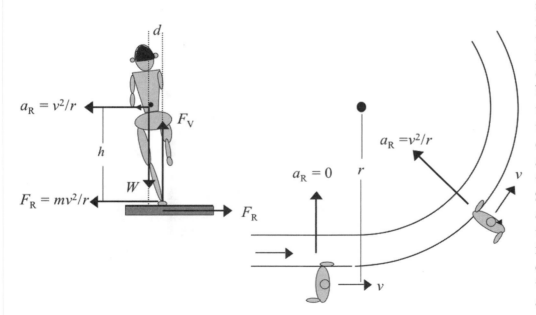

FIGURE 6.18 Although speed may be constant, its changing direction indicates acceleration during bend running.

force to produce this acceleration. Figure 6.18 illustrates the difference between velocity and speed.

On the right is shown a runner who requires a radial acceleration a_R of v^2/r toward the center of the circle. The force F_R required to produce this acceleration is acceleration multiplied by the runner's mass (or mv^2/r), and is provided by the frictional force at the foot. Remember how difficult it is to run in any other than a straight line on ice. Skates have sharp edges to allow this force to be applied without slipping. Sprinters use shoes with spikes in the sole for just this purpose. This centripetal force F_R creates a clockwise moment of $F_R h$ about an axis through the CM. For lateral toppling to be avoided, this moment must equal the counterclockwise moment $F_V d$ created by the upward ground reaction force F_V on the foot. Because F_R varies as v and r vary, d must change in order for toppling to be avoided. Sprinters must anticipate such change as they enter and leave a bend. The problem of bend running is even more complicated than that described in the foregoing quasistatic case. When airborne, the runner will travel in a straight line and will land with the CM moving in the wrong direction. So F_R must be greater than during midstance in order for the circular motion to be recaptured. Furthermore, the vertical ground reaction force F_V will vary, leading to changes in the counterclockwise moment about the CM. Consequently the degree of body lean will need to undergo many subtle changes in order for successful bend running to be performed. This complex motion would provide a suitable graduate topic using simulation and optimization. My apologies for any work of this nature that I have overlooked.

All circular motion, such as swinging a stone on a string overhead and the motion of our moon around us, requires a centripetal force. The former arises from tension in the string, and the latter arises from gravitational attraction. Also you feel the door of your automobile pressing on your shoulder as you make a sharp turn at speed unless the seat belt restrains you, as it should.

Most athletes who sprint between 200 and 800 m run part of their races in lanes that have different values for r. As we have deduced that the force F is inversely related to r, F is greatest on the inside lane if v is the same. It is a more difficult task to sprint on the inside lane, although most sprinters would choose to run in that lane if they could. Such a choice seems to relate to a psychological need to see the opposition in front of them in staggered starts. Unfortunately this choice is directly opposite to what would be recommended from a biomechanical viewpoint.

The human machine is not well designed to run around a bend from the point of view of lower limb length. If each lower limb has the same hip, knee, and ankle angles at foot contact, each hip joint will be at the same height above the ground at this instant. There are two strategies one can use to handle this situation. The necessary angle of lean of the body will require running with the line joining the hip joints horizontal using pelvic rotation. Thus this line through the pelvis will not be perpendicular to the long axis of the spine. The other strategy is to undergo a rocking motion of the pelvis from side to side with each step. In either case, continual practice at bend running may lead to some unbalancing of

■ **Key Point**

Radial or centripetal acceleration is least in the outside lane during bend running.

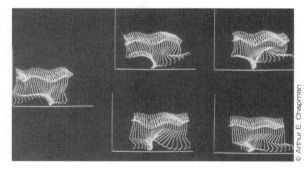

The essentials of different running styles can be visualized from cinematic images treated by computer.

the musculoskeletal system. The same potentially injurious effect applies to the knee and ankle joints, which are obviously subject to different force patterns than during running in a straight line. The solution would be to run every other race in the opposite direction, but the authorities do not allow this for obvious logistical reasons—some athletes would be running in the wrong direction.

There are many similarities between running and walking on sloping ground within the context of changes in potential energy as a function of the gradient of the slope. In running the forces are generally greater, since the momentum changes are greater. A significant feature of downhill running is weight acceptance by the lead lower limb. Upon landing, the downward velocity will tend to be great due to the accelerating effect of gravity. The knee and hip extensors will be required to create large moments at their respective joints. These muscles can easily produce such moments since their action in weight acceptance is eccentric. Therefore there is the possibility of very large forces induced in the tendons and across the joint surfaces. If weight acceptance also incorporates a definite heel strike, the anterior tibial muscles will also be subject to high eccentric forces as a means of avoiding heel shock. The potential for overuse injury, not to mention a single traumatic injury, is high. Running up a slope is far less dangerous since shock is not a factor. However, the ankle plantarflexors will be required to work within a changed range of shortening, with the same type of effect as seen in walking.

Enhancement of Running

Casual observation shows that some children are naturally faster runners than others. This difference persists with maturity, and it is remarkable that some people are much faster runners than others without any training whatsoever. While everyone can have his or her maximal sprinting speed increased, it is not possible to make a champion sprinter out of someone who is only an average sprinter to begin with. Biomechanical science appears not to have made any significant strides in explaining why some people are naturally fast runners. However, there are numerous biomechanical components to sprinting that can lead to increases in maximal sprinting speed.

One can achieve increases in maximal running speed by increasing stride length, stride frequency, or both simultaneously. Although not considered in the section on biomechanics, it is clear that the forces opposing the necessary range of motion of the lower limbs will counteract the work performed by muscles. Such forces can arise from connective tissue stretch. Mobility exercise is therefore required to increase the unimpeded range of motion of the hip joints. As a large part of the stride length occurs while the person is airborne, the horizontal velocity at takeoff must equal that at prior heel strike. This argues for specific muscular strength training to extend the force–velocity relationship of the muscle along the velocity axis. In other words, it is desirable to increase the maximal velocity of shortening of the muscles if this is possible. Increase in stride frequency necessitates strong muscles due to the accelerations required of the limb segments when one is changing direction rapidly. Therefore training to increase maximal muscular force will enhance sprinting. Although not part of the subject matter of this book, neuromuscular training is needed because of the rapid turning on and off of muscles.

Muscles of the upper limbs, particularly the shoulder flexors and extensors, require strength training for much the same reasons as discussed in the context of lower limb muscles. The need for strength training of the upper limb musculature is attributable to the fact the upper limbs provide a counteracting effect to that of the lower limbs in terms of angular momentum about the longitudinal axis. If the inertial forces produced by acceleration of the upper limb muscles are insufficient, the trunk will be subject to axial rotation. This will produce a rolling gait that will affect the motion of the pelvis and the length range within which the hip musculature acts. Another reason for strengthening the shoulder muscles is to increase the upper limb acceleration and thus provide an increased ground reaction force during sprint starts.

The relative importance of the various factors involved in sprinting has not been elucidated biomechanically. Therefore it is unclear where people should place the emphasis when training to enhance sprinting speed.

Running Safety

It is essential that modifications be made to the soles of running shoes in order to avoid sliding, and modifications over time have followed changes in the nature of track surfaces. Fifty years ago, tracks were variously constructed of clay, ash, or some similar particle material that was rolled to provide a hard surface. These surfaces didn't remain hard for long. In even earlier days, sprinters carried a trowel with which to dig holes for their feet at the starting line. The shoes were armed with long spikes to avoid slippage. Since then, composition tracks of a wide variety of chemically bound materials have been developed. Track shoes currently have a combination of a roughened, patterned sole and small, sharp spikes of a hard material that dig into the surface of the track.

Repetitive strain injuries are a potential feature of long distance running, and appropriate shoes are a necessity. Appropriateness is largely determined by the energy-dissipating properties of the shoes, but their construction also takes into account the nature of foot–ground contact of the individual.

Running is essentially safe, and the problems that arise are largely self-induced. Appropriate footwear for different types of running is very advanced in construction. Therefore the major safety problems are not biomechanically induced, but they have a biomechanical result. The problem appears to be that many of those who suffer injury in this activity utilize the advanced shoes not to run the same distance and avoid injury, but to run greater distances to reach the same injury level. This is not a biomechanical problem; it is psychological.

Practical Example 6.3

The lower limb shown in figure 6.19 is in the recovery phase of sprinting and under the influence of hip flexor and knee extensor forces.

The knee joint axis is disjointed for the purposes of illustration, and both hip and knee articulations are pin joints with their axes perpendicular to the vertical plane of progression. The data shown are typical of those one can obtain or calculate from a film or other optical device that registers movement of body markers on limb segments.

■ **Key Point**

People should train the muscles of the upper limbs for stability of the trunk and head during sprinting and also to increase the horizontal ground reaction force during acceleration.

■ **Key Point**

Running is safe if it is done in moderation with appropriate footwear on a plane surface with a clear visual field; alteration of any of these requirements spells biomechanical danger.

$$F_{RH} = ? \quad F_{RK} = ?$$
$$M_H = ? \quad M_K = ?$$
$$\theta_{RH} = ? \quad \theta_{RK} = ?$$
$$\beta_H = 5.0 \text{ rad} \quad \beta_L = 4.0 \text{ rad}$$
$$\omega_T = 2\pi \text{ rad/s} \quad \omega_L = \pi \text{ rad/s}$$
$$\alpha_T = 2.0 \text{ rad/s}^2 \quad \alpha_L = 3.0 \text{ rad/s}^2$$
$$v_H = 11.0 \text{ m/s} \quad v_K = 15 \text{ m/s}$$
$$\gamma_{VH} = 1.1 \text{ rad} \quad \gamma_{VK} = 1.2 \text{ rad}$$
$$a_T = 3 \text{ m.s}^{-2} \quad a_L = 5 \text{ m.s}^{-2}$$
$$\varphi_{aT} = 1.0 \text{ rad} \quad \varphi_{aL} = 0.5 \text{ rad}$$
$$m_T = 8 \text{ kg} \quad m_L = 4 \text{ kg}$$
$$I_{GT} = 0.24 \text{ kg.m}^2 \quad I_{GL} = 0.1 \text{ kg.m2}$$
$$r_T = 20 \text{ cm} \quad r_L = 25 \text{ cm} \quad L = 50 \text{ cm}$$

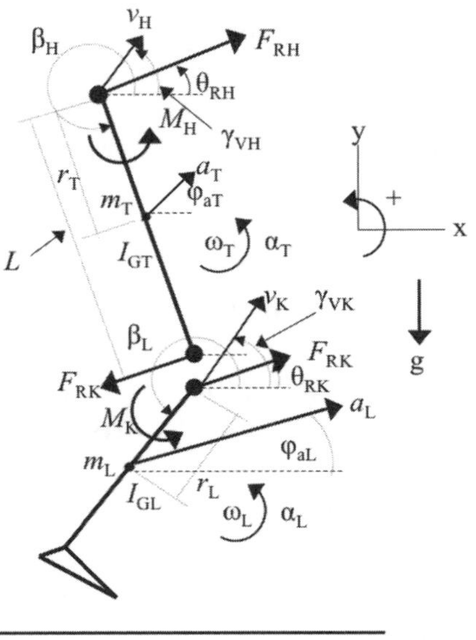

FIGURE 6.19 Forces and moments acting at one instance in the recovery phase of running.

The following questions require a process of inverse dynamic analysis beginning with motion of the leg-foot segment as the distal segment. This is necessary since the forces producing the movement are unknown.

Part 1

Calculate the resultant forces at the knee (F_{RK}) and the hip (F_{RH}) and the net moments at the knee (M_K) and hip (M_H) given the values shown.

Solution

F_{RK} and θ_{RK} are unknown, so we require two equations to solve for these two unknowns. We begin with resolution of forces in the x and y directions.

Resolving horizontally (x) and vertically (y),

$$F_{RK}\cos\theta_{RK} = m_L a_L \cos\varphi_{aL}; \quad F_{RK}\sin\theta_{RK} = m_L a_L \sin\varphi_{aL} + m_L g.$$

Dividing the right-hand equation by the left-hand one gives

$$\tan\theta_{RK} = (m_L a_L \sin\varphi_{aL} + m_L g) \,/\, m_L a_L \cos\varphi_{aL} = \tan\varphi_{aL} + (g \,/\, a_L \cos\varphi_{aL});$$
$$\theta_{RK} = \tan^{-1}[\tan\varphi_{aL} + (g \,/\, a_L \cos\varphi_{aL})];$$
$$\theta_{RK} = \tan^{-1}[0.9388] = 0.754 \text{ rad}.$$

Having obtained θ_{RK}, it can be substituted into one of the initial equations as follows:

$$F_{RK} = m_L a_L \cos\varphi_{aL} \,/\, \cos\theta_{RK} = 4 \times 5 \times \cos(0.5) \,/\, \cos(0.754) = 24.078 \text{ N}$$

F_{RK} now becomes an input force in an equal and opposite direction at the distal end of the thigh segment as follows:

$$-F_{RK}\cos\theta_{RK} + F_{RH}\cos\theta_{RH} = m_T a_T \cos\varphi_{aT}; \quad -F_{RK}\sin\theta_{RK} + F_{RH}\sin\theta_{RH} = m_T a_T \sin\varphi_{aT} + m_T g$$

Since F_{RK} and θ_{RK} are known, the equations can be reduced to

$$F_{RH}\cos\theta_{RH} = m_T a_T \cos\varphi_{aT} + (17.552); \ F_{RH}\sin\theta_{RH} = m_T a_T \sin\varphi_{aT} + m_T g + 16.963.$$

Again dividing the right-hand equation by the left gives

$$\tan\theta_{RH} = (m_T a_T \sin\varphi_{aT} + m_T g + 16.936) \ / \ (m_T a_T \cos\varphi_{aT} + 17.552);$$
$$\theta_{RH} = \tan^{-1}(115.611 \ / \ 30.519) = 1.313 \text{ rad.}$$

Substituting this result into the left-hand equation gives

$$F_{RH} = [m_T a_T \cos\varphi_{aT} + (17.552)] \ / \ \cos\theta_{RH};$$
$$F_{RH} = [8 \times 3 \times \cos(1) + 17.552] \ / \ \cos(1.313);$$
$$F_{RH} = 119.707 \text{ N.}$$

The moments M_K and M_H are obtained in a similar fashion based on the fact that the net moment acting about the knee joint on the leg has an equal and opposite effect on the thigh. The choice of the axis about which to calculate moments is a matter of convenience. The simplest is the joint axis rather than the one through the CM of the segment since the joint force, in passing through the joint, does not have a moment arm and therefore does not appear in the equations.

The equation of angular motion about the knee joint for the leg is

$$M_K - m_L g r_L \cos\beta_L = (I_{GL} + m_L r_L^2)\alpha_L;$$
$$M_K = [(0.1 + 4 \times 0.25^2) \times 3] + [4 \times 9.81 \times 0.25 \times \cos(4)];$$
$$M_K = -5.3622 \text{ Nm}$$

T_K now becomes an input moment in an equal and opposite direction at the distal end of the thigh segment as follows:

$$M_H - M_K - m_T g r_T \cos\beta_T = (I_{GT} + m_T r_T^2)\alpha_T;$$
$$M_H = [(0.24 + 8 \times 0.2^2) \times 2] + [-5.3622] + [(8 \times 9.81 \times 0.2) \times \cos(5)];$$
$$M_H = 0.2102 \text{ Nm}$$

The negative knee moment represents knee flexor muscle activity while the positive hip moment represents hip flexor activity, and it must be remembered that they are net moments due to the contraction of many muscles. They are mathematically generalized moments that can be applied to the segment at any point. Ascribing them as joint moments is reasonable since we know that the muscles producing them act across the joint. Some muscles actually cross both joints, so their effect requires a different form of analysis that cannot be performed from the data given here.

Part 2

Calculate the rate of change of kinetic energy of each segment due to force and moment at each segment's proximal end.

Solution

The rate of change of energy is the first differential of energy with respect to time. Since energy is Fs and $M\theta$, its rate of change is Fds/dt and $Md\theta/dt$, which can be rewritten as Fv and $M\omega$, respectively. These represent joint force power (JFP) and muscle moment power (MMP), respectively.

For the leg segment, JFP_L is

$$JFP_L = (F_{RK}\cos\theta_{RK} \times v_K\cos\gamma_{VK}) + (F_{RK}\sin\theta_{RK} \times v_K\sin\gamma_{VK});$$

$$JFP_L = F_{RK}v_K[(\cos\theta_{RK} \times \cos\gamma_{VK}) + (\sin\theta_{RK} \times \sin\gamma_{VK})] = F_{RK}v_K[\cos(\theta_{RK} - \gamma_{VK})];$$

$$JFP_L = 24.078 \times 15 \times \cos(0.754 - 1.2);$$

$$JFP_L = 24.078 \times 15 \times (0.902) = 325.84 \text{ W}.$$

For the leg segment, MMP_L is

$$MMP_L = M_K\omega_L = -.553 \times \pi = -1.737 \text{ W}.$$

The leg segment is gaining kinetic energy largely due to JFP, while the small MMP represents a loss of kinetic energy. The reason for the latter is that the moment and angular velocity are in opposite directions. The rate of change of energy of the leg segment due to MMP and JFP at its proximal end is

$$JFP_L + MMP_L = 325.84 - 1.737 = 324.103 \text{ W}.$$

For the thigh segment, JFP_H is

$$JFP_H = F_{RH}v_H \times \cos(\theta_{RH} - \gamma_{VH});$$

$$JFP_H = 119.707 \times 11 \times 0.977 = 1286.49 \text{ W}.$$

For the thigh segment, MMP_T is

$$MMP_T = M_H\omega_T = -0.323 \times 2\pi = -2.029 \text{ W}.$$

The rate of change of energy of the thigh segment due to MMP and JFP at its proximal end is

$$JFP_H + MMP_T = 1286.49 - 2.029 = 1284.461 \text{ W}.$$

Part 3

At what rate is energy being transferred across the knee joint by JFP and MMP?

Solution

Since the velocity of the proximal end of the leg segment is identical to that of the distal end of the thigh segment, the product Fv for one segment equals $-Fv$ for the connecting segment. Therefore the rate of gain of energy of the leg segment represents the rate of transfer of energy from the thigh segment. So the leg segment gains energy at a rate of 325.84 W. As the thigh segment loses energy through this mechanism at the same rate, the total rate of gain of energy of the thigh segment through JFP is $JFP_H - JFP_L$, which equals

$$JFP_H - JFP_L = 1286.49 - 325.84 = 960.65 \text{ W}.$$

In the case of transfer of energy due to MMP, the angular velocities of each segment are different. The rate of change of energy is then

$$M_K(\omega_T - \omega_L) = -0.553(2\pi - \pi) = -1.737 \text{ W}.$$

Since this value is negative, the thigh is gaining energy through MMP at the expense of the leg.

Practical Example 6.4

Two sprinters of mass = 80 kg (176 lb) are accelerating around the bend of the track in different lanes of radii 15 m (16 yd) (A) and 20 m (22 yd) (B) at identical tangential speeds and accelerations of 10 m/s and 0.5 m/s². Calculate the magnitude and direction of the force on the foot of each runner.

Solution

With reference to figure 6.18, tangential forces are equal for the two runners and are

$F_T = ma_T = 80 \times 0.5 = 40$ N.

The radial force for runners A and B are

$F_{RA} = mv^2 / r = 80 \times 10^2 / 15 = 533.33$ m/s and
$F_{RB} = mv^2 / r = 80 \times 10^2 / 20 = 400.00$ m/s.

For A, the resultant force F_A is

$F_A = (533.33^2 + 40^2)^{1/2} = 534.83$ N

at an angle to the radius of the lane of

$\theta_A = \tan^{-1}(40 / 533.33) = 4.29°$.

For B, the resultant force F_B is

$F_B = (400.00^2 + 40^2)^{1/2} = 402.00$ N

at an angle to the radius of the lane of

$\theta_B = \tan^{-1}(40 / 400.00) = 5.71°$.

Despite the fact that the kinematics determining success in the race are identical, the runner in the inside lane requires approximately 133 more newtons of force to achieve this equality.

The general assumption in this analysis is continuous motion, as if the runner were a wheel traveling around a bend. In fact, the runner spends time airborne during which the CM travels at a tangent to the lane. So the runner will always land at a greater distance from the center of the radius of the lane than when leaving the ground. In this case the radial acceleration will be greater than that calculated in the preceding example and correspondingly greater for the runner in the inside lane. Running in different lanes is equal in terms of distance traveled, but it is different in terms of average force and the associated mechanical impulse.

Summary

Walking is the most frequently used activity for moving around. Therefore it is essential that the energetic cost of walking be minimized. Energy is saved through use of a process of exchange from potential to kinetic forms by pendular motion of the lower and upper limbs. There is also evidence of transfer of kinetic energy between connected segments of the lower limb such as the thigh and leg. In this way the height of the nonstance foot is increased sufficiently to clear the ground during recovery. These exchanges are not sufficient to account for the total energy requirements of walking. Bursts of activity from muscles such as hip flexors and extensors are required to counter energy dissipation caused by joint and ground contact friction. Also there is activity in the knee extensor musculature in the lower limb that supports the body against gravity in the stance phase.

Normal walking over ground requires weight acceptance of the stance heel at the beginning of stance and a push-off at the end. For these events to be achieved, the ground must provide frictional resistance. When such friction is absent, as during walking on ice, the pattern of motion is quite unlike that of normal walking. Heel strike requires onset of activity in the weight acceptance muscles prior to contact, followed by strong eccentric contraction of the dorsiflexors to avoid the sole of the foot slapping too hard on the ground. At the end of stance there is a final push-off, with the great toe being the last point of contact with the ground. Between heel strike and toe-off, the center of pressure on the ground migrates from heel to toes along a path that avoids strain of the fascia and ligaments of the longitudinal and transverse arches of the foot.

Walking induces large forces in the feet. Therefore feet should be strong—a feature that people can achieve by walking without shoes when conditions are favorable. In this way the intrinsic muscles of the foot are exercised.

People can achieve safety in walking by observing the walking surface continually for areas of reduced friction. Hiking boots with firm soles should be worn to avoid high-pressure points during walking on uneven surfaces such as pebbles and soft sand.

Many biomechanical factors contribute to human running, and the relative importance of these factors depends on which type of running is being performed. Minimization of muscular work and minimization of peak forces that lead to injury are two major considerations during running at a constant speed. Muscular work can be minimized partly because stored elastic energy in the weight acceptance phase of stance is returned in the push-off phase. This is another example of the beneficial use of the stretch–shortening cycle of muscle. Furthermore, energy changes in the more distal limb segments result from distal transfer across joints of energy created by the strong proximal muscles (Chapman and Caldwell, 1983a). Mechanical efficiency is also enhanced by eccentric muscular contraction of the muscles during the recovery phase.

Acceleration requires maximal force from the extensor muscles of the hip, knee, and ankle. Acceleration of the arms enhances acceleration by increasing the ground reaction force. Sustained running requires the production of mechanical work at a rate that can be supplied by metabolic use of oxygen. Maximal running speed cannot be sustained for much longer than 4 s because the muscular work rate requirement outstrips the rate of supply of chemical energy resources.

Success in running depends largely on muscular capabilities, although body shape and proportion cannot be ignored. Individuals who have muscles with a large proportion of fast-twitch glycolytic fibers are suited to sprinting, while those with more slow-twitch oxidative fibers become good long distance runners. Whether fiber type can be changed by training is still a matter of conjecture, but sprinters cannot do any harm by training for high force-producing capabilities in their muscles. Equally, distance runners can stress the energy supply mechanism in training by running over long distances.

Safety in running depends largely upon the nature of force patterns at the feet. In the stance phase, the nature of ground reaction forces is determined by the type of foot strike employed, the footwear worn, and the nature of the running surface. These factors make repetitive strain injury more or less likely.

While much is known about the running mechanism, the relative contributions of many factors to success in different types of running have not been elucidated. Unfortunately, training for running must continue based on ideas of "what seems to work" rather than on reliable expectations of the outcome of training. A particularly useful source of information for those specifically interested in the biomechanics of running is Cavanagh, 1990.

RECOMMENDED READINGS

Alexander, R. McN. (1968). *Animal mechanics.* London: Sidgwick and Jackson.

Hay, J.G. (1978). *Biomechanics of sports techniques.* Englewood Cliffs, NJ: Prentice Hall.

McGinnis, P.M. (2005). *Biomechanics of sport and exercise.* Champaign, IL: Human Kinetics.

McNeill-Alexander, R. (1992). *The human machine.* New York: Columbia University Press.

Nigg, B.M., and Herzog, W. (Eds.) (1994). *Biomechanics of the musculo-skeletal system.* New York: Wiley.

Nigg, B.M., MacIntosh, B.R., and Mester, J. (2000). *Biomechanics and biology of movement.* Champaign, IL: Human Kinetics.

Robertson, D.G.H., Caldwell, G.E., Hamill, J., Kamen, G., and Whittlesey, S.N. (2004). *Research methods in biomechanics.* Champaign, IL: Human Kinetics.

Rose, J., and Gamble, J.G. (2006). *Human walking.* Philadelphia: Lippincott, Williams & Wilkins.

Vogel, S. (2001). *Prime mover.* New York: Norton.

CHAPTER 7

Jumping

Jumping refers to a conscious effort to get off the ground. It is done as an end in itself as in high and long jumps, as part of a more complex activity such as a basketball shot, and in order to avoid obstacles in our path. Our usual concept of jumping involves upward movement following foot-to-ground contact. But the concept can be extended to hand-to-ground contact, among others, since the same biomechanical process is invoked. So jumping is implicated in almost all of our movements other than walking. One way of getting off the ground is to pick up one's feet with a greater acceleration than the concomitant downward acceleration of one's center of mass (CM). However, this describes falling rather than jumping. When you are asked which muscles you use to jump, a reasonable answer would be the muscles of the hips and legs during foot-to-ground contact. More information would be contained in an answer that identified specific muscles. Most people would not consider the muscles of the arms to be contributors to jumping, but they would be wrong. In fact, all movable body segments can contribute to jumping performance.

Aim of Jumping

The primary aim of jumping is to raise the CM to a certain height following foot-to-ground contact. Therefore the problem is how to produce upward force at the feet that is greater than body weight in order to produce upward acceleration and to achieve an upward velocity at takeoff. Following this stage is a period of freefall in which the jumper cannot influence the path of the CM.

Mechanics of Jumping

The height to which the CM is raised in the air depends upon movements generating force at the feet to achieve a given velocity and height of the CM at takeoff, followed by the decelerating effect of gravity while the jumper is airborne.

The mechanical analysis begins with identifying the final mechanical aim, which is vertical displacement. Since we are airborne under the influence of constant downward gravitational acceleration following takeoff, we can apply an appropriate equation of uniform motion to determine the total vertical displacement of the CM. The appropriate equation must contain at least displacement, which we require, and acceleration, which we know to be that due to gravity. Equations 2.2 and 2.3 satisfy these conditions, but Equation 2.2 also requires time, which we do not know. Using Equation 2.3,

$$v_f^2 = v_i^2 + 2as,$$

and knowing that our final vertical component of airborne velocity is zero and gravitational acceleration acts downward or negatively, we obtain

Jumping may be a simple activity or part of a more complex activity such as a basketball shot.

© Human Kinetics/Sarah Ritz

$$0 = v_i^2 - 2gs \text{ or } s = v_i^2 / 2g \qquad \text{(Equation 7.1)}$$

where v_i = initial vertical component of velocity at takeoff, which we rename as v_{TO}, and s = vertical displacement while airborne. This equation tells us our vertical displacement while airborne. To obtain v_{TO}, work must be done in the preceding ground contact phase according to Equation 2.20:

$$\int Fds + mv_i^2 / 2 = mv_{TO}^2 / 2$$

By definition, v_i is zero at the beginning of the upward phase during ground contact, so we can rewrite this equation as

$$\int Fds = mv_{TO}^2 / 2 \text{ or } v_{TO}^2 = (2 / m) \int Fds.$$

Substituting v_{TO}^2 for v_i^2 in Equation 7.1 gives

$$s = (2 / m) \int Fds / 2g. \qquad \text{(Equation 7.2)}$$

Equation 7.2 describes the mechanical variables that contribute to the height achieved in jumping from the lowest position of the CM in the ground contact phase to the top of the airborne phase. The total height above the ground requires addition of the height of the CM when the jumper is in the lowest position during the ground contact. The only variable in Equation 7.2 is the integral of force with respect to displacement $\int Fds$ of the CM during the contact phase since m and g are constants. This integral will be maximized by maximizing both force and displacement of the body CM. The latter will be maximized if the jumper ensures that all body segments have their individual centers of mass as high as possible at takeoff. As a result of this mechanical analysis, we are in a position to advise biomechanically on how to maximize height jumped. It should be noted that there has been no attempt to analyze the mechanics of any particular jump that may require specific airborne motion.

Biomechanics of Jumping

Sir Isaac Newton penned three laws of motion that capture the essence of movement of bodies, human as well as inanimate. These laws have stood the test of time in our state of being close to the earth and traveling at modest speeds. One in particular, the third law, states that "action and reaction are equal and opposite." If you accelerate your arms upward, the muscular force required to do this has an equal and opposite reaction pushing the remainder of your body against the ground at your feet. The reaction to this force is that of the earth pushing up on you. Try this on your bathroom scales, and you will see your weight apparently increase and then decrease. The reason is that the bathroom scale is a force transducer that measures the ground reaction force. The subsequent apparent decrease in force is due to upward deceleration of the arms, which requires a downward force on the arms and an equal upward force on the remainder of the body. If your initial upward arm motion is sufficiently vigorous, the ground reaction force will go very high and subsequently drop to zero as you jump upward off your bathroom scale. This is jumping. There is evidence that the ancient Greeks knew about the arm effect at least on a practical basis. They used large rocks, known as *halteres*, that they held to increase the mass of the arms. It is probable that they

■ **Key Point**

Jumping to achieve a given height is determined by the amount of muscular work that can be performed during ground contact and the height of the CM at the instant of takeoff.

■ **Key Point**

Jumping is acceleration of body parts upward to increase the mutual force between us and the earth above the force of body weight.

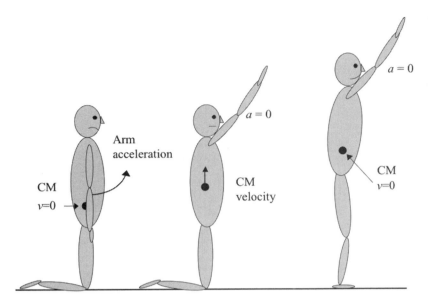

FIGURE 7.1 Jumping can be performed using the arms only.

were performing standing long jumps off two feet at the time, as the halteres would preclude fast running. We experience jumping as going upward because the earth is our reference for all things stable. In fact, an observer on the moon would see you and the earth moving apart—you a great deal more than the earth.

The key point about jumping being acceleration implies that we can accelerate upward any part of the body to satisfy the definition of jumping. This is indeed the case if we widen our perspective on jumping. It is possible with practice to leave the ground by upward acceleration of the arms alone, as seen in figure 7.1.

This strategy requires keeping the whole of the body rigid while vigorously accelerating and then decelerating the arms upward. In fact it is nearly possible to attain a standing posture from an initial kneeling posture by these means. The problem with this strategy is the relatively small mass of the arms. Since force is the product of mass and acceleration (F = ma), the force at the shoulders can be significant only if the arm acceleration is very large. Clearly the shoulder musculature cannot compete with that of the lower limbs as used in traditional jumping. Yet shoulder musculature can add to the effect at the feet.

The previous equation (Equation 7.2) shows how the integral of force with respect to displacement of the CM determines the height jumped. In this simple mechanical form we can visualize a force being applied at the CM. Of course this is misleading, as the force is applied at the feet and the displacement (and therefore acceleration) of the CM is a function of the sum of the displacements of all body segments. In fact the force at the feet remains equal to body weight when we are either stationary or moving our CM at a constant velocity, and it becomes of use in jumping only if we accelerate body segments. The key to maximizing ∫Fds is therefore to develop as much force in as many muscles as possible and to shorten these muscles as much as possible. This statement can be summarized as follows:

$$\int_{i=1}^{i=n} F_{cm}ds_{cm} = \sum \int_{i=1}^{i=n} F_i ds_i \qquad \text{(Equation 7.3)}$$

which means that the total work is the sum of work done by muscles 1 through n. Since the muscles produce a moment about a joint axis, Equation 7.3 can be rewritten as its rotational equivalent, Equation 7.4.

$$\text{Muscle work} = \sum \left(\int_{i=1}^{i=n} M_i d\theta_i \right) \qquad \text{(Equation 7.4)}$$

Readers should remember that Equations 7.3 and 7.4 are equivalent since they have the same dimensions of work (kg.m^2/s^2). It is this total muscle work that adds rotational and translational kinetic energy along with potential energy to the system. Not all muscles in the body are equally able to contribute to this sum of work. The most important contributors to work are those that can move significantly massive parts of the body over a considerable displacement. Here we are considering the limbs that have a large range of motion. A further consequence of using limbs not directly in contact with the ground is that at takeoff, the centers of mass of the limbs will be as high as possible, leading to the greatest height of the CM.

As in all skills, there is a question of appropriate timing of intersegmental motion. Should the reaction forces in the arms, for example, be too great for muscles below them to resist, the body mass below the arms will be displaced backward and the total displacement of the CM will be reduced. While it is unlikely that the arms can induce such an effect in jumping, practice is essential for ensuring that this scenario is avoided. Another question that arises is the manner in which the free body segments should be moved upward. If the arms are raised while the elbows are flexed, their moments of inertia about the shoulder joint will be small. The moment produced by the shoulder flexors will produce rapid angular acceleration of the arms; and as the velocity of shoulder flexion increases rapidly, the shoulder flexor moment will drop rapidly. But if the arms are swept upward with the elbows extended, the shoulder flexor force will remain relatively high. Therefore the work done will be high and will contribute more to the total work done than that during motion with the flexed elbows. The benefit to jumping of involving upward motion of free segments will be greater the more extended those segments are.

In the preceding description of the use of free segments, there are so many mechanical factors to appreciate that confusion is possible. Therefore computer simulation of jumping using the arms only was performed by your author. Only the output values of interest are given here since the manipulation of equations of motion is complex and difficult to follow. The simulated body comprises two segments: a single, rigid trunk, legs, and head block; and an arm block (representing both arms) free to rotate about the shoulder joint due to a constant moment M at the shoulder. The body began in a stationary position with the body block standing vertically and the arm block hanging vertically downward from the shoulder joint (see figure 7.2 for a pictorial representation of the motion). Two versions of the arm block were used: one with straight arms and the other with the arms fully flexed at the elbow. The work done by the shoulder flexor muscles was useful only while the feet were in ground contact. From the start to the time when ground contact was lost, the straight arms and the bent arms produced changes of energy of 527.5 J and 515.3 J, respectively. Each of these energy values was distributed as increased rotational and translational energy and increased potential energy of the system CM. While the straight-arm swing generated a greater amount of energy, these single values do not reveal the direction of the kinetic energy of the CM. A more detailed analysis using the impulse–momentum relationship is used for this purpose later.

This concludes discussion of the strictly biomechanical basis of jumping using the work–energy relationship. The following are some other observations that

■ Key Point

Jumping should be performed with upward motion of extended rather than flexed free limbs.

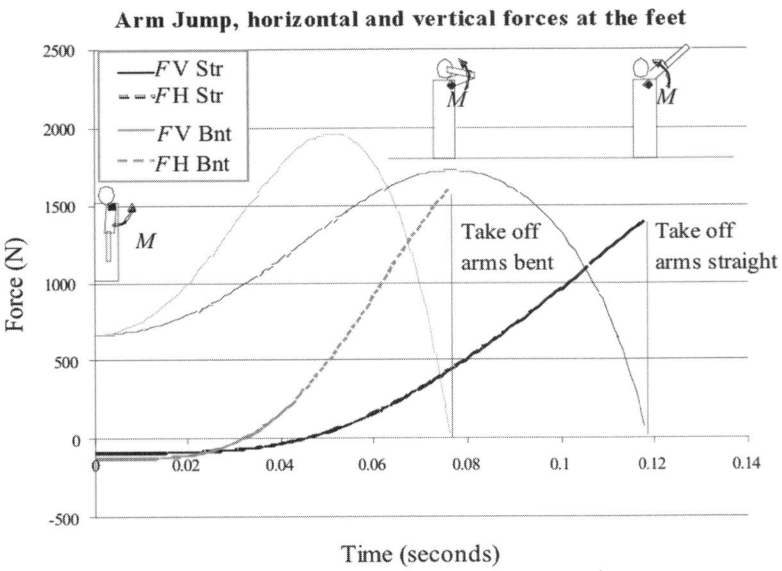

Arm Jump, horizontal and vertical forces at the feet

FIGURE 7.2 The contribution of the arms to jumping forces depends upon their configuration.

■ Key Point

Acceleration of body segments is of benefit only while some part of the body is in contact with the ground. If body segments accelerate while the person is airborne, the remainder of the body accelerates in the opposite direction.

are designed to enable the reader to view jumping from different biomechanical perspectives.

The simulation used to obtain work done with straight- and bent-arm swings also revealed the different effects from an impulse–momentum viewpoint. Figure 7.2 shows the results of the vertical force profiles at the feet (FV Str and FV Bnt).

The straight-arm swing generates vertical force above body weight for a longer period of time than the bent-arm swing. The area under the vertical force–time profile in each case represents the mechanical impulse $\int Fdt$, which is responsible for the vertical change in momentum of the system. As these are unequal, as can be seen by inspection, takeoff obviously occurs with different velocities of the combined CM of the system. These velocity profiles VH and VV are shown in figure 7.3. The major point in this figure is that the vertical takeoff velocity of the combined CM using the straight-arm swing (VV Str) is greater than for its bent-arm counterpart (VV Bnt). A minor point for consideration is that takeoff occurs at a velocity slightly less than the maximum achieved. The reason for this is that the vertical force profile spends some small time dropping body weight to zero, in which case gravity has an increasing effect in reducing vertical velocity. Symbolically this means that $\int (F - mg)dt$ is increasing negatively as F decreases to zero.

The two figures just discussed also demonstrate horizontal effects on the combined CM (CCM). The reason is that when the arm segment is swung upward, its CM has a horizontal component of motion in addition to the vertical component. At the beginning of the arm swing there is a reaction tending to rotate the body counterclockwise. The negative horizontal force (to the left) represents the force of friction that stops the positive sliding of the feet (to the right). As the positive horizontal component of acceleration of the arm CM drops, as in the case of such circular arm motion, the force becomes positive. This gives a net positive horizontal impulse to the combined CM. Figure 7.3 shows this effect, and takeoff is achieved with a substantial horizontal velocity that is greater with the straight-arm swing (VH Str).

What is being observed here is something like a horizontal standing broad jump using only the arms. If the simulation had been designed to produce pure vertical motion of the combined CM, the body would have had to begin the jump by leaning backward. Alternatively, some other segments of the body would require acceleration to counteract the horizontal effect. While jumping in this manner seems simple to us, it clearly requires some intricate interplay among the timing of segmental moments. Gymnasts use many varieties of nontraditional jumps, of which figure 7.4 is an example.

In this case the strong muscles of the buttocks generate a large acceleration of the relatively massive legs. The mutual force between the shoulders and the ground is therefore large; the takeoff velocity of the CM is therefore large, and gymnasts can land upon their feet. Added to this force are the parallel forces generated by the pushing arms. So in terms of arm and leg motion, figures 7.1 and 7.4 are mirror images.

A point of further interest in figure 7.4 is that angular velocity of the body accompanies linear velocity of the CM. The mechanics are as follows. Acceleration of the lower limbs has both linear and angular (rotational) components. The linear component increases the velocity of the CM at takeoff to raise us into the air. The angular component gives the lower limbs angular velocity at takeoff. If and when the lower limbs are stopped relative to the remainder of the body, the whole body gains angular velocity. This is common to a variety of jumps and is sometimes beneficial, as in figure 7.4, as well as something to be minimized in other jumps. Sometimes it is difficult to appreciate what effect a given velocity of the arms has on the remainder of the body. The events described can be otherwise explained through recourse to momentum. Momentum is the product of mass and velocity *(mv)*. The lower limbs in figure 7.4 have a given amount of momentum. When they stop in relation to the remainder of the body, this momentum is shared by the whole body. Since the lower limbs have smaller masses than the whole body, the velocity of the whole body will be less than that of the lower limbs. This effect is summarized in the following equation:

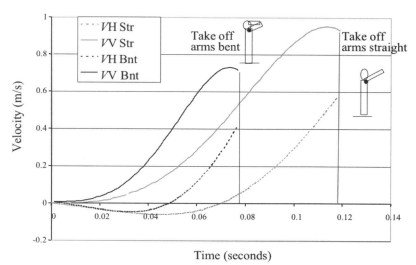

FIGURE 7.3 Arm-swing configuration determines components of takeoff velocity in jumping.

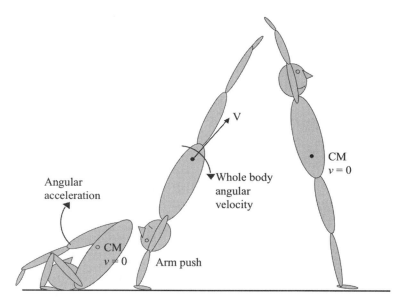

FIGURE 7.4 The back spring uses the same mechanical process as does jumping, but the roles of the arms and legs are reversed.

Momentum = m(legs) × v(legs) = m(whole body) × v(whole body);

v(whole body) = v(legs) × [m(legs) / m(whole body)]

As the value in the square brackets is less than one [<1], the velocity of the whole body will be less than the velocity of the lower limbs. The same equations could be written for angular momentum of the arms in figure 7.1 and would show that arresting motion of the arms in relation to the whole body results in a lower angular velocity of the whole body. Whether we examine rotation at takeoff through the concepts of either angular acceleration or angular momentum, the

Angular acceleration α = F x d / I

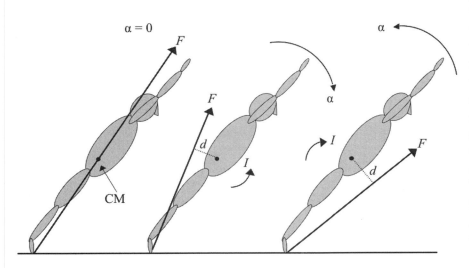

FIGURE 7.5 Segmental motion determines the direction of the force vector at the feet with subsequent effects upon angular motion.

phenomenon is due ultimately to the direction of the average force vector at the body–ground contact area. The term "area" is used rather than "point" because inevitably we have multiple points of contact with the ground in jumping. If a very expensive bathroom scale called a force plate is used, the average point of contact or center of pressure on the plate can be identified (see Robertson et al., 2004). Furthermore, the direction and position of the average force vector can be identified.

Whenever the force vector passes through the CM, pure translational acceleration of the body will be seen. When the force vector does not pass through the CM, angular acceleration α will occur as seen in figure 7.5.

In figure 7.5, I represents the rotational resistance to motion or moment of inertia, and it is the rotational equivalent of mass. Also d is the perpendicular distance between the CM and the force vector. Gravitational force has been omitted from this figure for the purposes of simplicity, but its inclusion would be necessary for a fully correct mechanical analysis. Unfortunately, this analysis cannot be done on your bathroom scale unless you are lucky enough to be able to afford a bathroom scale and allied equipment costing about $100,000.

The considerations just outlined have to be taken into account by any performer who jumps to execute rotations while airborne. This cannot be done at the conscious level, as the analysis is too complicated to undertake during the short duration of a jump. In addition, the direction and magnitude of the average force vector at the feet will change throughout the jump. Indeed there will be periods during which the force vector changes from a direction in front of the CM to behind it. Furthermore, it is possible to combine segmental accelerations to produce backward rotation while moving either forward or backward as seen in diving. What matters finally is the direction of the force vector in relation to the position of the CM averaged over time. Therefore performers are required to practice takeoff maneuvers by playing a tune on the relative motions of body segments.

Variations of Jumping

The act of vertical jumping has been our model for analysis so far. However, jumping technique can vary depending upon the required conditions either preceding or following the jump. In this sense jumping is commonly part of a continuum between running and airborne maneuvers.

■ **Key Point**

A jumper maximizes takeoff velocity by accelerating upward all body segments that are not in contact with the ground.

Jumping Following a Run

Jumping following a run is generally performed off one foot and usually produces a takeoff velocity that is greater than in a jump from stationary conditions, although the characteristics of the jumping phase are the same in the two types of jump. With a prior run we can generate a greater countermovement velocity than we can by simply dropping prior to a standing jump. The explanation of this strategy is detailed in the section on enhancement.

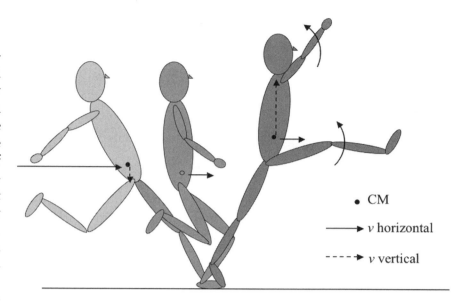

• CM

⟶ *v* horizontal

----⟶ *v* vertical

FIGURE 7.6 Arresting horizontal velocity by eccentric muscular contraction aids in generating vertical velocity in the high jump.

The large horizontal velocity of the CM is arrested by eccentric contraction of the jumping muscles. The CM not only loses velocity but also rotates about the foot until it passes over the foot. The conditions for going upward are now favorable, and the concentric force pushes the jumper upward as shown in figure 7.6.

An additional benefit is also the result of using upward acceleration of the nonjumping leg as well as the arms. The amount of horizontal velocity retained depends upon the airborne events that follow takeoff. A huge variety of airborne maneuvers follow jumps, depending on the sport. While these are discussed separately later, it is worthwhile pointing out now that the aim is generally one of two types. Either the maximal rise of the CM is the sole aim, or the time in the air is maximized to allow time for the bodily maneuvers to take place. These two aims are obviously connected. For now we shall confine ourselves to activities named jumps. Bar clearance in the high jump depends directly upon the height to which the CM is raised. The postural changes of the body during bar clearance are obviously important but secondary to the upward velocity generated at takeoff. However, the amount of horizontal velocity that must be retained is determined by the type of bar clearance chosen. Much less horizontal velocity is required after rather than prior to takeoff.

The long jump contrasts with the high jump in that a large horizontal velocity is necessary at takeoff. In a mechanical sense, success simply means having a large horizontal velocity and plenty of time in the air, because velocity (meters per second) multiplied by time (seconds) gives us distance (meters). Fortunately, plenty of time and plenty of distance are available during the approach run to build up a large horizontal velocity. All that is required in the takeoff phase is to direct force upward. There is a fine interplay between horizontal and vertical components of velocity because they determine the angle of projection of the CM at takeoff. This angle is another factor that appears in the rather frightening equation that determines the horizontal range of a projectile (the long jumper). However, it may be that a small amount of horizontal velocity can be given up to improving the vertical velocity so that the angle of projection is optimal. Whether

Pole vault

Stiff   Flexible

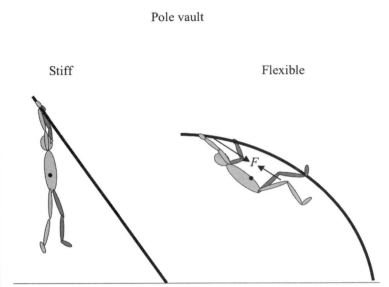

FIGURE 7.7 The flexible pole is an additional energy storage mechanism in jumping.

this is done will depend upon the capabilities of the long jumper. For example, a long jumper who is a top-class sprinter but not so good a vertical jumper may have to adopt the strategy described. Poorer sprinters will not be able to afford the luxury of reducing horizontal velocity during takeoff and may suffer from the lack of the beneficial effects of large muscle stretch. It is no surprise that good long jumpers are good to excellent sprinters, without exception.

Finally, a clue to the interplay of loss of horizontal velocity and gain in vertical velocity can be seen in the triple jump. Here the horizontal velocity is reduced successively with each takeoff. Yet the velocity of the final jump is not reduced so much that these jumpers couldn't beat 99.9% of the population who were performing a single long jump. The key to success is that each takeoff phase is identical in most respects as the arms and free leg are driven upward to maximize the force at the takeoff foot. In this case the maneuvers undertaken while the jumper is airborne between successive takeoffs prepare the arms and free leg segments for maximal acceleration during takeoff.

The last jump of interest is not so much a jump as a vault—the pole vault as seen in figure 7.7. Mechanically the pole-vault bears considerable resemblance to the high jump and long jump combined. The obvious difference is that there is a pole connecting the vaulter to the ground following takeoff. Prior to the 1960s, the pole was stiff in comparison with the poles of today. Even so, the pole afforded a means by which the kinetic energy of the vaulter at takeoff could be converted into potential energy at a later time. But as usual, gravity affects this conversion process so that not all the kinetic energy is converted into potential energy.

To understand this time-reduction technique, we must examine the vault from the point of view of angular momentum. It is obvious that at takeoff the vaulter rotates with the pole about an axis through the pole–ground contact point. We can say that the vaulter-pole system has angular momentum. As we know that angular momentum is moment of inertia multiplied by angular velocity, reduction of moment of inertia will lead to an increase in angular velocity. Therefore the time from takeoff to a vertical pole position will be decreased and gravity will act for a shorter period of time. The vaulter achieves this result by hanging in an extended posture from the hands for as long as possible to bring the body mass close to the axis of rotation. At some stage the trunk and hip flexors must contract to bring the feet above the head in preparation for bar clearance. The flexibility of the current poles allows bending, which brings the mass closer to the axis of rotation. Consequently the vaulter can hold the pole farther away from the point of rotation, which represents a greater starting height for final release of the hands when the pole straightens.

The flexible pole can also be thought of as an energy storage mechanism. As the kinetic energy of the vaulter following takeoff is being reduced by an opposing force from the pole, the force of the vaulter on the pole is increasing potential (strain) energy in the pole. This effect is large, since it is possible to store almost all of the body's kinetic energy ($mv^2/2$) as strain energy because of the considerable deformation that occurs in the pole. The vaulter also rocks backward and pulls the legs toward the hands by means of hip and trunk flexor forces F. The opposite of this force is on the pole, which gives further bend and therefore further stored strain energy. Following rotation the pole straightens, losing strain energy in the process and adding to the potential energy of the vaulter. What could happen to the high jump if we had a takeoff leg with the characteristics of a vaulting pole is an interesting but hypothetical thought.

Since the vaulter is in contact with the ground via the pole at all times until hand release, muscular force of the arms can be used to add to the potential energy. This ability somewhat compensates for loss of energy in the conversion process since the pole is not a perfect spring. The advantages gained by use of the flexible pole are impressive. The Olympic gold medal was won in 1956 with a height of 4.56 m. In 1960 the height increased by 14 cm. From 1960 to 1964, the height increase was 40 cm; and from 1960 to 2004 an increase of 125 cm was seen.

Trampolining and Diving

Repeated jumps on a trampoline or a diving board are an excellent example of energy exchange. On the first landing after an initial jump, the body has kinetic energy in a downward direction. Should the athlete simply land with straight legs, most of the kinetic energy will be stored as strain energy in the bed of the trampoline. Of course this is not to be recommended, as the force will rise rapidly and put stress on all joints in series such as the vertebrae. Yet some of this energy can be used to add to muscular work done in subsequent jumps. The athlete can therefore reach jumping heights much greater than that during repeated jumps on the ground. Such a mechanism suggests that the athlete could continue rising higher and higher. There are three major reasons why this cannot occur. The first is that the trampoline bed or diving board is not 100% efficient, so not all of the stored energy will be returned. The second is that there is a limit to the stress that the jumping muscles can withstand. With each landing there will be ever-increasing amounts of work to be done. Eventually the force-producing capabilities of the muscle or the available displacement of the body (or both) will be exceeded. Yet the effect is dramatic. The third reason is that diving rules allow only one bounce.

In some sports, such as basketball and volleyball, a two-footed takeoff follows a run. The run achieves the same result in terms of maximizing force at the feet through a countermovement as seen in the single-footed takeoffs described

Peak force at the feet when landing can be reduced by using the greatest amplitude of motion of the maximal number of segments of the body, including the arms.

previously. But in basketball and volleyball, the eventual aim is to achieve a vertical takeoff velocity. Should the athlete land in preparation for takeoff with the CM vertically above the feet, the frictional force will arrest foot motion and the athlete will rotate forward. The athlete must therefore land with the feet ahead of the CM so that the force vector at the feet passes in front of the CM. The resulting backward angular acceleration due to gravity will reduce the forward angular velocity to zero, and takeoff will occur vertically. Such a skill is easily learned; the main point is that the run provides a more effective countermovement than a simple downward motion can achieve.

There is one final type of jumping that is very impressive but not usually characterized as jumping—it is simply doing a handstand and then jumping using the arm muscles. While this activity is impressive, it is mechanically no different from "normal" jumping. The actual difference is that the arm and leg muscles have their roles reversed. Some circus performers are very good at this activity, and observation will illustrate how the lower limbs are accelerated in much the same manner as the arms in jumping off two feet. This is similar to the motion shown in figure 7.4 but done from a handstand position. In this manner performers can jump up a flight of stairs on their hands. Descending the stairs on the hands presents little difficulty to those who can ascend stairs in this way because of the eccentric nature of the descent. Skillful technique is required for this activity, but no amount of skill will work unless the arm and shoulder muscles are extremely strong and the lower limbs are used appropriately.

Enhancement of Jumping

In our context, enhancement means simply maximizing takeoff velocity, as opposed to subsequent airborne movements that may enhance specific types of jumps. We can be guided in this aim by the first key point, which requires us to produce force above body weight force in order to accelerate upward.

Small children begin jumping by using various combinations of hip, knee, and ankle extensors, usually in a rather uncoordinated manner. With practice and maturity they learn to use both lower limbs simultaneously. Such a strategy leads to improved jumping because the lower limbs are connected mechanically in parallel. Therefore the total force is the sum of the force in the two limbs. Use of the arms comes a little later and adds to the height we can reach. One might wonder why the high jump is not done off both legs simultaneously. The fact is that the rules don't allow it. If they did, we would see gymnasts winning gold medals on the high jump until high jumpers learned the necessary gymnastics skills.

In the following discussion we shall resort to the impulse–momentum relationship:

$$\int F dt = mv$$

Simply standing up has the characteristics of a jump, but we don't leave the ground because it takes a relatively long time. Again our bathroom scales will show force greater than body weight when we stand up, followed by a decrease below body weight as we come to a stop. The force profile in standing up is shown in figure 7.8, in which the shaded areas of the force curve above and below the line W are equal.

The first area gives us upward velocity, and the second, being equal but below the body weight line, reduces the velocity to zero when we achieve erect standing. This means that $\int(F - mg)dt = 0$. A greater magnitude of the force with a smaller duration is seen for jumping. In the jump from a stationary crouching posture, the force rises from W and the whole of the black area represents our momentum at takeoff (TO). This is not our usual technique, as we have learned that a countermovement from an erect posture

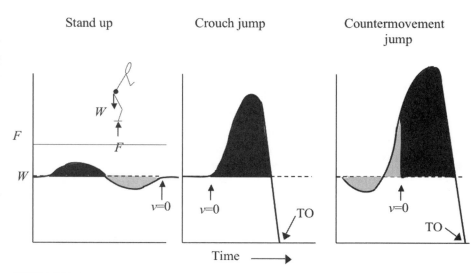

FIGURE 7.8 The countermovement enhances the vertical mechanical impulse over that obtained from a crouched position where velocity is zero.

gives us a better result. In the countermovement jump, the two initial shaded areas are equal and physically drop us to zero velocity. The significant facts are that after we have dropped initially and we begin our upward movement at $v = 0$, the force is already well above W and the subsequent black area gives us a much greater change in upward velocity at takeoff. Note the difference in the rate of rise of force in the countermovement jump, and also the position of zero velocity in relation to the force profiles. Try the two jumps and see the benefits of the countermovement.

The countermovement jump also involves a countermovement of the arms. This is done because the muscular force accelerating the arms upward is already well developed when their upward velocity begins. This well-developed force at zero velocity is the result of the biomechanical properties of muscle. In the countermovement, the muscles are being stretched and therefore are acting in the negative region of the force–velocity relationship (see the section "Use of Muscular Force" in chapter 3). Consequently the force developed is high. Secondly, the muscular contractile apparatus is enhanced by prior stretch, adding further benefit to the countermovement. Thirdly, strain energy is stored in elastic structures in series with the contractile apparatus of the muscle. This energy is given up to increasing the kinetic energy of the system in the final concentric jumping phase. Finally, the eccentric phase provides time for the contractile apparatus of the muscle to become fully active. In the examination of other activities in which maximization of velocity is the aim, we shall see the persistent use of countermovements.

It is clear that takeoff velocity in jumping is related to the strength of the muscles involved. So strength training is an essential part of a jumper's training. Clearly the extensors of the ankles, knees, and hips must be strong. But what jumpers need to remember is the role played by acceleration of body segments not normally considered to be essential to jumping. Strength training of the muscles of the shoulders is essential for enhancing jumping performance by virtue of the role played by the arms in jumping. If the jump is confined to a single-foot takeoff, it is the flexors of the hip of the nonjumping leg, rather than the extensors, that must be strong. The picture that is emerging here is that jumpers must have strong

■ Key Point

Takeoff velocity is maximized by a countermovement of body parts, which ensures that upward velocity begins with force at the feet well above body weight *W*.

muscles distributed throughout the body, and that a complete strength training program covering the whole body is essential to enhancement of performance.

The jumper must take into account the beneficial effects of the countermovement in jumping. This phenomenon cannot be applied indiscriminately. For example, a slow countermovement will allow the transitory force enhancement effect of prior muscle stretch to disappear before it can be used. Also a slow countermovement will allow dissipation of the force required to arrest downward motion. The subsequent upward phase of the jump will therefore begin at a level of force equal to body weight, as if no countermovement had been done. There are no numerical rules for the speed and magnitude of the countermovement, since individuals differ in terms of mechanical muscle capabilities. This aspect of enhancement of performance is treated in training only by trial and error, but performers should be aware that the timing of their countermovement will undoubtedly change as their muscles grow stronger.

Jumping Safety

The action of jumping is essentially safe unless it is done either in a room with a low ceiling or off a surface that is uneven. The inherent safety is attributable to the fact that as our upward velocity and consequent shortening velocity of the muscles increase, the ability of the muscles to produce force decreases. In this sense jumping is a self-defeating activity. Danger arises in the jumping phase when the surface has unforeseen characteristics such as low friction and hollows and bumps. In the latter case, the ankle and knee joints are likely to experience force vectors that do not pass through the center of the joint; an unusual load of a rotational nature is therefore likely to injure these joints.

Low foot–floor friction will have the same potential effects as described for walking. In many cases jumping is performed at a speed greater than walking pace. So the chances are high that the horizontal component of force at the feet will overcome limiting friction and produce sliding. Therefore it is essential that modifications be made to the soles of jumping shoes in order to prevent such sliding. This topic has been dealt with in the context of running. The addition in jumping is that the heel is the first part of the foot that makes contact with the ground. Therefore heel spikes are inserted into the jumping shoes to arrest potential sliding.

Apart from the initial countermovement, jumping up is a concentric muscular activity. We are working, so to speak, on the concentric or shortening part of the relationship between force and shortening velocity. One consequence of this fact is that we can never jump up to a height from which we cannot land safely. The reason for such a statement is simply the fact that landing is an eccentric activity in which we can produce more muscle force than we can in muscle shortening. The fact is that our ability to dissipate kinetic energy is much greater than our ability to generate it. This would seem to be a safety mechanism that does not allow us to generate more motion than we can handle. Whether this is an evolutionary result or a simple accident of physiology is open to question. In either case, it works for us. But this safety aspect applies only to the activity of jumping as high as we can and landing from that height. Climbing up stairs and landing from the 14th floor of a building will not work, no matter how fast your muscles are lengthening during landing. Yet humans can land from great heights without death, although usually with serious injury. Muscles can generate eccentric forces that are large enough to rupture tendons and to pull the tendon insertion off the bone, usually with a bit of the bone remaining attached to the tendon. The latter is known as an avulsion fracture.

Practical Example 7.1

Figure 7.9 shows a jumper of mass 70 kg (188 lb) taking off from both feet, having begun from a stationary crouched posture. An average force of $F = 1100$ N directed at an angle of $\theta = 1.4$ rad to the horizontal is applied for 0.8 s. The average perpendicular distance between the force vector and the CM is $d = 15$ cm (6 in.), and the average moment of inertia about the transverse axis is $I = 12.0$ kg.m^2.

Part 1

Calculate the magnitude and direction of the translational velocity of the CM.

Solution

Since mass, force, and time are given as constants, it is appropriate to use Equation 2.12, which is the impulse–momentum relationship:

$$Ft = mv_f - mv_i$$

As the question defines v_i as zero, the equation reduces to

$$Ft = mv_f, \text{ so } v_f = Ft / m.$$

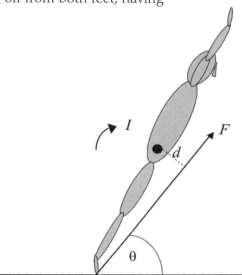

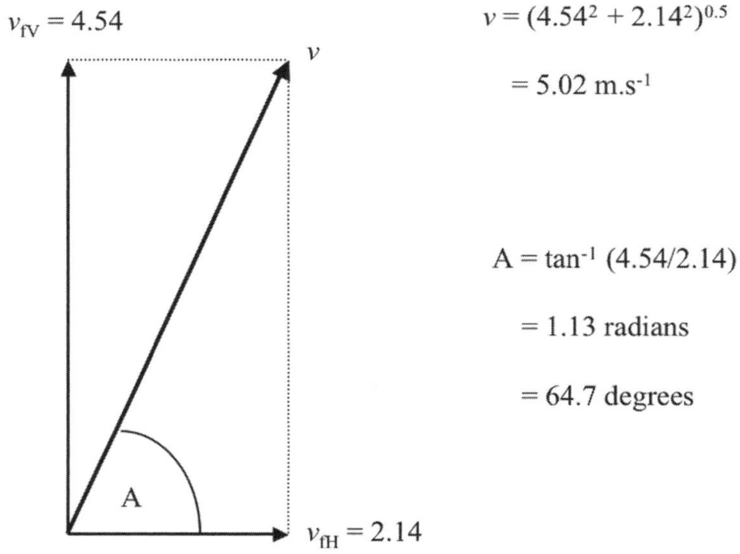

FIGURE 7.9 Average force vector in relation to the position of the CM in a jump.

$$v_{fV} = 4.54$$

$$v = (4.54^2 + 2.14^2)^{0.5}$$

$$= 5.02 \text{ m.s}^{-1}$$

$$A = \tan^{-1} (4.54/2.14)$$

$$= 1.13 \text{ radians}$$

$$= 64.7 \text{ degrees}$$

$$v_{fH} = 2.14$$

FIGURE 7.10 Resultant takeoff velocity vector of the CM.

The force vector must be resolved into its horizontal and vertical components, as the latter is opposed by gravity.

Horizontally

$$V_{fH} = F\cos\theta t \,/\, m$$
$$V_{fH} = \left[1100 \times \cos(1.4) \times 0.8\right]$$
$$V_{fH} = 2.14 \times ms^{-1}$$

Vertically

$$V_{fV} = \left(F\sin\theta - mg\right)t \,/\, m$$
$$V_{fV} = \left[1100 \times \sin(1.4) - 70 \times 981\right] \times 0.8 \,/\, 70$$
$$V_{fV} = 4.54\,ms^{-1}$$

The solution is depicted in figure 7.10.

Part 2

Calculate angular velocity about the transverse axis through the CM at takeoff.

Solution

In this case gravity does not have to be considered since it does not create a moment about an axis through the CM. Since moment of inertia, force and time, and perpendicular distance are given as constants, it is appropriate to use a simplified version of Equation 2.17, which is the angular impulse–momentum relationship:

$$Mt + I\,\omega_i = I\,\omega_f$$

As the question defines ω_i as zero, the equation reduces to

$$Mt = I\,\omega_f, \text{ so } \omega_f = Mt \,/\, I.$$

The moment M is due to the force acting at some perpendicular distance from the CM, so

$$M = Fd \text{ and } \omega_f = Fd \times t \,/\, I;$$
$$\omega_f = 1100 \times 0.15 \times 0.8 \,/\, 12 = 11 \text{ rad.s}^{-1} = 630.3°.\text{s}^{-1} = 1.75 \text{ rev.s}^{-1}.$$

This example represents a very simplified depiction of the real situation. In fact all of the constant values given will vary during the jump. The force will change as seen in figure 7.8, and its direction of application will vary. The values d and I will change due to changes in the configuration of the body. A deeper analysis of this jump is impractical in this book, as a substantial computer program would be required to capture the changes in variables and their resulting contributions to the jumping performance.

Practical Example 7.2

A jumper of mass 90 kg (241 lb) is performing a standing long jump for maximal horizontal distance. A load of 10 kg (27 lb) is held in each hand and used to increase the mechanical impulse at the feet. The velocity of the CM of the system is 5 m/s directed at an angle of 45° to the horizontal, and the CM is 0.6 m (24 in.) above its final level during landing. The loads are thrown backward from a position of 0.2 m (8 in.) vertically below the system CM with a velocity of 2 m/s relative to the ground on two separate jumps. Jumps I and II have load release when the velocity vector of the system is horizontal and when it is 20° below the horizontal, respectively. Jump III is a jump during which no load is carried. Calculate the total horizontal displacement D of the CM after takeoff for each jump.

General Solution

Such problems and many others are best solved with the aid of a diagram. In this case we can deal with the system as a point mass and with the person and loads as point masses after their release as seen in figure 7.11. Kinematics can be analyzed using the equations of uniform motion because there is no acceleration horizontally and only gravitational acceleration downward occurs following takeoff. The second consideration is knowing the velocity of the body when the loads are released. We determine this by using the principle of conservation of momentum. The vertical and horizontal displacement from takeoff to the point at which the loads are released is required, as are those displacements following release to landing.

Solution to Jump I

From takeoff to load release, the horizontal component of velocity is unchanged and equal to $v\cos(45)$. The initial vertical component of velocity is $v\sin(45)$ and the final value is zero. The horizontal displacement requires the time during this period, which is calculated as follows:

Vertically up $v_f = v_i - gt$; so $t = [v\sin(45) - 0] / 9.81$, $t = 0.36$ s.

Horizontally displacement is $v\cos(45) \times t = 5\cos(45) \times 0.36 = 1.273$ m.

The height h of the system CM above its takeoff level is obtained as follows:

Vertically up $v_f^2 = v_i^2 - 2gh$; so $h = (v_i^2 - v_f^2) / (2 \times 9.81)$.

$h = [v^2\sin^2(45) - 0] / 19.62 = 0.637$ m.

The next requirement is to calculate both the height and velocity of the body CM immediately after release of the loads. The system CM (total mass 110 kg or 295 lb) lies on a line joining the body (90 kg or 241 lb) and load (20 kg or 54 lb) CM and is closer to that of the body since the body is more massive than the load. If we designate L as the distance between these centers of mass, the body CM and load CM lie at distances of (20/110)L and (90/110)L, respectively, from the system CM.

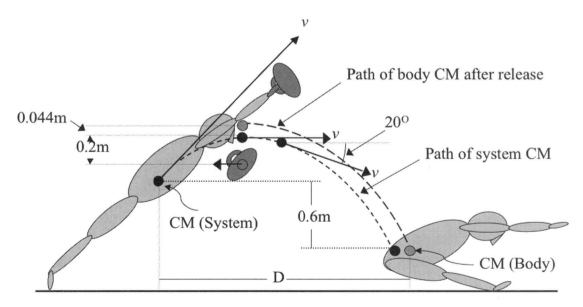

FIGURE 7.11 The ancient Greeks knew that projecting masses backward increased the horizontal displacement of the standing broad jump.

We are given that the load's CM is 0.2 m (8 in.) below the system CM at release, which yields the following:

(90 / 110)L = 0.2; so L = (0.2 / 0.8181) = 0.24444 m

Therefore the body CM lies at (20/110)L or 0.0444 m above the system CM. The system CM is 0.637 m above its takeoff level. The takeoff level is 0.6 m above the final level on landing. So the starting height of the body CM above its height on landing for the second phase of the jump is

h = 0.0444 + 0.637 + 0.6 = 1.2814 m.

To obtain the time of the fall, we use the following equation and its quadratic solution:

Vertically down $s = v_i t + gt^2 / 2$; so $gt^2 / 2 + v_i t - s = 0$

The solution to the quadratic equation $at^2 + bt + c = 0$ is $t = [-b \pm (b^2 - 4ac)^{1/2}] / 2a$, which in terms of the current symbols becomes

$t = [-v_i \pm (v_i^2 + 2gh)^{1/2}] / g$;
$t = [-0 + (0^2 + 2 \times 9.81 \times 1.2814)^{1/2}] / 9.81 = 0.5111$ s.

We now require the horizontal velocity of the CM of the body, which we obtain through the concept of conservation of momentum. The total horizontal component of momentum ρ relative to the ground prior to release is the product of the total mass and initial horizontal component of velocity as follows:

ρ = 110 kg × 5cos(45) = 388.909 kg.m/s

The load travels backward relative to the ground with a momentum of ρ_L = 20 kg × 2 m/s = 40 kg.m/s. Therefore the body momentum ρ_B is obtained as follows:

ρ_B = 388.909 + 40 = 428.909 kg.m/s

The horizontal velocity of the body is therefore 428.909 / 90 = 4.766 m/s. The horizontal distance covered following release of the loads is therefore 4.766 × 0.5111 = 2.436 m. When this value is added to the horizontal displacement before release, we obtain

D = 2.436 + 1.273 = 3.709 m.

Solution to Jump II

The following analysis for Jump II is performed in the same process, using the same equations, and using values applicable to this jump. The written explanations for the mathematics of Jump I are not included except when such explanation is necessary.

Vertically up $v_f = v_i - gt$; so $t = \{[v\sin(45)] [-v\sin(20)]\} / 9.81$, $t = 0.535$ s.
Horizontally displacement is $v\cos(45) \times t = 5\cos(45) \times 0.535 = 1.891$ m.

The height h of the system CM above its takeoff level is obtained as follows:

Vertically up $v_f^2 = v_i^2 - 2gh$; so $h = (v_i^2 - v_f^2) / (2 \times 9.81)$
$h = \{[v^2\sin^2(45)] - [-v\sin(20)]^2\} / 19.62 = 0.488$ m

The starting height of the body CM above its height on landing for the second phase of the jump is

h = 0.0444 + 0.488 + 0.6 = 1.1324 m.

To obtain the time of the fall, we use the following equation and its quadratic solution:

Vertically downward $s = v_it + at^2/2$; so $at^2/2 + v_it - s = 0$

where $s = h$, $v_i = v\sin(20)$ and $a = g$; $t = \{-v\sin(20) + [v^2\sin^2(20) + 4gh/2]^{1/2}\}/g =$ −0.6855 s or +0.3368 s.

The correct value for the time of the fall of the last stage is +0.3368 s. With a horizontal component of velocity of the body of 4.766 m/s and a time of flight of 0.1684 s, the displacement in the final stage is 4.766 m/s × 0.3368 s = 1.605 m. When this value is added to the horizontal displacement before release, we obtain

D = 1.605 + 1.891 = 3.496 m.

The earlier release of the loads in Jump I gave greater total horizontal displacement (3.709 m) than the later release seen in Jump II (3.496 m). The reason is that the earlier release gives a greater airborne time over which the enhanced component of horizontal velocity acts. This enhanced component is due to the greater momentum gained by the body, equal to the loss of momentum of the loads when they are thrown backward. Relative movements of body segments will not achieve this result, as the combined CM of the body will not change its path unless the arms fly off backward.

Solution to Jump III

Here there is no mass to release, and the initial vertical and horizontal components of velocity along with the vertical displacement are sufficient to solve the problem.

Vertically downward $s = v_it + at^2/2$; so $at^2/2 + v_it - s = 0$

where $s = 0.6$ m, $v_i = -v\sin(45)$, and $a = g$.

$t = \{v\sin(45) + [v^2\sin^2(45) + 2g \times 0.6]^{1/2}\}/g = -0.4846$ s or +0.8626 s

The horizontal range is the product of this time and the horizontal component of velocity as follows:

D = $v\cos(45)$ × 0.8626 = 3.05 m

This value is much less than those obtained when loads are thrown backward, since no transfer of momentum can be used to enhance the horizontal component of velocity.

Summary

The height to which the CM is raised in jumping depends upon the height of the CM at takeoff and its velocity. The former is maximized when the jumper has all parts of the body as high as possible. The velocity of the CM depends upon the work done and the mechanical impulse generated against the ground. Both impulse and work can be maximized through use of as many muscles as possible to accelerate all free body segments upward. This is why it is essential to swing the arms upward vigorously. In general, free segments should be swung upward in an extended configuration. The jumper can increase the mechanical impulse due to such actions by ensuring that it begins with as high a force as possible. This in turn results from arresting a downward motion of the body in

■ **Key Point**

Increase in both upper and lower body muscular strength will enhance jumping.

the manner of a countermovement. Not only will the force be high at zero velocity in the countermovement due to arresting of the downward momentum; also the properties of increasing force as muscle lengthens, temporary enhancement of the contractile apparatus, and storage and release of strain energy in elastic components will add to this effect. In some jumping activities, a run-up will serve to provide the countermovement effect. In this case the jumping muscles reduce the run-up velocity by eccentric contraction prior to their shortening. Pole-vaulting uses the pole as an elastic storage mechanism intervening between the vaulter and the ground. As all segments of the body should be recruited for maximal jumping, training for jumping should include increasing the strength of all muscles, not those of the lower limbs alone.

RECOMMENDED READINGS

Hay, J.G. (1978). *Biomechanics of sports techniques.* Englewood Cliffs, NJ: Prentice Hall.

Ivancevic, V.C., and Ivancevic, T.T. (2006). *Human-like biomechanics: A unified mathematical approach to human biomechanics and humanoid robotics.* Dordrecht (Great Britain): Springer.

McGinnis, P.M. (2005). *Biomechanics of sport and exercise.* Champaign, IL: Human Kinetics.

Yamaguchi, G.T. (2001). *Dynamic modeling of musculoskeletal motion.* Boston: Kluwer Academic.

CHAPTER 8

Object Manipulation

We manipulate many and varied objects, including our own and other persons' body masses, in addition to external inert objects. Objects are pulled and pushed, lifted and lowered, and carried. A common factor in these activities is the mechanical interface between us and the object. This interface is usually, but not always, the hands. This chapter begins with one aspect of the interface, which is gripping.

Why can we accomplish such diverse activities as hanging from a bar while sustaining much more than body weight and also write in very fine script with minuscule forces? Some of our genetically related cousins can hang from a bar with their feet but are not much good at writing. Even very young human infants have sufficient grip strength to hang from their parents' (or more likely their grandparents') fingers. The hand is a marvelous tool in its variety of capabilities, which are under the control of the human brain. One of the wonderful manipulations that the hand can perform is gripping (see MacKenzie and Iberall, 1994; Tubiana, 1981).

Pushing and pulling are means by which we can move objects around. Usually the objects to be moved are inert and therefore predictable in their behavior; they move only as a result of gravity and forces applied by us. The reaction to our applied forces is frictional force. Friction was recognized by our forebears, who minimized it by using logs on which to roll objects rather than dragging them along the ground. This was followed by the development of the wheel, which in effect can be carried along with the load. This is yet another example of human ingenuity preceding the formalism of mechanics.

For most of us lifting is a necessary, occasional activity, but for some people it is their occupation. Lifting can be a source of serious injury to an extremely sensitive part of the body, the back. We may suffer a variety of pain, from mild aches and pains to severe pain shooting down the legs and chronic spasm of the muscles of the back. Injuries such as hernias can be the result of lifting. These are only some of the disabling conditions that can arise in back injury. It appears that the untrained casual lifter is in considerably more danger than the trained person who lifts for a living. The danger to the body, apart from dropping something on one's toes, arises from the complexity of the architecture of the back.

People carry objects from one place to another occasionally in the home or garden, and frequently as an occupation. For small objects, the only inherent danger is in carrying the mass of the body on two lower limbs, and in this respect carrying is no different than walking. But objects come in various shapes and sizes, including suitcases, television sets, and small children who wriggle. If all objects were the same shape, size, and mass, we could learn correct carrying practices and repeat them without thinking. But since they are not, avoidance of injury to the carrier or to the small child requires thought regarding technique. In the developed world, carrying generally is performed using the hands, arms, shoulders, back, and legs. In other societies the head is often used to advantage (for carrying rather than thinking about carrying), as we shall see.

Aim of Gripping

When we grip an object, the aim is to produce force by one part of the hand against an opposing force by another part. In some cases this is the sole aim, while in others it is a means of producing a force normal to an object that can then be twisted using friction. People also use gripping to suspend themselves from external objects such as bars in gymnastics and tree branches, as when children steal fruit, and also to suspend objects like suitcases. Sometimes it is knowing when to release the grip that is important.

Mechanics of Gripping

A simple representation of gripping by the hand is shown in figure 8.1.

Readers will appreciate that the fingers and thumb are each represented here by one less phalanx than exist in reality for the purpose of simplicity. The pincer grip gives two opposing forces at two respective points of contact with the object. The friction grip gives numerous forces at numerous points of contact. The source of these forces is muscle force, which, in creating moments about the joint axes, produces external forces against the object gripped (see figure 8.2).

As the number of joints involved increases, more muscles can be active (F_{M1}, F_{M2} . . .) and more moments will be generated ($F_{M1}d_1$, $F_{M2}d_2$ . . .). At the most distal joint, the force F_{M2} creates a moment $F_{M2}d_2$ that is responsible for generating a moment $F_{R2}l_2$. Force F_{M1} creates a moment $F_{M1}d_1$, which is responsible for generating a moment $F_{R1}l_1$ plus a reaction force at the most distal joint. This reaction force arises from the compressional effect that F_{M2} has on the most distal joint. The compressional force vector is difficult to include in this figure, but its presence must be remembered. Finally, muscle force F also generates a moment $F_R l$ and sustains the effects at the distal joints, so it would be expected to be the greatest muscle force. This analysis also applies in the pincer grip where a single force is applied to the object. However, the individual muscle forces must exist to stop the finger and thumb joints from extending.

When gripping while rotating an object we require friction. The frictional force arises from the relative sliding

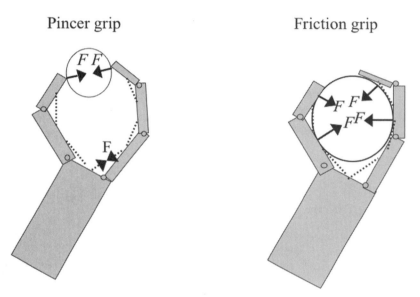

FIGURE 8.1 Pincer gripping requires two opposing forces, while the friction grip has many points of force application.

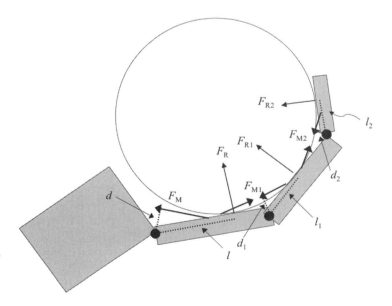

FIGURE 8.2 Moments created by muscle F_M result in normal forces F_R on the object.

Friction grip

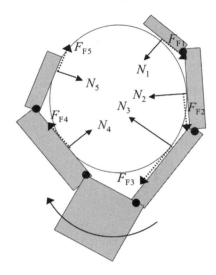

FIGURE 8.3 Gripping produces normal forces *N*, and additional rotation of the wrist or forearm produces frictional forces F_F.

motion between the surface of the object and the points of contact with the skin. Such an effect is shown in figure 8.3.

In this case the forces from the fingers become the normal forces (*N*), which are related to the frictional forces (F_F) acting at a tangent to the object's surface as follows:

$$F_F = \mu N$$

The total frictional force tending to rotate the grasped object clockwise is

$$\Sigma F_F = F_{F1} + F_{F2} + F_{F3} + F_{F4} + F_{F5},$$

which becomes

$$\Sigma F_F = \mu N_1 + \mu N_2 + \mu N_3 + \mu N_4 + \mu N_5. \qquad \text{(Equation 8.1)}$$

As these forces act at a tangent to the surface of the object, each will contribute to the total moment applied to the object by the hand. Should any normal force be reduced, the total frictional force will be reduced correspondingly.

■ Key Point

The pincer grip induces a reaction force on the object that is largely perpendicular to its surface; the friction grip produces a normal force and also a force parallel to the surface.

Biomechanics of Gripping

The mechanical situations just depicted are very simplified versions of the real thing. This is the case because the flexibility of the hand lies in its complex architecture, which comprises 14 bones in the fingers, five in the palm, and eight in the wrist. Such a complex architecture is usable only because of the many and varied muscles attached to these bones. Figure 8.4 gives a simplified single view of the arrangement of some muscles.

Muscles that move the fingers are located in the forearm and the hand. They flex (e.g., flexor digitorum) and extend the wrist, hand, and fingers and rotate the wrist laterally (abduction and adduction). Other small intrinsic muscles such as the lumbricals and interossei open (abduction) and close (adduction) the fingers laterally. Other muscles cross both elbow and wrist joints so that their forces create moments about both of these joints. The muscles are situated at various locations from the underlying radial and ulnar bones, giving rise to names including superficialis and profundus, which are close to the surface and deep within the forearm, respectively. The tendons of many of the muscles divide and insert in a number of other bones. The deep muscle flexor digitorum profundus is an example. Each tendon is surrounded by a synovial sheath. There are many more muscles than bones in the wrist and hand than are shown. Such an arrangement enables control of an extremely large number of intersegmental configurations.

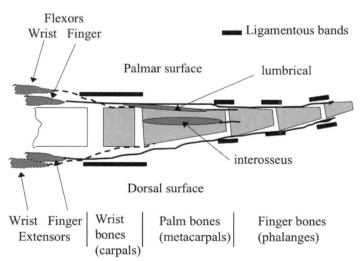

FIGURE 8.4 Some important muscle groups involved in gripping.

There is a significant mechanical benefit to the manner in which these muscles are located in the upper limb. The moment of inertia of the forearm about an axis through the elbow joint is reduced by the proximal location of the greater part of the muscle mass. Moment of inertia of the whole upper limb about the shoulder axis is therefore similarly reduced. The second benefit is that range of movement of the wrist joint is not opposed by large amounts of bulky muscle.

A band of ligamentous connective tissue or retinaculum surrounding the wrist serves to keep the tendons from moving away from the underlying radius, ulna, and carpal bones when the wrist is in flexion or extension. In a similar manner there are ligamentous bands surrounding the joints between the phalanges of the fingers. These ligamentous bundles are aligned transversely and obliquely with respect to the long axes of the phalanges. A complete anatomical diagram of the structure is very complex and is unnecessary here. The movements of the bones of the hand, wrist, and forearm are too numerous to discuss in any single book other than one devoted entirely to the hand. Readers interested in fine detail on this topic should consult a book on musculoskeletal anatomy.

Gripping requires flexor moments at the metacarpophalangeal and interphalangeal joints. Such moments produce high fingertip forces by well-defined muscular recruitment patterns in adults (Valero-Cuevas et al., 1998). Other more complex tasks such as piano playing will involve recruitment patterns that must be learned individually. Whatever the activity, the joints must be lubricated. This appears to occur by means of the rolling action of one articular surface on another, which tends to pump or redistribute the synovial fluid in the joint capsule (Sagowski and Piekarski, 1997). Sagowski and Piekarski also demonstrated a complex sliding as well as rolling motion in the metacarpophalangeal joints, changes in the position of the joint center of rotation, and moments that are dependent on joint angle.

A particular problem in gripping is the "bowstring effect." This results from disruption of the ligaments surrounding the joint such that the tendon moves away from the joint as shown in figure 8.5.

This figure shows the forces that are active in one metacarpophalangeal joint. In the intact case, the flexor tendon force F_T attempts to straig1hten the tendon as in the case of a piece of string being pulled on at each end. The ligaments oppose such an action and restrain the tendon in its appropriate location, having forces F_L applied to them. The ligaments are under a constant threat of rupture by overuse of the gripping hand. Also external trauma such as slicing by a knife can be responsible for ligament rupture. Consequently the tendon is straightened and pulled away

Key Point

The musculoskeletal arrangement of the wrist and hand allows an enormous number of degrees of freedom of the gripping system.

Key Point

The proximal location in the forearm of the bulk of muscles that control the wrist and fingers reduces the moment of inertia of the forearm-hand segment about the elbow and allows unrestricted movement.

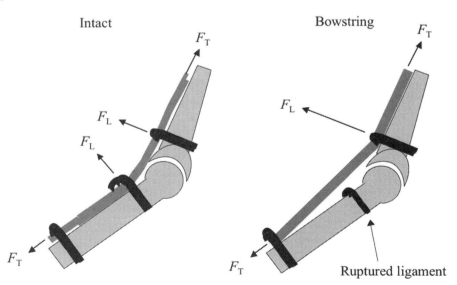

FIGURE 8.5 Tendons are held in situ by ligamentous bands unless rupture allows the bowstring effect.

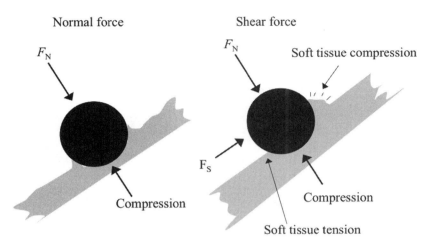

FIGURE 8.6 Friction force leads to shear force in soft tissues.

from its normal location as in the bow-string diagram. Then there is only one ligament capable of restraining the tendon, so the force applied to it will increase. This ligament is therefore more likely to rupture under the increased value of F_L. This scenario occurs when the flexor muscles are highly active in both the pincer and friction grips.

A problem with the friction grip is shearing of the soft skin and underlying subcutaneous fat. Figure 8.6 demonstrates this effect by contrasting a simple normal force F_N with a combination of normal F_N and shear forces F_S.

With the normal force, a compression of soft tissues will occur, as well as a certain amount of stretching where the tissue is displaced near the area of application of the force. In the case of shear, there will be greater compression on one side of the application area while the other side will experience significant stretching. Such frictional force application describes the biomechanics of bruising of tissues and actual rupture in very severe cases.

Another dangerous mechanism is experienced if limiting friction is overcome and sliding friction is present. In this case the friction is a dissipative force and heat is generated during sliding. The obvious consequence is burning of the skin on the palmar surface of the hand.

A further painful condition is experienced when the shape of the articular surfaces of the fingers is modified by trauma or arthritis. In such cases individual finger joints can lock and may require an external force to straighten them.

Variations of Gripping

The variety of gripping tasks can be classified into those that require large finger forces and those in which force is minimal. Limitations to the ability to perform the latter are more neuromuscular than biomechanical. Biomechanical factors provide the limitations in the high-force gripping tasks.

High finger flexor forces are required when one works with hand tools to produce either a large pincer force using a pipe wrench or high grip friction using a screwdriver. There are also muscular limitations to carrying heavy objects and to hanging and swinging from bars. The limitations in these cases vary from the maximal isometric force that the flexor muscles can produce to the ultimate stress and strain that the tissues can withstand in shear.

Throwing requires use of the finger and wrist flexors in combination. Such a combination differs from what is seen in gripping in that the wrist and finger extensors cannot be used. It might be argued that finger extension is required to release the grip on a ball that is being thrown. There little evidence for such an argument, for two reasons. One is that only gentle grips are needed to hold most balls. The other is that acceleration of the ball is usual in throwing, so that the

■ **Key Point**

Tendons of the finger flexors are kept in position by ligamentous bands surrounding the bones and tendon; this arrangement prevents the "bowstring effect."

■ **Key Point**

Shearing of skin and underlying tissue is a potential problem inherent in the use of the friction grip.

■ **Key Point**

In some cases, muscles can induce frictional force in gripping that is greater than the shear strength of the underlying skin.

inertia of the ball presses on the fingers, which then need only small amounts of force for control. This effect is seen dramatically in shot-putting: The 16 lb (6 kg) mass induces extreme wrist extension that puts the wrist flexors in considerable stretch prior to their shortening in the final phase of putting.

Enhancement and Safety of Gripping

The ability to grip is enhanced through an increase in the maximal isometric force of finger flexors. While this enhances gripping, it also enhances the shear stress in soft tissues in friction gripping. People can enhance safety in this case by wearing gloves that provide good frictional resistance but that also are sufficiently sturdy to avoid shear strain. Gymnasts use a leather pad that is located on the palmar surface of the hand and fixed over two middle fingers and the wrist to offset frictional shear in activities involving gripping a bar. Even with the use of such protective pads, athletes develop calluses because the normal force N and therefore its frictional counterpart $F = \mu N$ induced during swinging are large. Increase in maximal muscle force is also likely to increase the bowstring effect on the ligaments of the finger joints that support the tendons. This problem is not one that can be avoided by the wearing of gloves; it is preferable to use some other mechanical device intervening between the hand and the object.

Handles of an appropriate shape and size are required for carrying heavy objects. Very thin handles will induce high pressure on the soft tissue of the fingers. This can lead to pressure pain, reduced blood supply, and early onset of muscular fatigue. Alternatively, handles with a large radius can lead to slipping and loss of the load.

Tendons that cross the wrist are susceptible to dangerous loading of a frictional nature. The carpal (wrist) tunnel is a space bounded by ligamentous sheets or bands surrounding the forearm. It contains flexor and extensor tendons, bones of the wrist and forearm, and numerous nerves and blood vessels. Forming a tight fist reveals some individual flexor tendons that disappear under this sheet in the direction of the palm. The tendons are tightly packed together and have a tendency to rub on the underlying bones, despite the fact that they are enclosed in lubricating sheaths. The avoidance of excessive wrist flexion and extension lessens the chances of this repetitive strain injury. Many other wonderful actions that the hand can perform are better explained by recourse to neurophysiological and motor behavior analysis than by biomechanical analysis of its machinery alone.

The hand is a beautiful machine, and it should be treated with care. Whenever there is any doubt as to the safety of the hand, people should use an appropriate mechanical device.

Practical Example 8.1

Figure 8.7 shows the ends of fingers gripping the screw-top of a can. The top will unscrew when the moment $M = 10$ N.m. If the fingers share the load equally, what is the normal grip force N required to open the can if the coefficient of limiting friction between the fingers and screw-top is 0.65?

■ Key Point

Continued use of a frictional grip necessitates the wearing of protective gloves to avoid tissue shear and burning of the skin.

■ Key Point

The design of handles should ensure no local pressure points in gripping.

■ Key Point

Gripping should be performed with the wrist in its neutral position to avoid excessive frictional contact between tendon sheaths and restraining ligaments.

$r = 5\text{cm}$

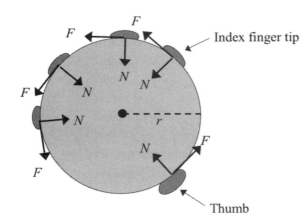

FIGURE 8.7 Normal and frictional forces produced by the finger pads in opening a twist-top can.

Solution

The product of the force F applied by each finger and the value r will need to equal $M / 5$ N.m. Using the expression $F / N = \mu$ will give the value N.

$$F \times r = M / 5; \text{ so } F = (10 / 5) / 0.05 = 40 \text{ N}$$

Since $F / N = \mu$,

$$N = F / \mu; N = 40 / 0.65 = 61.54 \text{ N}.$$

One point apparent in this problem is that increasing the radius of the screw-top will reduce the required value of F and therefore N. This leads to the design of attachments that fit over the top of cans to be opened by those whose muscular strength is compromised by age or by pain as in the case of people with arthritis. Another point is that dry hands have a greater value μ than do wet hands. Furthermore, when gloves are worn, they should exhibit high frictional qualities and be dry.

Aim of Pulling and Pushing

The most frequent aim in pulling and pushing is to produce force parallel to the surface on which a load is being moved. Since moving an object from rest includes acceleration of the load and overcoming inertia and friction, the greatest force in pulling and pushing occurs at the beginning of the movement. When a given speed is reached, the only resistance is rolling or sliding friction. The aim necessarily includes motion of the body. This implicates safety and also the manner in which body motion can influence the force applied to the load. Other aims are to topple an object and to slide an object that does not have wheels.

Mechanics of Pulling and Pushing

There are two stages to pushing and pulling of movable objects: getting the object started and continuing its motion. In either case the problem is to apply force in the most economical manner with the least effort and least risk of injury. In starting the motion of a cart, we are working against static friction of the wheels and the inertia of the cart. Figure 8.8 illustrates the competing forces at the handle and at the wheels.

Friction will be overcome when

$$F_C > \Sigma F_{FC} \qquad \text{(Equation 8.2)}$$

where ΣF_{FC} is the sum of the frictional forces at the wheels. Since the coefficient of static (or limiting) friction μ is related to force by $F / N = \mu$, where $N = mg$, motion in this case will occur when

$$F_C > N\mu \text{ or } F_C > mg\,\mu. \qquad \text{(Equation 8.3)}$$

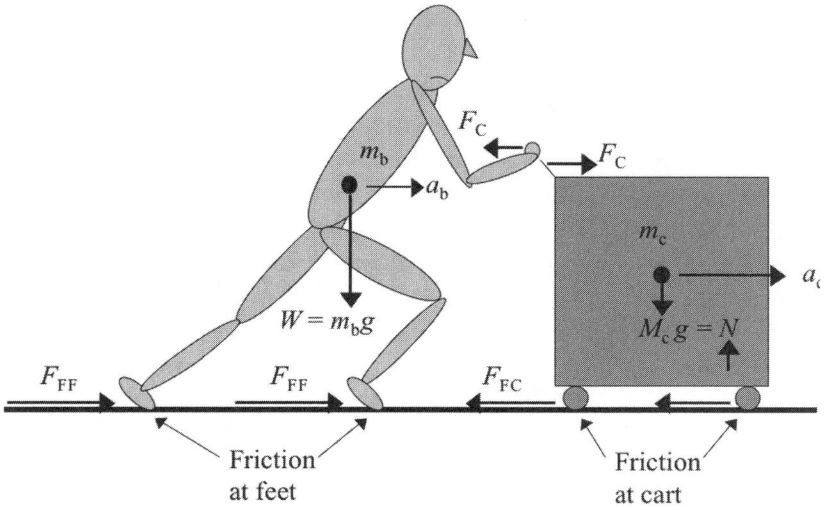

FIGURE 8.8 Friction at the feet aids force production, while friction at the wheels retards cart motion.

This leads to the conclusion that the greater the mass of the cart and coefficient of friction, the greater will be the force required to begin motion. Since μ is greater for an object to be slid on the floor than for one on wheels, the force to move the former will be greater. Correspondingly, more massive carts will be more difficult to move as shown by Equation 8.3 and also due to the greater inertia resisting acceleration.

In the period of horizontal acceleration the human operator will also accelerate, and the appropriate equations are

$$\Sigma F_{FF} - F_C = m_b a_b \text{ and } F_C - \Sigma F_{FC} = m_c a_c;$$

$$\Sigma F_{FF} = m_b a_b + m_c a_c + \Sigma F_{FC}. \qquad \text{(Equation 8.4)}$$

Therefore the frictional force at the feet will be greater in the acceleration phase because the masses of both the body and the cart require acceleration and the frictional force ΣF_{FC} must be overcome. As the dynamic frictional coefficient decreases with an increase in velocity, the frictional force ΣF_{FC} will decrease during acceleration. When the steady state velocity of pushing or pulling is achieved and acceleration is zero, the only resisting force will be ΣF_{FC}. This dynamic friction will be lower than the corresponding static friction, and ΣF_{FF} will be less than in the static and acceleration phases. The elements in Equation 8.4 illustrate that motion of the body segments, inasmuch as they modify acceleration of the body center of mass (CM), is a means of modifying the relationship between ΣF_{FF} and ΣF_{FC}.

Biomechanics of Pulling and Pushing

The initial stage of the pull or push requires increase in isometric muscular activity of numerous muscles to overcome friction. Whether the muscular activity is flexor or extensor will depend on whether a pull or a push is being performed and also on the posture adopted as shown in figure 8.9.

The directions of the moments shown are therefore approximate, particularly in the case of shoulder and elbow moments. The moments that differ between

■ Key Point

The coefficient of friction limiting initiation of movement is generally greater than that during subsequent movement.

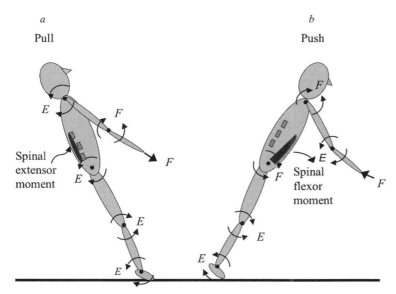

a
Pull

b
Push

FIGURE 8.9 Approximate directions of moments in pulling and pushing.

the depicted pull and push occur at the hip, shoulder, and elbow joints. Pulling and pushing require activity of the spinal extensors and flexors, respectively. Moments at the wrist will be very much determined by the wrist joint angle and the direction of the reaction force on the hand.

One might consider the optimal posture to adopt in the isometric case to be that in which maximal moment can be obtained at each joint. In this case no single joint moment need be maximal and therefore a potential limiting factor. If the reaction force at the hand is large, it is likely that the moments generated at the shoulder, elbow, and wrist joints will be unable to resist displacement of the body CM induced by high ankle, knee, and hip moments. There are two solutions to this problem. In pushing, the first is to remove arm moments from the linkage by applying force with the upper torso directly against the load as shown in figure 8.10*a*.

Pushing with the back against the load is preferable since the force is distributed over a large area of the back, leading to reduced pressure on any single part (figure 8.10*b*). Pulling is easier than pushing from a biomechanical viewpoint, mainly due to the difference in behavior of the arm musculature in the two activities. Contraction of the leg and trunk muscles produces forces at the shoulder joints that must be transmitted to the cart by the arms. Figure 8.11*a* shows that during pushing, the elbow and wrist joints need to be kept at fixed angles by moments generated by muscles on one side of each joint.

With the application of high forces from the feet, the elbow extensors and wrist flexors must work strongly, so there is considerable danger of the elbows collapsing into flexion and the wrist into extension. When pulling, the joints of

a

Forward push

b

Backward push

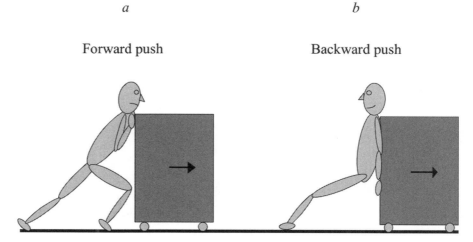

FIGURE 8.10 Removing the arm linkage by pushing with the torso can facilitate larger forces.

the upper limb are automatically straightened, and both flexors and extensors can
reduce stress in elbow ligaments, which can
also contribute tensile force (figure 8.11*b*).
A similar situation exists at the shoulder
joints.

A second solution for getting a large load
moving against a high frictional force is
to allow changes in segmental configuration. Beginning with the arms straight in
pushing and developing large moments
at the ankle, knee, and hip joints will allow
acceleration of the body CM. The lower
limbs will straighten and the upper limbs

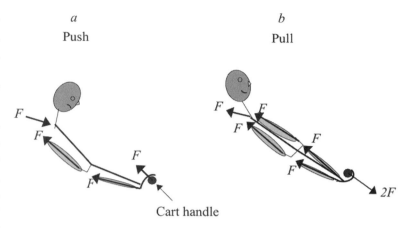

FIGURE 8.11 Pulling is easier on the wrist than pushing.

will flex, leading to development of horizontal momentum of the body CM. This
momentum can be arrested by the arm musculature working in its much better
force-producing capacity of eccentric contraction. In this way the force applied to
the load as a reaction to decreasing the momentum of the body can be increased
as a means of overcoming limiting friction. The same effect can be applied in pulling. This effect is not so much a true transfer of momentum since the load does
not move initially, but more a means of raising F_C to overcome F_{FC} (see figure 8.8).
Since the motion is horizontal, it may seem that we do not have to fight against
gravitational force. In fact, the tendency of the body to rotate about the ankle
joints gives a gravitational component that aids us a little. We experience such
an effect when we lean against an object that is unexpectedly free to move.

In the case in which $\Sigma F_{FF} > \Sigma F_{FC}$, both $m_b a_b$ and $m_c a_c$ can be positive (Equation
8.4), but acceleration of the body CM and acceleration of the load will generally
not be the same value if the segmented body changes its configuration. Such
changes can be performed to allow the joint angles to be maintained close to their
maximal moment-producing capacity.

The coefficient of friction gradually decreases in the approach to a desired
constant load velocity. The force at the feet can be reduced gradually and appropriately so that the load can be maintained at or near constant velocity. Should
the person take large steps, large variations of ΣF_{FF} will lead to large variations of
ΣF_C with consequent acceleration and deceleration of the load. One can avoid this
inertial effect by taking small steps to decrease variation in ΣF_{FF} so that muscular
effort is needed only to counteract friction. The likelihood of injury is lessened as
a consequence unless one encounters some unseen obstruction. Therefore people
should attempt movement of objects only when they have clear vision of what
is coming next.

Variations of Pulling and Pushing

The biomechanical difference between pulling and pushing a load uphill is significant. Figure 8.12 depicts the forces that are likely to be applied when moving
a load uphill using the postures shown in figure 8.9.

In figure 8.12, the push (*a*) and pull (*b*) are treated and considered separately
using the solid black lines. The angle of the slope to the horizontal is θ. The push

■ Key Point

Postural changes
can aid in reducing
weak links in
the kinetic chain
between the floor
and the load during
pushing and pulling.

■ Key Point

Transfer of body
momentum to the
load is an aid to
getting motion of the
load started.

■ Key Point

When one is pulling
or pushing a large
load, the force
applied will fluctuate
less if one takes
small steps rather
than large steps.

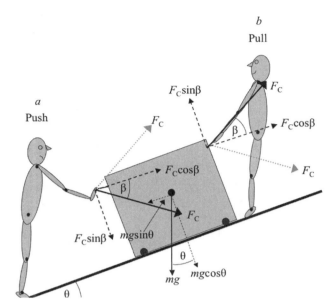

FIGURE 8.12 Pulling uphill reduces the gravitational effect.

(*a*) is conducted with a force F_C acting at an angle β to a line parallel to the slope. When F_C is resolved into its components, parallel and perpendicular to the slope, the pushing force has a component acting toward and perpendicular to the slope surface of $F_C\sin\beta$. This adds to the cart's normal force $mg\cos\theta$ to yield a total force N normal to the slope of

$$N = F_C\sin\beta + mg\cos\theta.$$

The force (F_p) parallel to and up the slope is

$$F_p = F_C\cos\beta - mg\sin\theta.$$

Movement will take place when $F_p > N\mu$ or

$$F_C\cos\beta - mg\sin\theta > (F_C\sin\beta + mg\cos\theta)\,\mu. \qquad \text{(Equation 8.5)}$$

Rearranging Equation 8.5 gives the force at which the limiting static friction will be overcome as

$$F_C > mg(\sin\theta + \mu\cos\theta) \, / \, (\cos\beta - \mu\sin\beta) \text{ (Push)}. \qquad \text{(Equation 8.6)}$$

The same manipulation can be made in the case of pulling as follows:

$$F_C > mg(\sin\theta + \mu\cos\theta) \, / \, (\cos\beta + \mu\sin\beta) \text{ (Pull)}. \qquad \text{(Equation 8.7)}$$

All factors are the same in Equations 8.6 and 8.7 except that $\mu\sin\beta$ is negative in Equation 8.6 and positive in Equation 8.7. These equations indicate that for equal values of θ and β, the denominator in pulling will be greater than that in pushing. Therefore the force F_C required for movement will be less in pulling than in pushing. Consequently it is generally preferable to pull a load uphill rather than push it. This is not universally true, particularly if the forces are directed as indicated by the dotted lines labeled F_C in figure 8.12. However, the postures required to generate these forces are extreme and are unlikely to be controllable by the cart operator. In order to avoid such postures, carts and other movable objects should have numerous projections where force can be applied. The result will be maintenance of favorable force-producing postures leading to enhanced performance and safety.

The preceding analysis also can be applied to moving objects downhill. In this case, gravity is helpful and may produce a force that is greater than frictional force. The smallest push may be all that is necessary to initiate movement in this case, and further force application will be an

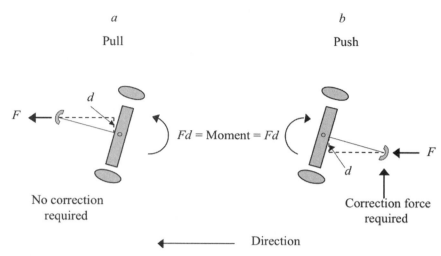

a
Pull

b
Push

$Fd = \text{Moment} = Fd$

No correction required

Correction force required

Direction

FIGURE 8.13 Pulling tends to correct for deviations from the intended line of motion.

upward pull to control downhill motion. In this case the operator would do well to face down the slope while pulling upward against gravity and also moving forward. Letting a load down the slope by facing up the slope and pushing to counter the effect of gravity would not be recommended, as unusual floor conditions will be unseen.

If the operator is moving a cart that has two wheels side by side, such as a golf cart, the simplest method of control is to pull. Figure 8.13 shows how pulling has a self-correcting effect on changes to the direction of motion.

When the path of the cart deviates from the direction of the applied force, pulling produces a moment Fd tending to return it to the original path. Pushing tends to exacerbate the deviation, and sideways force has to be applied to right the motion to correct the deviation. However, there are reasons for a preference for pushing that involve the obviation of back and shoulder pain.

Enhancement and Safety of Pulling and Pushing

Carts or trolleys are moved around in many occupations. Therefore workers encounter a wide variety of floor conditions such that frictional resistance to pushing a cart can be highly variable. In low-frictional conditions, forces at the wheels will be small, perhaps because the wheels are running on a hard, smooth floor. To achieve a given velocity under these conditions we require a small force at the feet, but one that will need to be greater if the cart has a larger mass. A "harder push" is also required if the cart is on a thickly carpeted floor. In these situations the body lean tends more to the horizontal to allow our push to be directed horizontally, thus reducing the normal force. The apparent advantage of this strategy is counteracted somewhat by the fact that the frictional force at the feet will necessarily be greater. So a major requirement is a sufficiently high coefficient of friction at the feet to avoid slipping. There is no guarantee that a high coefficient of friction between the load and the floor will equate to a correspondingly high coefficient between the shoes and the floor. It is therefore essential that in occupations in which pushing and pulling objects around are part of the job, attention be directed toward appropriate footwear and flooring

■ **Key Point**

Pulling uphill is preferential to pushing, particularly if the pulling force has a vertical component that acts against gravitational force.

■ **Key Point**

Downhill motion of a load is best achieved by means of an eccentric pulling force.

■ **Key Point**

All sources of frictional force must be reduced when objects are pushed or pulled, but not at the expense of sufficient friction between the floor and footwear.

and the upkeep of both. In the household case, people need to judge these factors before attempting movement. It is in starting the push that the forces are highest and the effects of a mistake are most serious.

Some people complain of back pain when pulling rather than pushing, although the reverse is common also. Undoubtedly the difference is due to individual back construction and perhaps to some preexisting back disorder. During pushing there is a reaction force at the shoulders that tends to extend the spine (see figure 8.9). To avoid extension, the flexors of the trunk, namely rectus abdominis, must be contracted. Back muscle involvement is likely to be restricted to the deep muscles that maintain the spatial orientation of the vertebrae. Absence of force by the superficial back muscles probably reduces effects such as compression and pressure on the intervertebral discs and articular surfaces, respectively. Therefore the chance of back strain is lessened. During pulling, the reaction force tends to flex the spine, and the large extensor muscles of the spine, namely erectors spinae, must contract. The loads on the discs and articular surfaces are therefore increased. Because of individual characteristics, no universal preference can be advised for pushing as opposed to pulling. The statement "If it hurts, don't do it" remains a good guide to preference of one technique over another. A more detailed treatment of the role of the back muscles is given in the section on lifting and lowering later in this chapter.

The use of multiple handles to which force can be applied will enhance performance and improve safety. Handles at different heights from the ground will allow the operator to change the direction of the applied force so the force in the pushing direction is maximized and the normal force is minimized. Such changes will also affect the frictional requirements between the floor and footwear. People can achieve a further modification to force application through the use of a harness of some type. Use of the relatively weak muscles of the arms and shoulders can be obviated by means of wide strapping or a harness or yoke surrounding the shoulder girdle. An additional advantage of such a device is pressure reduction on any specific area of the shoulder girdle. In this case the harness should be wide and should cover a large area; a rope will not provide the same pressure-reducing effect. The negative aspect of using a harness is that fine control of a cart will be minimized as the hands and arm muscles are removed from the linkage.

© Human Kinetics/Neil Bernstein

Pushing using the back eliminates the weaker link of the arms and reduces the pressure on the tissues that is induced by the force.

Historically humans have been able to plough fields with such a device, and a yoke is the traditional means by which horses pull carts.

Pulling a load while moving backward has its advantages in terms of enhancing performance, but it also has the disadvantage of potential contact between the cart and the feet and legs. Therefore people should carefully consider the space available between body and cart before pulling a load.

Safety will also be a consideration if there is variability of the coefficient of friction among the wheels of the cart. Not only will a high-friction wheel increase the overall coefficient of friction; it will also induce rotation of the cart about an axis perpendicular to the plane of the wheels. All readers will be familiar with the problems produced by sticky wheels in stores and airports. The fact is that carts should be serviced sufficiently and frequently. A young, strong person may well be prepared to wrestle with a sticky cart, but older persons may suffer substantial injury if they try to do so.

Pulling and pushing are involved when people use manual tools such as wrenches. Here we are using a force at the feet applied through the hands to generate a moment, when unscrewing a rusted nut, for example. The major problem occurs when the nut becomes loosened suddenly while a large force is being applied. The consequence is often a fall, which can increase in severity as the force applied is increased. There are two major contributing factors to the fall. One is that when applying a large force, the muscles (probably of the lower limbs) are in a high state of activity. Upon release of the resisting moment, the reflex instructing the muscles to shut off takes time to develop and the muscle itself requires time to deactivate. Therefore there will be force accelerating the body until the muscles are completely deactivated. A second factor is that the operator may be using a component of gravity to add to the force by leaning with the CM outside of the base of support. The operator can reduce both of these factors by decreasing the amount of applied force that is necessary. A simple means of achieving this objective is to increase the lever arm of the implement. If the moment at which release will occur is M, one can produce it by applying a force F at some perpendicular distance d such that $M = Fd$. Increasing the length of the lever arm will mean an increase in d so that F can be reduced. Hand tools that require application of large force should allow for increase of the length of the lever arm. For anyone who has changed an automobile tire manually and while doing so found that the securing nuts had corroded over time, it should come as no surprise that power tools are used in shops where this is done repetitively.

There are many ways of pulling and pushing, and one can learn how to enhance performance and safety of these activities. The ultimate safety skill, however, is to know when to use a machine rather than one's body to push and pull things around.

Practical Example 8.2

Figure 8.14 shows a person supporting a load statically to keep it from rolling downhill. This example illustrates the biomechanical analysis required to determine how the cart can remain stationary when the operator applies the force F_C in different directions.

Part 1

Calculate the minimal coefficient of friction μ_F at the feet that will be required to avoid movement given the following:

$\theta = 30°$, $\beta = 25°$, $m_c = 25$ kg, $m_b = 80$ kg, $\mu_c = 0.15$

Solution

We approach the solution by resolving forces parallel and perpendicular to the inclined surface along which frictional force acts. To obtain μ_F, we need the parallel

■ **Key Point**

Wheels on carts should be maintained so that friction is minimal and is close to equal among the wheels.

■ **Key Point**

Tools that help in producing moment of force should be of sufficient length to minimize muscular force.

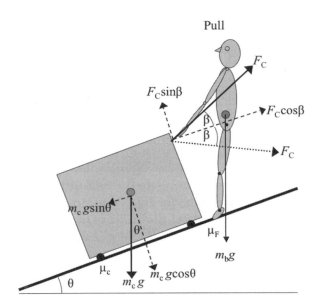

FIGURE 8.14 Two different directions of static force supporting a cart on a slope.

and perpendicular forces at the feet. We can obtain these only by knowing the force F_C, which the person is applying to the cart. Dealing first with the tendency of the cart to roll down the slope and using the general relationship $F = \mu N$, the cart will remain stationary if the sum of forces applied to it is zero. The following equation applies:

$$m_c g \sin\theta - \mu_C (m_c g \cos\theta - F_C \sin\beta) - F_C \cos\beta = 0;$$

$$m_c g \sin\theta - \mu_C (m_c g \cos\theta) = F_C (\cos\beta - \mu_C \sin\beta); \qquad \text{(Equation 8.8)}$$

$$122.63 - 31.86 = F_C \times 0.843;$$

$$F_C = 107.67 \text{ N}$$

F_C represents the mutual force between the person and the cart and is now considered the force tending to slide the person down the slope. The person will tend to slide according to the following equation:

$$F_C \cos\beta > \mu_F (-F_C \sin\beta + m_b g \cos\theta);$$

$$\mu_F < F_C \cos\beta \,/\, (-F_C \sin\beta + m_b g \cos\theta); \qquad \text{(Equation 8.9)}$$

$$\mu_F < 97.58 \,/\, 634.14;$$

$$\mu_F < 0.154$$

The feet will slide when the coefficient of static friction between the shoes and the floor is less than 0.154. This value of limiting friction is similar to that seen between the cart and the floor.

Part 2

Calculate the value of μ_F if the force is applied parallel to the slope.

Solution

The question implies that β is zero. The result of Equation 8.8 would then be

$$F_C = 90.77 \text{ N}.$$

The result of Equation 8.9 would be

$\mu_F < 0.134.$

Should the force produced by the person be applied so that β is zero, the value of F_C would be reduced according to Equation 8.8 (90.77 N vs. 107.67 N). Correspondingly the feet will slide when the coefficient of static friction between the shoes and the floor is reduced (0.13 vs. 0.15). An alternative interpretation is that not only is the force required less, but also a lower coefficient of friction at the feet can be tolerated when the force is applied parallel to the slope.

Part 3

Calculate the minimal coefficient of friction μ_F at the feet if the force F_C is directed at an angle $-\beta$ as shown by the dotted force vector F_C in figure 8.14.

Solution

From Equation 8.8 the force required will be

$F_C = 93.61$ N.

The result of Equation 8.9 would be

$\mu_F < 0.118.$

In this case $(-\beta)$, the force is increased above that obtained when the force is parallel to the slope ($\beta = 0$), but not above that when the force is applied at the same angle ($+\beta$) away from the slope (93.61 N vs. 90.77 N and 107.67 N, respectively). However, the frictional requirements at the feet are greatly reduced when the direction of the pull has a component toward the surface ($-\beta$) compared with parallel and upward applied forces. This example indicates that pulling with a component toward the surface is the best solution for keeping a cart from moving down a slope given the present value for the coefficient of limiting friction between the cart and the slope. Readers may wish to vary this value of β to see whether this statement is universally correct.

Practical Example 8.3

A worker is attempting unsuccessfully to topple a cabinet by pushing with a constant, horizontal maximal force of F_1.

Part 1

What is the minimal work required by a partner who pulls on a rope from the other side of the cabinet with a variable force of F_2 given the values shown in figure 8.15?

Solution

The minimal total work required to topple the cabinet is equal to its gain in potential energy when its CM reaches a vertical line through the point of rotation as shown by the broken outline of the cabinet. The assumption is that the cabinet has negligent kinetic energy at this point. When the cabinet is at the toppling position, its CM has risen by $b(1 - \cos\mu)$ where

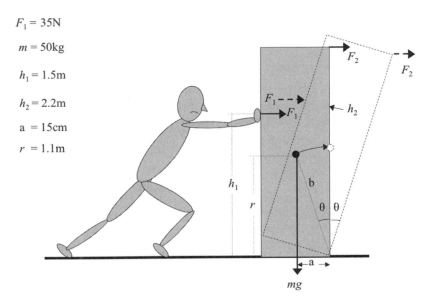

$F_1 = 35N$

$m = 50kg$

$h_1 = 1.5m$

$h_2 = 2.2m$

$a = 15cm$

$r = 1.1m$

FIGURE 8.15 Toppling a cabinet by means of two forces F_1 and F_2 applied horizontally.

$b = (a^2 + r^2)^{1/2} = (0.15^2 + 1.1^2)^{1/2} = 1.2325$ and

$\theta = \tan^{-1}(a / r) = \tan^{-1}(0.15 / 1.1) = 7.765°$.

The gain in potential energy is

$mgb(1 - \cos\theta) = 50 \times 9.81 \times 1.2325[1 - \cos(7.765)] = 5.54$ J.

The force F_1 in acting through a horizontal displacement of "a" gives the work done as

$F_1 \times a = 35 \times 0.15 = 5.25$ J.

Therefore the work to be done by the second worker is the difference between the potential energy required and the work of which F_1 is capable, which is $5.54 - 5.25 = 0.29$ J.

Part 2

What is the minimal value of this variable force F_2?

Solution

There is a question whether the cabinet can be moved from its initial position, where gravity is exerting the greatest moment upon it. If F_2 was constant, its least value applied over a displacement of $h_2\sin\theta$ between the initial and toppling positions would be

$F_2 = 0.29$ (joules) $/ h_2\sin\theta$ (meters) $= 0.29 / [2.2 \times \sin(7.765)] = 0.976$ N.

The moment created by gravity is

$mga = 50 \times 9.81 \times 0.15 = 73.575$ N.m.

The sum of the moments from both forces, assuming that the constant value of F_2 is used, is

$F_1h_1 + F_2h_2 = (35 \times 1.5) + (0.976 \times 2.2) = 54.636$ N.m.

As this total moment is less than gravitational moment on the load, the load will not move. The moment due to force F_2 has to be considerably larger and equal to the difference between gravitational moment and that due to F_1 as follows:

Moment due to $F_2 = mga - F_1 h_1 = 73.575 - (35 \times 1.5) = 21.075$ N.m.

Therefore F_2 must be at least $(21.075 / h_2) = 21.075 / 2.2 = 9.58$ N.

Since the initial value of F_2 is greater than the average value obtained previously, it is possible that F_2 would need to be negative to maintain constant angular velocity. It is possible to develop an expression for the value of F_2 required to allow constant angular velocity of the cabinet, but such an expression is rather too complicated for inclusion here. A simulation program of the activity would best represent the kinetics involved.

A general application of the preceding analysis would lie in the area of man–machine interface in repetitive tasks.

Aim of Lifting and Lowering

The aim of lifting is to raise an object from one vertical position to another. In this sense raising the mass of the upper body after stooping to observe something on the floor is classified as lifting. We can lift with many combinations of segments from one finger to the whole body. Single lifts are part of daily life and are done with little danger to the body. Problems arise when the load lifted is large or if the same lift is done repetitively and the lifter is untrained for the lifting task. So the initial aim of lifting can be rephrased to include raising an object from one vertical position to another safely. The aim of lowering a load is to reduce its height without danger to either the person or the load. So the aim of lowering can be different when one is handling a block of concrete versus a packing case containing a fragile and valuable Ming vase. Biomechanically, lifting and lowering are reverse operations done with the same muscles.

Mechanics of Lifting and Lowering

The simple mechanics of lifting indicates that work must be done in order to give a mass a greater potential energy as depicted in figure 8.16. During the lift, the load will have both kinetic energy $(mv^2/2)$ and potential energy (mgs) due to the work $(\int F ds)$ done by the force F. At the end of the lift when the velocity is zero, only potential energy will be present. At this stage the work done represents the total gain in potential energy as follows:

$$\int F ds = mgh \tag{Equation 8.10}$$

where h is the height through which the load is lifted. It should be remembered that h is the height from the initial starting height and not always the height measured from the ground. The initial value of F must be greater than mg in order for a lift to be accomplished. When a load is lowered, the initial value of F must be less than mg but not zero unless the load is being dropped in an unsupported manner. Once the load is moving either upward or downward, it will move at a constant velocity if $F = mg$. This analysis deals only with vertical motion, for that

■ Key Point

The essence of lifting is the performance of muscular work to raise the potential energy of a load and the mass of the body.

At any position *s*, work done by *F* is:

$$\text{WORK} = \text{KE} + \text{PE}$$

$$\int_0^s Fds = mv^2/2 + mgs$$

For *F* = constant, h_1 & h_2 initial and
final heights at rest, $s = h_2 - h_1$:

$$\text{Work} = \Delta\text{PE}$$

$$Fs = mg(h_2 - h_1)$$

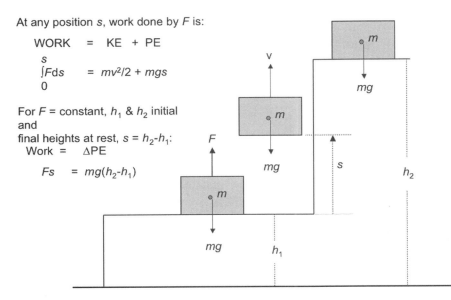

FIGURE 8.16 Exchanges of potential energy and kinetic energy in load lifting.

is how potential energy is measured in this context. Of necessity there will be horizontal motion to consider if a load is to be placed upon a table.

Biomechanics of Lifting and Lowering

Lifting and lowering are biomechanically reverse actions. The former requires concentric muscular contraction to increase the potential energy of the system, while the latter is eccentric and serves to decrease or dissipate energy.

The biomechanical expression of lifting is to increase the potential energy of an object. If during such a lift we raise the body CM by 0.3 m (12 in.) and our body mass is 70 kg (154 lb), the potential energy produced is about 206 J per lift (Equation 8.10). This represents the energy produced to increase the potential energy of the body alone whether we are lifting 1 kg (2.2 lb) or 100 kg (220 lb). Assuming that the 100 kg load is lifted through 0.5 m (20 in.), the energy produced to move the load would be 100 times 9.81 times 0.5, which equals 490 J. So a single lift of 100 kg would require a total of 696 J. However, it may be dangerous to lift a mass of 100 kg due to the possibility of injuring tissues such as muscles, tendons, and articular surfaces, and particularly intervertebral discs. Alternatively it would make little sense to lift a total of 100 kg in 1 kg packets, as the energy produced would be 100 times 206 J or 20,600 J for a total energy production of 21,090 J. Most of the energy in lifting small packets the way we lift large loads is due to having to lift the body. Clearly a compromise is required depending upon the magnitude of the load to be lifted. Some loads are too small to require a full knee-bend lift, and some are too large to lift safely at all. This is a question of judgment, which is much more accurate if the lifter is practiced at the task. Two basic types of lift are the stoop lift, using hip joint extension while keeping the knees extended, and the crouch lift, in which the trunk is kept as vertical as possible and knee extension produces the lift.

■ Key Point

The stoop lift saves energy but may be injurious to the back; in comparison the crouch lift requires greater energy but is better for the back.

Stoop Lift

Lifting primarily with hip extension is shown in figure 8.17. The masses of the load m_L and upper body m_B (including trunk, arm, and head) produce the following clockwise moments:

$$m_L gr\cos\theta \text{ and } m_B gp\cos\theta = (m_L r + m_B p)g\cos\theta$$

Lengths *r* and *p* are measured from the axis of rotation to the shoulder and body CM, respectively. In this figure we are using an axis of rotation located well

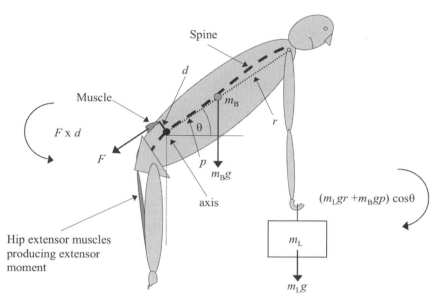

FIGURE 8.17   Forces in sustaining a load in the stooped posture.

down the lower back, where many injuries occur. Traditionally the axis used is one that is transverse and passes through the intervertebral disc between the fifth lumbar and the first sacral vertebrae (L5/S1 joint). However, we could use an axis located anywhere along the spine between the shoulders and the pelvis should our interest be specific to a particular intervertebral joint of the spine. Moving the axis up the spine would require appropriate changes in m_B, p, and r. Moments about the chosen axis of rotation in the static condition equate as follows:

$$Fd = (m_L r + m_B p)g\cos\theta \qquad \text{(Equation 8.11)}$$

The source of F is the muscles, which need to increase their force if the load $m_L g$ is increased or the value θ is decreased. The value d depends upon the architecture of the spine, and extension or flexion of the lumbar spine can respectively increase or decrease d by a small amount. As θ decreases to zero, $\cos\theta$ increases to a value of 1. So the magnitude of Fd increases to its maximal value when the trunk is horizontal. One can appreciate the magnitude of the task by using some reasonable values of m_L = 20 kg, m_B = 50 kg, r = 0.5 m, p = 0.3 m, θ = 45°, and d = 0.06 m. The value of F then becomes 2890 N (295 kg or 648 lb) of force produced by the extensor muscles of the lumbar spine. While this force may not seem high in comparison with the Achilles tendon force calculated in chapter 4, "Balance," it remains a large force for the intricate structure of the back to withstand. It certainly is a large value for the untrained lifter. Fortunately there are many extensor muscles of the spine; and in bodybuilders, weightlifters, and other "power" athletes these muscles are large and well defined. An additional muscular requirement is that of the hip extensors, which must stop the pelvis from rotating clockwise in this figure.

When acceleration occurs, there is an inertial effect due to angular acceleration of the trunk, arms, and head. The appropriate equation in this case is

$$Fd = (m_L r + m_B p)g\cos\theta + (I_G + m_B p^2)\alpha \qquad \text{(Equation 8.12)}$$

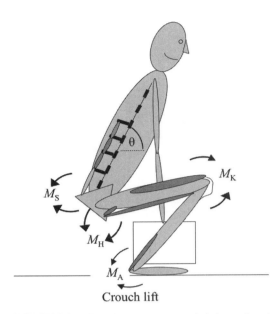

FIGURE 8.18 Moments in sustaining a load in the crouched posture.

where α = angular acceleration or second differential of θ or $d^2θ/dt^2$. Note that α is not equivalent to acceleration of the lifted load m_L since vertical displacement of the load is not linearly related to θ. While it is possible to relate α to acceleration of the load, it is sufficient to understand that acceleration and deceleration of the load will require greater and lesser muscle force F, respectively.

Equation 8.11 shows how the muscular force required increases as the angle of the trunk in the stoop lift becomes more horizontal. An interesting observation using electromyographic techniques is that the erector spinae muscles shut off when the trunk becomes horizontal in the unloaded case. This posture is exactly the one in which the activity would be expected to be greatest. The fact is that the ligaments and inactive muscles posterior to the axis of rotation within the vertebral column provide sufficient force to bear the load. One can perform this test using body mass alone, but attempts to shut off these muscles while bearing an external load should be avoided.

Crouch Lift

The crouch lift uses extension of ankle, knee, and hip joints as shown in figure 8.18.

Extensor moments M at the ankle, knee, hip, and spinal joints illustrate the greater muscular contribution over that seen in the stoop lift. The simple fact is that the CM of the body is raised through a greater vertical displacement in this type of lift. Therefore a greater part of the muscular energy requirement contributes to lifting the mass of the body as governed by the ds component in Equation 8.10. The greatest load with respect to body mass is borne by the ankle extensor moment M_A. As we progress up the body from joint to joint, the mass supported by each successive moment decreases.

The major benefit of the crouch lift is that υ can be kept close to 90°, which will reduce M_S significantly. The inertial effect of angular acceleration of the body mass shown in Equation 8.12 will also be minimized by this technique. Even though angular acceleration of the upper body may be eliminated, the mass of the upper body needs to be accelerated vertically. This poses a problem for the knee extensors. The ability to produce M_K is very much dependent upon the mechanical advantage of the joint and the force–length characteristics of the knee extensors. A simple attempt to raise the body from the position shown to a standing posture, even without a load in the hands, will lead to experiencing a position that feels weak. This can be a factor limiting the load that can be lifted in this manner. The danger in attempting to overcome this problem is stooping over to get the knees more quickly to an extended angle. The objective of the crouch lift is therefore defeated, and the problems of the stoop lift reappear. A further problem is that the knees have to be spread well apart to accommodate a load of any significant size. Young hips may be able to cope with this, but older, potentially arthritic hips will not. Lifting is complicated and potentially dangerous.

The vertebral column is shown as a straight structure in figure 8.18. This is far from accurate, as the vertebral column has both thoracic and lumbar curvatures.

■ Key Point

For any given load held in the hands, the force in the extensor muscles of the back increases as the body leans forward.

In the stoop lift, the lumbar curvature is reduced, whereas it remains present in a large range of the crouch lift. Such anatomical changes affect the mechanics of the lift and require inclusion in a detailed analysis of the lift and therefore the potential for tissue injury (McGill, 2002).

Lowering a load is governed by the same equations as is lifting, but the implications for force production by the spinal erector muscles are somewhat different. Equation 8.12 can be rewritten so that for simplicity the last term is named an inertial component:

$$Fd = (m_L r + m_B p)g\cos\theta + \text{Inertial component}$$

In lifting, the inertial component will be positive at the beginning of the lift and negative at the end where the upward acceleration is negative. So force F will be greater at the start and less at the end. The exact opposite applies in lowering a load once it is unsupported. Some implications for safety arise from these biomechanics. A load can be judged too great at the beginning of a lift where the force requirement is high. Therefore a decision to abandon the lift can be made early and without risk of the load falling from a substantial height. The positive inertial component in the late stage of lowering a load cannot be judged a priori and may be a surprise to the operator of the task. The only safe manner in which this increased force can be obviated is to drop the load, with unknown consequences of damage to the operator or the load. The preceding discussion implies that lifting and lowering are equally dangerous. However, lifting is a concentric activity involving muscle shortening, while lowering is eccentric. We have seen in chapter 3 that our ability to develop force eccentrically is greater than when muscles work concentrically. So we are in a better position to handle higher force at the end of lowering than we might expect. Furthermore, the energy cost of eccentric work done by a given force applied over a given displacement is less in eccentric contraction. On balance it appears that lowering is easier than lifting a load. This is a simple qualitative statement that will be untrue in many cases because of the number of variables that constitute lifting and lowering and depending on which criteria we use to judge what is "easier."

The basic mechanics of moving a load upward were introduced using the work–energy relationship. The majority of the biomechanical analysis is done by relating moment produced by muscle to opposing moments due to a load and a rotational inertial component. This is an example of the need to understand a range of mechanical concepts in order to gain an appreciation of the biomechanical nature of an activity.

Forces in the Vertebral Column

An apparently unfortunate aspect of the construction of the human machine is that its control system (the brain) resides at some distance from the muscles that are controlled. The brain sends signals to the muscles via nerves emerging from the bony spine that protects the spinal cord. The problem is that the bony spine is broken into segments called vertebrae so that it can flex, extend, and rotate axially and laterally. So the intervertebral structures represent those of a joint through which the nerves emerge.

If we move inward to examine the internal forces, we can identify some of the sources of danger to the back during lifting, including the dreaded slipped (herniated)

■ Key Point

The load in the crouch lift is limited by the strength of the knee extensor muscles.

■ Key Point

The inertial component in lifting increases as its acceleration increases.

Cross section   Longitudinal section

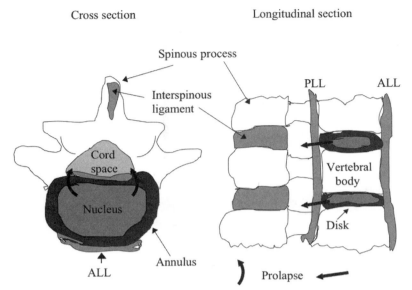

FIGURE 8.19 Cross (transverse plane) and longitudinal (sagittal plane) sections of the bones and ligaments and disc of the vertebral column.

■ **Key Point**

In disc compression it is the gelatinous nucleus of the intervertebral disc that is forced out of place, not the whole disc.

disc. The complex nature of the spinal intervertebral joint is only partly captured in figure 8.19. The basic architecture is vertebrae linked by the intervertebral discs. The bumps and knobs on the vertebrae are sites of insertion of tendons and ligaments, although the tendons are very short. The projections of bone increase the leverage of the muscles and ligaments, as we have seen previously in the knee joint. Human beings are not unique in this respect because most vertebrates show almost identical projections, and for the same reasons.

A significant point is that the disc is not like a uniform sponge and does not slip out of position as is often mentioned in casual conversation. In fact, the disc comprises a ring of fibrous cartilage called the annulus fibrosis with a gelatinous center called the nucleus pulposus. The disc is bounded on the top and bottom by a relatively smooth surface of the main body of the vertebrae. If the vertebrae are pushed toward each other, the pressure in the nucleus increases and the annulus tends to bulge outward in all directions. Fortunately there are longitudinal ligaments (anterior, ALL, and posterior, PLL) that aid in resisting bulging of the annulus, but they do not completely surround the annulus. Thus there are sites, shown by the bold arrows in figure 8.19, where the annulus can split if the intradiscal pressure is high enough. The result of this is intrusion of the nucleus into the space where the spinal cord is located. This intrusion can press on the cord or the spinal nerves that emerge from the cord. Irritation of the nerve leads to pain that feels as though it is located along the course of the nerve and not necessarily in the spinal cord. Sciatica is a typical condition of this kind. The biomechanical question is how the disc gets compressed.

The forces on the disc are shown in figure 8.20. While this figure is rather complicated, it shows only a fraction of the complex architecture of the real thing. The compressive force C on the disc is increased by increased muscle force F and by a component of the body weight acting longitudinally down the trunk. As the trunk bends forward, F increases and W_I decreases. However, these two effects do not change equally to cancel out each other at different trunk angles. The fact is that the greatest compressive force C is produced with the trunk horizontal. There is another force that helps to reduce compression of the disc. The force P results from an increase in the pressure in the abdomen, caused by contraction of the diaphragm, the muscles of the pelvic floor, and the concentric muscles surrounding the abdomen. It acts in all directions, but it has a longitudinal component that acts parallel to the spine. This component reduces the value of C. Evidence of the use of the diaphragm comes from our tendency to expel air in a grunt in the later stages of a lift as the diaphragm relaxes. Evidence of the use of the muscles of the pelvic floor comes from the fact that if we do not contract them we defecate—an embarrassment if our only intention is to lift something. The pressure in the abdomen can become quite high, so high that a small amount of abdominal tissue can be forced through the inguinal ligament in the groin as a hernia.

The disc may also be disrupted by a shearing force S due to a transverse component of body weight W_T. Such a force would tend to produce sliding of one vertebral body on another. That this does not normally occur is probably due to the strength of the disc and increased contact pressure between the posterior articular processes (not shown), which are designed to allow one vertebra to rotate on another in a rather complex manner. However, if a spine that is presently under compression and shear is given a twisting component, the shear pattern will be both increased and changed in direction and the chances of injury will be magnified significantly. A comprehensive study of the biomechanics of lifting is presented by McGill (2002).

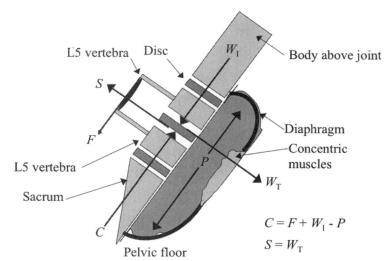

FIGURE 8.20 Compressive and shear forces acting on the intervertebral disc during forward leaning.

$$C = F + W_I - P$$
$$S = W_T$$

Variations of Lifting and Lowering

Prior analysis has dealt with lifting and lowering to a final position with an erect stance and straight arms. In many cases, such a final posture will not suffice if we are required to lift an object onto a table. The obvious modification to the lifting strategy in this case will be to flex the arms at the elbow and shoulders, stand on one's toes, or both. The problem with these variations in technique concerns the ability of the elbow flexors and ankle plantarflexors to generate the required joint moments. The elbow flexors are relatively weak in comparison with the muscles that have done the lifting. So we are faced with an inability to control finally a load that may have been easy to lift initially. Such a situation presents a danger, as does the possibility of not being able to produce sufficient ankle joint moment because the plantarflexors are at a short length.

In relation to the many and varied types of lifting and lowering that can be performed, the sport of weightlifting provides an indication of our ultimate lifting capacity. Two types of lift—the clean and jerk (CJ) and the snatch (S)—are depicted in figure 8.21. The basic biomechanical problem common to the two types of lift is to raise the load from position *a* to position *b* to allow the body to move under the load and secure it in position *c*. In the CJ, the load is secured at the sternoclavicular joint and the athlete stands up to complete the clean at position *c1*. Then the lifter must move the load overhead by first flexing the ankle, knee, and hip joints to *c2* and then extending them to give the load upward momentum at *c3*. This is followed by flexing the lower limb joints and simultaneously flexing the shoulder joints and extending the elbow joints to complete the jerk at *c4*. The lifter finally extends the lower limb joints to stand up (position *d*, if he or she can). The S lift appears less complicated because having achieved position *c*, the athlete then simply stands up to position *d*.

■ **Key Point**

It is the compressive force on a disc that forces the nucleus out of place, and this compression increases as the trunk is bent forward.

■ **Key Point**

The worst scenario for rupture of an intervertebral disc is a combination of compression and shear forces; intra-abdominal pressure can alleviate this effect somewhat.

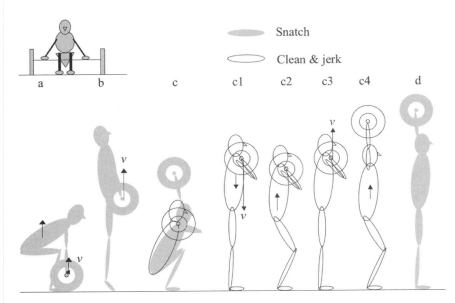

FIGURE 8.21 The single movement of the snatch lift and the double movement of the clean and jerk.

■ **Key Point**

A lift comprising two stages with rest between them may be easier than a one-stage lift from the point of view of both muscular strength and control.

■ **Key Point**

Maximal lifting should be performed in as short a time as possible so that the opposing mechanical impulse due to gravity is minimized.

The exercise of doing muscular work to raise the potential energy of the system is made a little easier if the lifter adopts the grip shown in the small inset. Both activities begin with the hands spread well apart, which means that holding the bar at arm's length requires less vertical displacement and less gain in potential energy than if the arms were vertical at the end of the lift.

The initial lifting action is the same for both types of lift, in which the aim is to raise the barbell to a height that allows time for the athlete to descend below it and support it against gravity. One way of overcoming this problem is to apply the maximal vertical force over the greatest vertical displacement to achieve the greatest potential energy of the bar. However, the vertical displacement is limited by the anatomy of the lifter. It is appropriate at this stage to view the lifts from the force–acceleration viewpoint and the attendant impulse–momentum relationship in addition to the work–energy relationship. Maximizing vertical force to achieve the greatest vertical momentum is the essence of these types of lifts. The greatest acceleration achieves two objectives. The first is to minimize the magnitude of the opposing gravitational impulse by reducing the time taken to reach maximal pulling height. The second is to achieve the greatest upward velocity of the bar at the end point of the pull. When the upward force drops to the level of mg (close to b), upward acceleration of the bar becomes negative, but the load has upward momentum. At this time the attempt must be made to maneuver the body under the bar.

The lifter can drop the body under the bar by ensuring that the force at the feet is zero so that gravity accelerates the body downward. A way to aid in this is to continue to apply an upward force on the bar to retain as much momentum as possible. Such a force results from contraction of shoulder musculature, which is relatively weak. Despite the unfavorable biomechanical motion for creating mutual hands-bar force, any small amount will help the body downward and the bar upward according to Newton's third law. In this manner, the time to allow the body to drop under the bar will be maximized. At this stage the aims of the CJ and the S differ. The CJ requires securing the bar against the torso at the level of the sternoclavicular joint, while the S requires the bar to be overhead with the upper limbs straight and the elbow joints locked (c). The lowest position of the body when these aims are achieved requires the bar to be higher in the S than in the CJ. If the same mass were lifted in each technique, more potential energy would appear in the S lift than in the CJ lift. Since there is a maximal amount of work that can be done in the upward pull, mgh will be the same in the two types of lift. With a greater h required of the S lift, the consequence is that a smaller value of m can be lifted. This fact explains the difference in world records for the two types of lifts, the record for the CJ being greater than that for the S.

Following position c, the problems inherent in the two types of lift differ. In the S lift, the athlete simply stands up to position d. The only joints involved in the motion are those of the lower limbs. The success of this lift therefore depends upon the work-producing capabilities of the muscles crossing those joints. There is a certain amount of strategy necessary in this task due to the moment–angle relationships exhibited at each joint. Here we are essentially dealing with isometric moment–angle relationships, since the movement is very slow under such loads. If all three joints (ankle, knee, and hip) extend at the same angular velocity, there is no guarantee that they will all be capable of generating the maximal moment at each joint position simultaneously. In fact, it is likely that one of the joint's moments will limit the motion, with the other two perfectly capable of performing the task. The suitable strategy is to stop motion of the capable joints while the problem joint gets past its weak position in the moment–angle relationship. It is even possible that the opposite motion of one of the capable joints (say hip flexion) is allowed, for then the extensor joint moment will decrease and the load placed upon the knee extensors, for example, will not exceed their capabilities. Large loads can be lifted with this nonuniform intersegmental motion in which all extensors do not move at the same uniform angular velocity simultaneously. Unfortunately, the chance of failure using this strategy has to do with the need to maintain the vertical line through the CM of the system within the base of support. Additionally, there is the need to maintain the vertical line through the CM of the line coincident with the axis of the shoulder joint. If these two provisos are not met, gravity will create moments that are too great to recover from.

The CJ lift following position c requires the athlete to stand up while retaining the bar at the level of the sternoclavicular joint to finish at position $c1$. The strategy adopted and the biomechanics are the same in principle as described for the S lift. Following position $c1$, the athlete lowers the body to position $c2$ and rapidly extends the lower limb joints in one continuous movement. When position $c3$ is achieved, the body-load system has upward momentum. The athlete then reduces or eliminates the lower limb moments while creating a mutual force between the hands and the load. Newton's third law comes into play again, pushing the load upward and the athlete downward to reach the stationary position $c4$. The final action is to produce extensor moments in the lower limb muscles to achieve position d. The two stages of the CJ exhibit the same biomechanical phenomena but with different bodily motions involved. It should be noted that motions from $c1$ to $c4$ and then to d do not involve knee bends as deep as that seen in the initial lifting stage *(a to c)*. What is required in the CJ is sufficient momentum of the load upward to allow the upper limbs to be straightened. They cannot do this by themselves because the shoulder flexors and elbow extensors are too weak to push such loads upward.

The CJ lift is more effective than the S lift in raising a large load overhead. Current world records (as of April 2007) for athletes with a mass under 56 kg (123 lb) are 168 kg (370 lb, CJ) and 138 kg (303 lb, S). For athletes under 105 kg (231 lb), the records are 242 kg (532 lb, CJ) and 199 kg (438 lb, S). Reasons for these values from a purely mechanical point of view were given earlier. However, the skill component of this activity must not be minimized, for it is a difficult task to maintain one's balance while performing activities that require such high levels of muscular activation.

■ **Key Point**

Different kinematics may be seen at different joints in lifting to compensate for the different mechanical advantages of the joints.

One further interesting point arises when the ratio of load lifted to body mass is compared among weight classes of lifters. The values given in the preceding paragraph show that the smallest lifters (56 kg class) achieve ratios of 3 (CJ) and 2.46 (S) while the largest lifters produce ratios of 2.3 and 1.9, respectively. These latter two values are not necessarily representative of the larger lifters, since most in this class have masses much greater than 105 kg. This means that smaller athletes are relatively stronger than larger athletes but not absolutely stronger. Why this is the case is open to question. One view of this question involves comparison of the dimensional contributions to mass and muscular force. If we consider a cylinder with a uniform density as a gross oversimplification of a body segment, the volume, which will determine the segmental mass, will be proportional to the product of length and cross-sectional area or $\pi r^2 l$. The muscular force produced will be loosely proportional to cross-sectional area alone. The ratio of force produced to segment mass will be related to $\pi r^2 / \pi r^2 l$. If for a given segment the length is doubled, the ratio will be halved. If the length is halved, the ratio will increase. This indicates that for a segment with a given cross-sectional area, a shorter segment will be more beneficial for lifting than a longer one. Changing the cross-sectional area has no effect on the ratio. The fact that smaller lifters have shorter segments implies that this analysis is a possible contributing factor to the relationships between size, shape, and strength. The author makes no claims that this is true or that it is the sole factor explaining the large differences in the load per body mass ratio in weightlifters. This area appears to be worthy of investigation.

Enhancement and Safety of Lifting and Lowering

When the desk-bound executive decides to build a wall around the garden, he realizes quickly that lifting the concrete blocks "correctly" is energy consuming. So he cheats and bends at the waist for each lift to minimize energy expenditure. At the very best he wakes up the next day with a sore back. At the worst he is carried away from the garden on a stretcher. Why might this unfortunate incident occur?

The intervertebral joint is beautifully constructed to afford us spinal flexibility in many complex ways. With reasonable use it can perform without error for many years. With misuse it can wear inappropriately or break down seriously. If we have to lift something we should be aware of the dangers. It should now be apparent that the greater the load and the more forward the lean, the greater the chance of injury. It is better to lift a series of small loads and accept the increased energy cost of repetition than to lift one big load. It is better to lift a small object that can be kept close to the trunk than one that is bulky. Lifting two small boxes is better than lifting one large box of the same total weight. The more movements we combine during a lift, the greater the chance of injury. So people should perform straight lifting before rotating in order to position an object in a different place; they should rotate by re-placing the feet, not with axial rotation of the trunk.

Lifting injuries do not solely affect the spine. If in doubt, employ a stronger machine than yourself, human or otherwise. A typical stronger machine would be an Olympic weightlifter. These athletes lift loads in the order of a few hundred kilograms, and not just to waist height. Weightlifters do sustain injury, but not to the extent we might predict from the loads lifted. The point is that tissues adapt to training by getting stronger. Such strength in combination with correct technique keeps us out of danger.

Is there any hope for the puny casual lifter? We can reduce the required moment in the sagittal plane at any intervertebral joint by keeping the trunk more upright and the load close to the body. This strategy is the current conventional wisdom on lifting heavy loads. Most household lifting injuries occur through incorrect judgment of how much we can lift and also through poor technique.

A two-person lift is particularly hazardous unless each person is confident that his or her muscular capabilities are sufficient for the task and the two plan the lift beforehand and communicate verbally. The primary danger arises if one person loses the ability to sustain the load, leading to the imposition of a much larger load on the partner.

■ **Key Point**

Lifting is hazardous to the spine and other body parts and should be performed with caution and foresight regarding potential hazards; any doubt about your lifting capacity should stop the lifting attempt.

Practical Example 8.4

Figure 8.22 represents a particular position during a quasistatic lift. This simply means that the inertial effects can be ignored and the lift can be considered static.

Part 1

Develop an expression for the horizontal position (X) of the CM that ensures that the lifter will not fall over.

Solution

The required expression must ensure that the horizontal position X of the combined CM m remains within the base of length H such that

$$0 < X < H.$$

Such an expression must include the dimensions and angles shown, and taking clockwise moments segment by segment about 0 and dividing by g gives

$$mX = m_L r_L \cos\theta_L$$
$$+ m_T(l_L \cos\theta_L + (l_T - r_T)\cos\theta_T)$$
$$+ m_B(l_L \cos\theta_L + l_T \cos\theta_T + r_B \cos\theta_B)$$
$$+ m_A(l_L \cos\theta_L + l_T \cos\theta_T + l_B \cos\theta_B),$$

which can be rearranged to yield

$$0 < \{\cos\theta_L [r_L m_L + l_L(m_T + m_B + m_A)]$$
$$+ \cos\theta_T [(l_T - r_T)m_T + l_T(m_B + m_A)]$$
$$+ \cos\theta_B [(r_B m_B + l_B m_A)] / m\} < H.$$

Either of these expressions can be used to determine whether certain postures are safe from the point of view of stability in lifting, as is examined in Part 2.

m_L =	7.0kg
m_T =	16.4kg
m_B =	46.4kg
r_L =	24.0cm
r_T =	15.1cm
r_B =	36.0cm
l_L =	36.0cm
l_T =	40.2cm
l_B =	60.8cm
θ_L =	1.0rad
θ_T =	2.8rad
θ_B =	1.0rad
H =	30.5cm

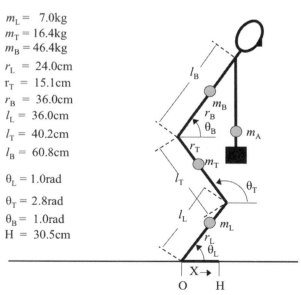

Part 2

What are the maximal and minimal values of m_A (load and arms) that can be sustained given the values shown in figure 8.22?

FIGURE 8.22 Definitions of symbols necessary to perform analysis of load lifting.

Solution

All values required in the developed expressions are given with the exception of the mass in question. Substitution and rearrangement of the expressions will yield the answer as follows:

$$X = \{0.54[0.24 \times 7 + 0.36(16.4 + 46.4 + m_A)]$$
$$-0.94[(0.402 - 0.151)16.4 + 0.402(46.4 + m_A)]$$
$$+0.54(0.36 \times 46.4 + 0.608 \times 46.4)] / 69.8\};$$
$$X = [13.12 - 3.87 + 24.25 + m_A(0.194 - 0.378)] / 69.8;$$
$$X = (-0.184m_A + 33.35) / 69.8;$$
$$X = 0.48 - 0.0025m_A;$$
$$0 < 0.48 - 0.0025m_A < 0.305;$$

X will be 0 when m_A = 0.48 / 0.0025 = 122.19 kg;

X will be H when m_A = (0.48 − 0.305) / 0.0025 = 70.0 kg.

When the maximal value (122.19 kg) is supported, the combined CM will be vertically above the rearmost point of foot support, labeled O. Alternatively, a much smaller load (70.0 kg) can be sustained when the combined CM is located at the foremost point of support, labeled H. These results apply only to the posture shown. There are other postures that can be adopted to allow even greater loads to be supported. In these cases it may not be possible to lift these loads because the load may impose physical constraints owing to its dimensions. This possibility brings into question the relationship between the moment that the muscles can produce about a joint axis and the joint angle. This question is dealt with in Part 3.

Part 3

What is the maximal, sustainable value of m_A when passing through the posture shown given the following relationship between knee joint extensor moment (M_K) and knee joint extensor angular velocity (ω_K)? Assume that any inertial forces due to acceleration are negligible.

$$\omega_K = 0.5 \text{ rad/s};$$
$$M_K = 200(e^{-\omega K}) = 200/ (e^{\omega K}), \text{ where } e = 2.71828$$

This expression for M_K tells us that there is an exponential decrease of M_K as the angular velocity of knee extension increases. This is only an approximation of the relation between joint moment and joint angular velocity that results from the force–velocity relationship of muscle.

Solution

The convention for the kinematics and kinetics used is positive upward and to the right for translational motion and positive for counterclockwise angular motion. As M_K will appear in the solution, we can equate moments about the knee joint and it will not be necessary to know the force at this joint. We must calculate the moment and force at the hip joint, as these will affect the total moment applied to the thigh segment. The upward vertical hip joint force F supporting the body, arms, and load in this quasistatic analysis is

$$F = (m_B + m_A)g.$$

This force acts in the downward direction on the thigh segment and creates a moment about the knee joint of

$$-Fl_T\cos\theta_T = -(m_B + m_A)gl_T\cos\theta_T.$$

The moment M_H at the hip that stops rotation of the trunk and arms is

$$M_H = (r_B m_B\cos\theta_B + l_B m_A\cos\theta_B)g.$$

M_H represents an equal and opposite moment applied to the thigh segment and thus will have a negative sign. The total extensor (clockwise) moment M_K at the knee applied to the thigh opposes the sum of the previous two moments and is

$$M_K = -(m_B + m_A)gl_T\cos\theta_T - M_H;$$
$$M_K = -(m_B + m_A)l_T\cos\theta_T g - (r_B m_B\cos\theta_B + l_B m_A\cos\theta_B)g;$$
$$M_K = m_B g(-l_T\cos\theta_T - r_B\cos\theta_B) + m_A g(-l_T\cos\theta_T - l_B\cos\theta_B);$$

$$m_A = [M_K - m_B g(-l_T\cos\theta_T - r_B\cos\theta_B)] / [g(-l_T\cos\theta_T - l_B\cos\theta_B)]. \qquad \text{(Equation 8.13)}$$

All that remains is to obtain the value of M_K for the given knee joint angular velocity as follows:

$$M_K = 200 / (e^{\omega K}) = 200 / e^{0.5} = 121.31 \text{ N.m}$$

Inserting appropriate values into Equation 8.13 yields:

$$m_A = [121.31 - 46.4 \times 9.81(0.379 - 0.195)] / [9.81(0.379 - 0.329)];$$
$$m_A = 76.56 \text{ kg}$$

This is a significant load (168 lb) that could be lifted by only a small percentage of the population. This analysis has only one joint moment as a limiting factor but could be extended to numerous joint moments to investigate which are likely to be the true limiting moments. Furthermore, this analysis used only angular velocity as a factor determining the joint moment, whereas joint angle will be a significant determinant, particularly for the knee joint. It is evident that there are numerous factors to take into account when we are analyzing the ability to lift a load in a safe manner.

Aim of Carrying

The aim in carrying is to move a load without injury and also in many cases to rotate the body about the longitudinal axis. The vertical aspect of manual handling was covered in the section on lifting and lowering, and the problem of gripping a load was covered in the section on gripping. Here we are concerned with supporting the load during moving, which raises the issues of safety and fatigue.

Mechanics of Carrying

The mechanical pathway from the load through the hands to the feet can be considered as a chain. Therefore it is only as strong as its weakest link. If one is carrying a large load, the back may prove to be the weakest link. If one is carrying a moderate load for a long distance, the grip may prove to be the weakest link. If

The lifter's back is dangerously extended because the load is too large and its center of mass cannot be located any closer to the lifter's CM.

the dimensions of the load are such that the elbow joints must be kept flexed at 90° at the elbow, the elbow flexors might be the weakest link. One other proviso that may affect the mechanics is that the position of the load must not obscure vision.

The gross mechanics of carrying are relatively simple and do not require equations other than those used in other chapters. The problem is to support the load in such a position that its CM creates the least moment about the joint centers in the segmental chain. We usually achieve this by keeping the load as close to the outer limit of the body as possible. Other postures that might be adopted may position the vertical through the load CM closer to the vertical through the body CM. Yet this modification may produce a particularly large moment about a particular joint center even though other moments may be minimized.

Since axial rotation of the spine frequently begins or ends a carrying task, the effect of the load on the spinal configuration must be considered. Axial rotation of the spine results from relative motion between the posterior articular facets of adjacent vertebral bodies in the thoracic region. In the lumbar region such rotation occurs about an axis through the vertebral body. Compression of these facets limits rotation in the thoracic region and throws a shearing load upon the intervertebral discs in the lumbar region. Therefore a compressively loaded spine has decreased rotational amplitude and increased chance of injury in rotation.

Biomechanics of Carrying

■ Key Point
A load should be carried as close to the vertical through the body CM as possible; this reduces muscle force and consequent compression on joint surfaces.

As a load carried either is increased or has its CM moved farther laterally from a vertical line through the body CM, the moment about any joint increases. Therefore the muscle forces that oppose the moment will increase as shown in figure 8.23. A consequence of this is increased compressional force F_C in the joint. This figure assumes that support is on one lower limb, possibly during walking, with only one foot on the ground. The muscle is intended to represent the combined action of all hip abductors that can contribute to the task.

With the previously explained mechanical requirements in mind, a number of carrying tasks are illustrated in figure 8.24. Method a is a common task in which the load is close to the body. The hands supply a force along a vertical line directly coincident with the gravitational attraction on the load. The feet are well placed to walk while avoiding toppling. The upper limbs are straight, which enables elbow flexors and extensors to avoid distraction of the elbow joint in concert with ligaments. The only suspect element of this figure is that there are no handles with which to support the load. Therefore there will be 90° of flexion at the metacarpophalangeal joints with the load supported on the palmar surface of the fingers as shown in the inset for *a*. If this finger configuration is maintained for too long a period of time, finger flexor fatigue is inevitable. Figure *b* is similar to *a* except that the dimensions of the load bring the CM of the load far in front of the body. An attempt to obviate this effect is leaning backward. Unfortunately, such a maneuver requires extension of the lumbar spine as shown in the inset for *b*. The effect is to bring the posterior articular facets together with consequent limitation of axial rotation of the lower spine.

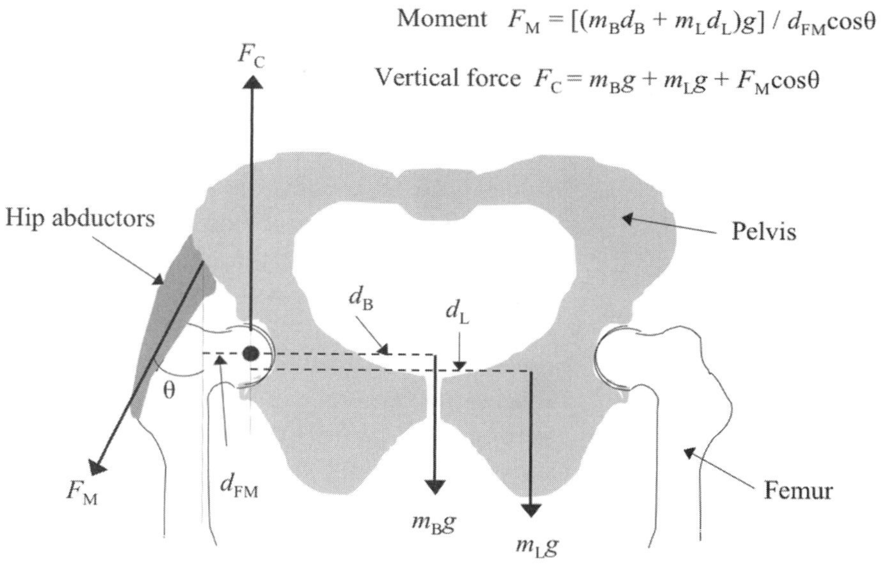

Figure *c* shows an example of asymmetrical carrying in which the load creates a clockwise moment about a number of joint axes, notably those of the spine. This moment is opposed by contraction of the lateral flexors on the left side, most noticeably the erector spinae muscles. Should the spine be bent further to the left, the lateral flexors on the right side of the body would be required to be active. Inset *c* shows lateral flexion to the left, a drop of the left upper articular process in relation to the lower one, and the reverse effect on the right. The intertransverse ligament on the left experiences reduced tension (C) while that on the right is strained significantly (T). Contraction of the left lateral flexor muscles will increase the compressive force on the hip joint over and above that required to support the load as shown in figure 8.23. A mechanical analysis of the intervertebral force occurring in *c* is very complex given the number of structures involved.

Figure *d* shows a child being carried in a sack or basket by means of a halter passing around the carrier's forehead. In addition to producing compression on the frontal bone of the cranium, this requires contraction of the neck flexors (e.g., sternocleidomastoid muscle in the inset for *d*). Again, such contraction will increase the compressive force on the cervical articular facets and intervertebral discs. The advantage of carrying a child in this manner, or in its modern-day equivalent in front of the body, is that hands are free to perform manual tasks such as gathering crops or using a cellular telephone.

$$\text{Moment} \quad F_M = [(m_B d_B + m_L d_L)g] / d_{FM}\cos\theta$$

$$\text{Vertical force} \quad F_C = m_B g + m_L g + F_M \cos\theta$$

FIGURE 8.23 Muscle force and gravitational force lead to a force of compression F_C at the hip joint.

■ Key Point

Handles should be available whenever anything is carried.

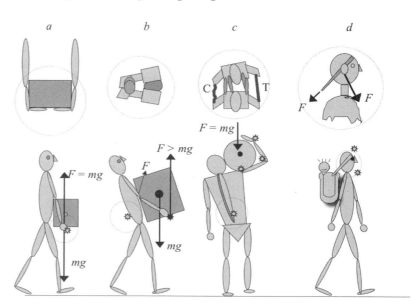

FIGURE 8.24 Insets show regions of potential mechanical problems in various carrying tasks.

Apart from the objective of keeping the load as close to the body as possible, the question of lateral symmetry should be examined. The back can suffer if we lean sideways under a load for a sustained period. This posture is characteristic of carrying a suitcase. Muscles on one side of the back are lengthened and on the other side they are shortened. The capability of the muscles for producing force is therefore different on the two sides of the spine. The same applies to the supporting ligaments, resulting in lateral wedging of the disc with the annulus stressed differently on the two sides. The possibility of injury is obvious. Fortunately, after years of toiling with one-handled suitcases, the public have finally been given wheels for their loads. This improvement not only removes the effect of gravity but also significantly reduces asymmetry in the spine. Why this improvement has taken so long is a mystery, since flight crews have been walking pain free through airports for years. No reasonable case can be made for carrying a suitcase when wheels can be used. Wheels have been around for even longer than flight crews.

A similar development has taken place with regard to carrying objects in bags slung over the shoulders. Again symmetry can be achieved with the use of a strap over each shoulder. While this seems obvious, there are still golf caddies who use only one shoulder, with potentially serious results. They do not even change shoulders during the round.

Variations of Carrying

While injury avoidance in individual carrying is the responsibility of one person, the carrying of an object by two or more people can result in injury to a nonguilty partner. There are many ways in which such injuries can occur. A primary cause is the situation in which one of the people is faced with an unexpectedly large load. This can happen when one of the two does something unexpected. The other person can lose footing, or be unable to provide his or her share of the load, or make a movement that the partner could not have anticipated. The mechanical consequence of such uncoordinated activity is a load that suddenly increases rapidly. Tissues of the body can be ruptured not only by large loads but also by the rapidity with which the load changes. Carrying therefore requires considerable planning and firm expectation regarding the moves to be made. If one of the people wishes to change his or her mind in the course of carrying, the task should be stopped. Exactly the same strategy should be applied to the initial lift of the load.

A particularly difficult task is for two people to carry a large sheet of material such as plywood that is oriented in a vertical plane. Inevitably the individuals are located on opposite sides of the load and cannot see each other. One senses the load toppling about a horizontal axis and applies a force transversely, which the other individual senses as toppling of the load in the other direction. The second individual reacts by pushing in the opposite direction. The first individual then senses the load toppling more in the original direction, and the process continues in the manner of a feedforward system. Eventually both individuals are pushing hard laterally in opposite directions. Should one of them cease to push, the whole load experiences angular acceleration and various dangerous situations can arise. This argues for one individual to be given the responsibility for main-

◼ **Key Point**

Asymmetry and axial rotation of the trunk should be avoided when loads are carried.

◼ **Key Point**

Hazards that suddenly increase a load must be avoided, particularly during lifting in partnership.

taining the vertical orientation of the load while the other is simply concerned with supporting the load vertically.

Enhancement and Safety of Carrying

Enhancement and safety in carrying are increased by training to increase muscular strength. While this should be a requirement in industrial situations, the casual carrier does not train for the enormous varieties of carrying activities. In both situations, planning and thought about the possible dangerous outcomes of lifting will both enhance and ensure safety.

The initial act should be to assess the dimensions and mass of the load. If clear vision cannot be maintained, carrying should be abandoned. If the mass of the load is great, carrying should be abandoned. Both of these assessments can be made before movement occurs in the carrying direction. Providing that these assessments are positive, the next strategy is to observe the route to be taken with reference to obstacles, distance to cover, and duration of the carry. Obstacles must be removed; otherwise a different route must be planned. What might appear to be within the capabilities of the carrier in terms of the magnitude of the load may be too great if the duration requires great muscular endurance. If the CM of the load does not lie at its geometrical center, it should be rotated prior to movement to keep the CM as low as possible. This strategy will ensure minimal moment due to gravity as the CM of the system is moved during walking.

A secure handhold must accompany carrying. It should be secure in terms of friction, minimization of pressure points, strength, and muscular endurance. To judge that one can "just make it" to the end of the route is to court danger. A good question to ask is whether one can perform the carry over twice the length or duration of the route.

Axial rotation of the spine is to be discouraged during carrying for reasons given previously. Inevitably, rotations to face from one direction to another will be required. It is preferable if the carrier does this by moving the feet in small steps rather than swinging the shoulders in a horizontal plane relative to the pelvis. So the strategy is to lift the load and shuffle the feet to face in the required direction before commencing forward motion.

When carrying is performed by two or more people, the plan should be discussed initially. During the carry there should be verbal communication. For example, it must be possible for one person to indicate that something is going wrong with his or her role. This should allow the other(s) to compensate.

Part of the initial planning should concern how to abandon the carry should something unforeseen occur. Verbal communication will enhance the ability to abandon the carry, or at least stop further motion.

■ Key Point

In most cases the use of a trolley or some other form of transport is preferable to using the human body to move objects.

Practical Example 8.5

A person has the opportunity to carry either one or two suitcases of the same total mass (m_S) as shown in figure 8.25. Calculate the muscular loading (F_M) and force (F_J) at the hip joint when the person is in midstance phase while walking. Assume

$\theta = 20^\circ$
$r = 18cm.$
$p = 8cm$
$l = 30cm.$
$h = 45cm.$
$w = 20cm.$
$m = 50kg.$
$m_S = 25kg.$

$\quad a \qquad b$
$\beta_a = 90^\circ \quad \beta_b = 95^\circ$

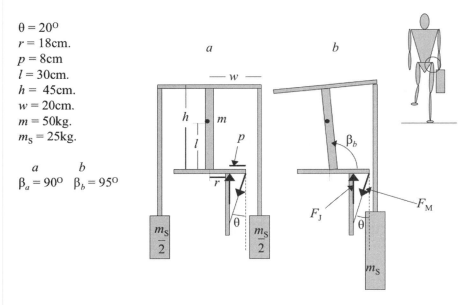

FIGURE 8.25 Stylized representation and postural adjustments required during the carrying of one load versus two equal loads of the same total mass.

that the motion in the frontal plane is quasistatic and that the combined CM of the system is not located vertically above the hip joint.

Solution

The forces contributing to F_J are mg and m_Sg acting vertically downward plus F_M acting at an angle θ to the vertical in cases a and b. F_M creates a clockwise moment about the hip joint to counteract the opposing moments due to mg and m_Sg, which are different in cases a and b. F_M is likely to be different in each case and must be calculated first in order for F_J to be calculated. Equating moments about the hip joint axis gives the following:

Case a:

$F_M p\cos\theta = (r - l\cos\beta_a)mg$
$+ (r - h\cos\beta_a + w\sin\beta_a)m_Sg / 2$
$+ (r - h\cos\beta_a - w\sin\beta_a)m_Sg / 2;$
$F_M p\cos\theta = (r - l\cos\beta_a)mg + (r - h\cos\beta_a)m_Sg;$
$F_M = [(r - l\cos\beta_a)m + (r - h\cos\beta_a)m_S]g / p\cos\theta;$
$F_M = [(0.18 - 0.3\cos90)50 + (0.18 - 0.45\cos90)25]9.81 / (0.08 \times \cos20);$
Case a: $F_M = 1761$ N

Case b:

$F_M p\cos\theta = (r - l\cos\beta_b)mg + (r - h\cos\beta_b - w\sin\beta_b)m_Sg;$
$F_M = [(r - l\cos\beta_b)m + (r - h\cos\beta_b - w\sin\beta_b) m_S]g / p\cos\theta;$
$F_M = [(0.18 - 0.3\cos95)50 + (0.18 - 0.45\cos95 - 0.2\sin95)25]9.81 / (0.08 \times \cos20);$
Case b: $F_M = 1410$ N

The adjustment made to body configuration from a to b reduces the muscle force required. In both cases, leaning the trunk more clockwise would bring the combined CM closer to the vertical through the hip joint and muscle force would be reduced further.

To calculate F_J, forces must be resolved into vertical and horizontal components:

Case a:

Vertically $F_{JV} = (m + m_S)g + F_M\cos\theta$
$F_{JV} = (50 + 25) 9.81 + 1761\cos20$
$F_{JV} = 2390$ N
Horizontally $F_{JH} = F_M\sin\theta$
$F_{JH} = 1761\sin20$
$F_{JH} = 602$ N

Case b:

Vertically $F_{JV} = (m + m_S)g + F_M\cos\theta$

$F_{JV} = (50 + 25)\ 9.81 + 1410\cos20$

$F_{JV} = 2061$ N

Horizontally $F_{JH} = F_M\sin\theta$

$F_{JH} = 1410\sin20$

$F_{JH} = 482$ N

While the vertical force is large as expected, the horizontal component is also in the order of hundreds of newtons of force or tens of kilograms of force (2.2 times that amount in pounds). It should also be recognized from the equations calculating F_{JV} in both cases that the compressional force at the joint is dominated by the muscle force. Since this force is present only because it has to counteract the moments due to the load and body mass, it is of considerable importance that the centers of mass of these loads be kept as close as possible to a vertical line above the joint center during carrying. If the analysis were performed for the left hip joint, the single mass m_S would create an extremely large clockwise moment. This would raise the muscle force and concomitant joint force to extremely large levels.

Summary

The hand can grip in the manner of pincers or of a frictional device. Each type of grip is produced by force in finger flexors that are located in the forearm or hand. Long tendons transmit the force from the forearm muscles to the fingers. The path of these tendons is secured by transverse ligamentous bands near and at the finger joints. High tendon forces with flexed finger joints induce large forces in the ligamentous bands. This is known as a "bowstring effect." Frictional gripping induces shear forces in the skin and underlying tissues of the fingers. These forces can injure tissues, and every effort must be made to obviate these effects by use of protective gloves or the use of machinery that intervenes between the hand and the object to be gripped. In this sense, enhancement and safety of gripping performance are closely allied.

Pushing and pulling on a horizontal surface both occur in two phases. The first phase is acceleration of the object from zero velocity to some constant value. In this case not only the inertia of the object but also the force of static friction need to be overcome. The second phase involves keeping the object moving at a constant speed, in which case dynamic friction is the only opposing force. During pushing uphill and downhill, gravity opposes and aids motion, respectively.

Pushing and pulling forces can be aided by body lean, which in creating a moment about the feet produces a component of force parallel to the pushing surface. In these cases, force produced by the flexor and extensor muscles of the trunk will require modification and may induce back strain.

Pulling is preferable to pushing in terms of fatigue of the upper limb musculature. When pushing, the muscles crossing the shoulder and elbow joints must contract to avoid shoulder joint motion in general and elbow flexion specifically. When pulling, muscles crossing both sides of these joints can contribute to force transmission and also help in avoiding joint distraction. Pulling uphill is preferable to pushing, since the direction of application of force tends to reduce the opposing frictional component.

Pulling and pushing can be enhanced through the use of some form of harness, which reduces the load on upper limb muscles.

The major safety requirement is to ensure that the floor provides sufficient frictional resistance to prevent slipping of the feet but does not have so high a coefficient of static friction that moving the object becomes difficult.

Lifting and lowering are mechanically opposite actions; the former represents generation of energy and the latter dissipation. Biomechanically these are quite different tasks due to the properties of muscle. At any given force and velocity during lifting, the concentric force available will represent a greater fraction of the maximal force than will the eccentric force during lowering. Consequently lowering is more economical than lifting from a muscular energetic point of view. A simple test of this is to lift a given load repeatedly at a given frequency until fatigue stops further lifting, then to repeat the action at the same frequency while lowering. The reader-experimenter can be the judge of the outcome; but if back pain is induced, the experiment should be abandoned.

The sources of back stress are numerous and include the compressive loads induced in the intervertebral discs and facet joints. Compression in the stoop lift arises from the muscular force in the posterior vertebral muscles and increases with the angle of forward lean. In the crouch lift, compression is due to body weight, which has a decreasing effect with forward lean. An additional shear force of the intervertebral disc is induced in the stoop lift. There is interplay between the magnitude of the lifted load and the number of repetitions required in lifting. The reason is that each lift requires the lifting of body mass also. Lifting is hazardous to one's health, mainly because of loading of the complex architecture of the spinal column. Casual household lifting is particularly hazardous because of the likelihood of untrained muscles and lack of skill. Lifting in pairs is much more hazardous. When in doubt, the lift should not be attempted and a machine should be used if available.

Many potential hazards are associated with carrying objects. The most serious would appear to be injury to the back. Therefore people must make every effort to minimize spinal asymmetry by keeping the load as close to the CM of the body as possible. Carrying half the total load in each hand rather than the whole load in one hand keeps the CM of the combined load close to the center line of the body. But such a strategy requires careful lateral flexion of the trunk to place the combined CM of the trunk and the load vertically above one or the other hip joint during walking. However, during walking, when loading is applied to each hip joint successively, lateral motion will be minimized with the two-hand carry. People in some societies have solved this problem by carrying the load on the head.

Asymmetrical carrying can amplify forces at the articular surfaces of a joint through the compression produced by muscular contraction. These contractions are greater if the moment due to gravity is increased by the asymmetry. The spinal joints and intervertebral discs are particularly susceptible to asymmetry, especially if rotation of the spine is performed during carrying.

Carrying should be planned beforehand with respect to the duration, presence of obstacles on the route, and the ability to grip securely. When the load is borne initially, the presence of any significant stress in the linkage between the load and the feet should indicate that carrying should be abandoned. This significant stress can only increase during the duration of the carry.

RECOMMENDED READINGS

There are many objects to manipulate, so one source of information is inadequate. The following sources cover a number of fields.

Bartel, D.L., Davy, D.T., and Keaveny, T.M. (2006). *Orthopaedic biomechanics: Mechanics and design in musculoskeletal systems.* Upper Saddle River, NJ: Pearson/Prentice Hall.

Chaffin, D.B., Andersson, G.B.J., and Martin, B.J. (2006). *Occupational biomechanics.* Hoboken, NJ: Wiley.

Salvendy, G. (2006). *Handbook of human factors and ergonomics.* Hoboken, NJ: Wiley.

Schmitt, K-U., Niederer, P.F., and Walz, F. (2004). *Trauma biomechanics: Introduction to accidental injury.* New York: Springer.

Tichauer, E.R. (1978). *The biomechanical basis of ergonomics: Anatomy applied to the decision of work situations.* New York: Wiley.

Vogel, S. (2001). *Prime mover.* New York: Norton.

White A.A., and Panjabi, M.M. (1978). *Clinical biomechanics of the spine.* Philadelphia: Lippincott, Williams & Wilkins.

Whiting, W.C., and Zernicke, R.F. (1998). *Biomechanics of musculoskeletal injury.* Champaign, IL: Human Kinetics.

Throwing, Striking, and Catching

Throwing and striking involve imparting velocity to an object so that it flies through the air. These activities are involved in casual pursuits, hunting for food, sport, and war. They rank highly as the main method, after lifting and carrying, of moving objects around. The human machine is very good at these activities and far superior in terms of accuracy and adaptability to any device that has as yet been engineered.

When someone throws a ball, inevitably someone wants to catch it. So is catching the opposite of throwing? The answer is yes, but catching is easier biomechanically. Much of the section on catching in this chapter is concerned with catching balls, but in the majority of examples other objects can be substituted for the ball. Such substitution does not include catching a desk that is falling down the stairs.

Aim of Throwing and Striking

The aims of throwing and striking are many and varied. A projectile is generally thrown from the hand, but sometimes from a device intervening between the hand and the projectile; a slingshot is an example. Striking is done with a variety of implements. The common features are that the object should be released with appropriate kinematics of position and velocity. These two factors determine the subsequent path of the airborne projectile. Striking requires the same kinematics when the object loses contact with the implement. However, the mechanical interaction between the implement and the object during contact has a significant effect upon the kinematics following release. Frequently, maximization of acceleration of the hand or implement is of primary importance during contact; but when contact ceases, further attempts at acceleration are useless. These kinematic requirements also apply in striking in the workplace.

Mechanics of Throwing and Striking

We throw for distance, speed, and accuracy, although mechanically any one of these primary aims includes at least one of the others as a secondary requirement. We cannot throw for maximal distance without optimizing the combination of magnitude and direction of the velocity vector. Equation 9.1 relates velocity (v), angle of projection from the horizontal (θ), and height of release (h) of a projectile to the horizontal range (R) from release to landing.

$$R = [v^2 \sin\theta\cos\theta + v\cos\theta \sqrt{(v^2 \sin^2\theta + 2gh)}] / g \qquad \text{(Equation 9.1)}$$

Interpretation of this equation is as follows. Range increases if both v and h are increased. Increasing v from 1 to 1.1 (10%) manifests its effect as an increase in range from 1 to 1.21 (21%) because v is squared. So increasing the magnitude of the velocity is of paramount importance in throwing far. Increasing h is limited to a large extent unless we climb to the top of a tall building to throw. In any case we benefit range only by the square root of an increase in h. The implications of changing θ are a little more complicated. The product $\sin\theta\cos\theta$ is maximized when θ is 45°. If we throw from ground level to land at ground level such that h is zero, the equation reduces to R = $2v^2\sin\theta\cos\theta$ / g or $v^2\sin(2\theta)$ / g. With θ at

45°, 2θ = 90° and sin2θ reaches its maximal value of 1. Therefore 45° is the optimal angle for achieving the greatest range when the throw starts and ends at the same horizontal level. In fact this value truly applies only to motion in a vacuum, since air resistance produces a force that alters the velocity of the projectile. Consequently some fine-tuning of the relationship between θ, v, and h is needed to achieve maximal range. Finally, there is nothing much we can do about gravity apart from throwing and striking at different places on the earth's surface (minor effect) or on the moon (extremely large effect but expensive). The question of air resistance belongs to the study of fluid dynamics and is considered further here only if it affects the biomechanics of throwing and striking.

Equation 9.1 also applies to our attempts to be accurate in throwing and striking. However, maximization of range can be sacrificed if accuracy means only hitting a specific target, and in this respect many combinations of v and θ can be used. But when accuracy means hitting a specific target at a specific speed in a specific direction, there will be very severe restrictions on the combination of v and θ. Control over this combination is at the very root of throwing and striking skills.

In previous chapters we have seen how velocity is imparted to an object (or mass). Force is the main factor, and the two relationships of interest are impulse–momentum ($\int F dt = \Delta mv$) and work–energy ($\int F ds = \Delta mv^2 / 2$). So we can maximize v by maximizing the product of muscular force and the time during which it acts, and also by maximizing muscular force and the displacement over which it acts. In throwing, force is the sole responsibility of the muscles. The same applies to striking, although the force applied to the mass depends on how the mechanical properties of the mass interact with those of the implement. This interaction appears at first sight to be simply a question of pure mechanics, but we shall see that it is important in determining how striking skills are performed biomechanically.

The problem at hand is how to apply force in an effective manner to move a mass or a striking implement that is connected to a distal (or peripheral) body segment. Since there are numerous segments involved in throwing, each one connected to adjoining segments by muscles or generators of force, there is a complex interaction of forces contributing to end-point velocity of the most distal segment. A simple model that will suffice to replicate the mechanics of a two-segment mechanism is shown in figure 9.1.

Masses $m_1 = 4$ kg (8.8 lb) and $m_2 = 2$ kg (4.4 lb) are connected by a spring of spring constant K2 = 5 N/m, and m_1 is connected to the external environment by a spring K1 = 10 N/m. The spring behavior is similar to that of muscle in that the shorter the spring, the less force it develops. In addition, the relative strengths of K1 and K2 replicate the greater force-producing

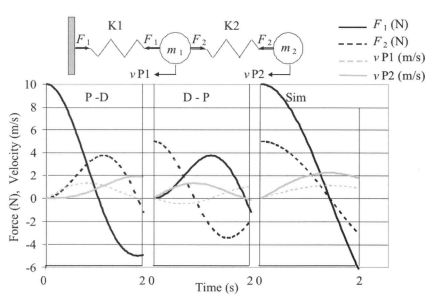

FIGURE 9.1 An analogy of muscle action in which a proximal-to-distal sequence of activation of force generators maximizes the velocity of the most distal mass.

capacity of the more proximal muscles. The left panel labeled P-D represents results when the muscles are activated in a proximal-to-distal sequence. At zero time, K1 is stretched by 1 m, and K2 is unstretched and so produces no force. The sequence of events is that the initially stretched K1 shortens, producing a force F_1 accelerating m_1 that simultaneously begins to stretch K2 against the inertia of m_2. Therefore K2 develops F_2, which accelerates m_2 and opposes acceleration of m_1. This interplay continues for a period of about 1 s, when the force F_1 becomes negative, analogously to an antagonist muscle acting against m_1. As m_1 displaces faster than m_2, the force F_2 rises, and as m_2 begins to catch up with m_1, F_2 falls and eventually becomes negative. The figure demonstrates how the early rise in velocity of m_1 (vP1) gives way to a later and greater rise in velocity of m_2 (vP2). The analogy here is that m_1 is a proximal segment with bigger muscles than the distal segment m_2. We shall see later how human throwers and strikers approximate this action.

When this simulation (D-P) is performed with K2, the distal spring, initially stretched and K1 at rest length, vP2 rises initially but decreases to zero at 2 s. At this time vP1 is positive; but since it is vP2 that is of importance, our aim is not achieved. Therefore the distal-to-proximal sequence of force onset is not beneficial for throwing when the distal muscles begin in an active state and the proximal muscles are inactive. A further simulation was performed with both K1 and K2 active at zero time, in other words having a simultaneous onset (Sim). In this situation, vP2 had a peak value that exceeded the value seen in the initial proximal-to-distal onset, but after 2 s it proved to be no greater than that seen in the initial proximal-to-distal sequence. Therefore m_2 would have to be released from the distal segment at about 1.5 s in this simulation. The reason for the greater value of vP1 is the greater total amount of work obtained from K1 and K2. As a greater range of motion was required to provide this result, the question arises whether there would be sufficient anatomical range of movement to allow simultaneous onset.

The implications for the interplay of forces can be seen in profiles of the momentum and kinetic energy of each mass. Kinetic energy is the value produced as $mv^2/2$ and is the result of work done by K1 and K2. Figure 9.2 shows how K1 does most of the work in the early stages of motion of the proximal-to-distal sequence. However, K2 subsequently becomes increasingly stretched. Therefore the work done by K1 in giving m_1 kinetic energy or KE (KE1) is being transferred to K2. This aids in increasing the KE of m_2 (KE2). It should be stressed that the total energy (TKE) as represented by KE of m_1 plus KE of m_2 is present when the KE of m_1 is maximal. Total energy then stays relatively constant as the KE of m_1 is transferred to m_2. The total work done by the two muscle-like elements is greater than could be done by one muscle alone, and the interplay reveals that almost all of the work done by muscle K1 eventually appears as KE in segment m_2. The distal-to-proximal sequence does not allow K2 to do sufficient work to transfer to m_2. The Sim sequence is very costly in terms of energy; the total energy peaks at a value greater than those of the previous sequences but shows no transfer of energy. However, this type of sequence may create problems of anatomical range as suggested earlier. Naturally the same story applies to profiles of momentum, since calculations of both momentum and energy use only velocity and mass.

The model just described was designed to illustrate how two segments joined by force generators having springlike behavior can interact to maximize the output of the distal segment. It was restricted to two masses and two muscle elements having force–length relationships only. The sequence of events is rather difficult to follow, which is why simulation is used to reveal the behavior of a model. The real model of the human is much more complex and includes the force–velocity relationship of muscle. The problem faced by human throwers and strikers is to fine-tune the interplay among numerous body segments using a large array of muscles, always keeping in mind that the proximal-to-distal sequence of motion (or onset of muscular contraction) appears to be the best guiding principle.

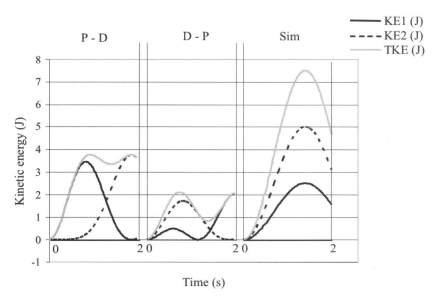

FIGURE 9.2 Proximal-to-distal sequence of activation of force generators maximizes the KE of the most distal mass.

Biomechanics of Throwing and Striking

The section on mechanics has shown how a proximal-to-distal sequence of onset of muscular contraction appears to be a guiding principle in increasing the velocity of the distal end point of a series of segments. It is impossible to know the relative temporal onsets of muscular activation by observing throwing and striking skills. Yet the effects can be seen in the relative motion of the body segments, which clearly demonstrates early motion of proximal segments followed in sequence by that of the distal segments. Close inspection of a golf drive, a baseball swing, and a soccer kick shows the commonality, and in fact universality, of this sequence.

The problem the human faces is how to time the onset of muscular recruitment in a multisegmented system in order to produce a proximal-to-distal sequence of motion. There are many body segments and therefore many degrees of freedom in the system between the toes and the fingers. We shall see how implementation of the proximal-to-distal sequence is partly due to the mechanical properties of muscle.

Intersegmental Throwing Patterns

The real system comprises bones linked by joints that allow articulation under the influence of muscular force. Maximizing release speed of a projectile implies acceleration, which implies force, which implies muscular contraction. We know that greater acceleration occurs with greater force and that bigger muscles produce greater force. But the bigger muscles are centrally located, while we wish to build up the speed of the object in the hand. So the problem is how to coordinate the muscles to throw successfully. In this aim the concept of muscular

■ Key Point

The velocity of the distal point of a segmented series of interconnected bodies is maximized by a proximal-to-distal sequence of onset of force generators.

■ Key Point

One can judge the proximal-to-distal nature of throwing and striking by observing either onset of motion in the required direction or peaking of segmental angular velocity, and by feeling the onset of muscular effort.

force is unhelpful, since obtaining a sum of instantaneous forces from a number of muscles in series has little meaning. The concepts of total muscular work and mechanical impulse are preferable because they are directly correlated, respectively, with changes in KE and momentum of the system. The basic principles are that muscular force multiplied by muscular shortening is work done and muscular force multiplied by time is mechanical impulse. This work changes the KE of the system, and impulse changes the momentum. In either case, it is velocity or angular velocity of body segments that is changed. Our aim therefore is to coordinate muscular contraction to maximize work and impulse, leading to maximal velocity of the distal end of the most distal segment.

There is little we can do about the force–length (or moment–angle) relationship of a given joint. Although the force–length relationship may identify maximal force at a given muscle length and therefore a given joint angle, the maximal moment may not occur at that joint angle due to the joint architecture (see chapter 3). But in our efforts to maximize muscular work done ($\int F ds$), it behooves us to begin segmental motion at a joint angle that gives us the greatest range of motion.

As force decreases with an increase in shortening velocity, the greatest force will be realized at zero shortening velocity. This implies that relative motion between adjacent segments should begin with zero angular velocity when muscle activation begins. The problem encountered in this respect is that as soon as activation begins there will be muscle force tending to accelerate one segment relative to its adjoining other. The movement that follows will occur before muscle activation is maximal. What is required is to have muscle activation maximal throughout the full range of movement. The simple solution is to allow time for muscle activation to rise fully when the muscle is lengthening. One can do this by the strategy of allowing gravity to act on a relaxed limb segment if there is a substantial vertical component to the movement. Another strategy is to use antagonist muscles to draw the segment in a direction opposite to the desired movement. Either or both strategies need be performed only to stretch the muscles attached between the stationary base and the proximal segment. In biomechanical terms this action is known as the stretch–shortening cycle of muscle. Readers will find such intersegmental motion easier to understand if they refer to the section on intersegmental motion within chapter 3. An additional benefit of either strategy is force enhancement of the contractile mechanism of muscle, which is greatest immediately after the cessation of stretch but unfortunately declines with time.

Events subsequent to the initial stretching of the most proximal muscle are depicted in figure 9.3.

This figure depicts a simple overhead throw in a vertical plane. The curved black arrows represent the magnitude and direction of angular velocity of the upper limb segments, with a circle representing zero angular velocity. Broken lines show intersegmental forces, and gray lines (smooth and wavy) represent muscles (active and inactive, respectively). The sequence of intersegmental motion is as follows. From a posture shown in *a*, the upper arm moves clockwise (anatomically named shoulder extension). The force created at the elbow joint produces counterclockwise rotation of the forearm and hand provided that the elbow extensor muscles are relaxed. With the shoulder extensors remaining active, the elbow extensors are activated to arrest elbow extension as seen in *b*. Clockwise rotation of the forearm induces counterclockwise rotation of the hand until wrist flexors then

■ Key Point

Muscular work will maximize the kinetic energy of a projectile if maximal muscle shortening is produced.

stop rotation of the hand *(c)* and begin its clockwise rotation. Finally the shoulder flexors slow down the upper arm, which aids clockwise rotation of the upper arm, and the wrist flexors produce a rapid flexion of the wrist *(d)*. There is a good reason for the considerable number of changes in direction of rotation of the limb segments. What we see is elbow extensors and wrist flexors beginning to contract (produce force) while they are in the process of being stretched.

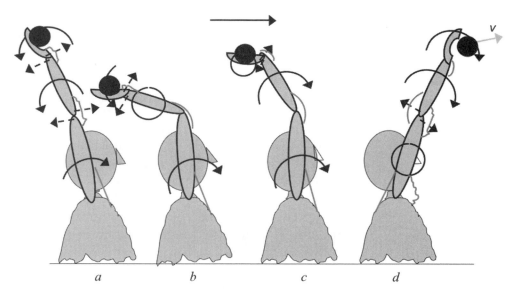

FIGURE 9.3 Overhead planar throwing illustrating active (smooth lines) and inactive (wavy lines) muscles (gray lines), segmental angular velocities (black curves), and approximate intersegmental forces (broken lines).

Data calculated from a planar overhead throw are shown in figure 9.4. The first point to note is that zero moment at each joint occurs when the joint is undergoing motion in a direction opposite to its final motion. These events are shown by the leftmost vertical lines in each joint moment and angular velocity profile. The second point is that zero angular velocity of each joint occurs when the moment is well established and at or near its maximum. The rightmost vertical lines show these points in time. A throw not only benefits from the stretch–shortening cycle of active muscle, but also does so in a proximal-to-distal temporal and spatial sequence.

■ **Key Point**

A "backswing" will allow muscular activation to rise during the eccentric contraction prior to the concentric phase.

Joint forces (broken lines) are implicated in the transfer of energy between segments in a proximal-to-distal direction. The fact that the product of force and velocity is power or rate of change of energy has been discussed previously. The joint forces are equal and in opposite directions, but the velocity of the joint center is in only one direction as it represents a point. The proximal segment therefore experiences $-Fv$ while the distal adjoining segment experiences $+Fv$ of equal magnitude. These powers represent the rate of transfer of energy from segment to segment. This is how work done by proximally located muscles can appear as KE in a ball held in a distal segment. An analogy to this motion is cracking a whip. It is no accident that a whip tapers in size along its length in much the same way the upper limb exhibits smaller segments as we move distally. The same transfer of energy is seen from the proximal to the distal parts of the whip, but in this case the tip of the whip cracks because it reaches the speed of

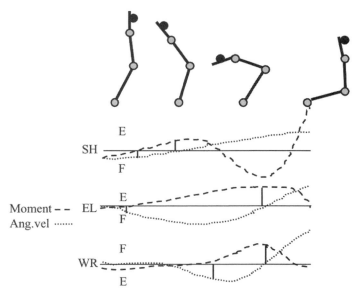

FIGURE 9.4 Data calculated from planar overhead throwing. F and E represent, respectively, flexor and extensor directional moment and angular velocities at the shoulder (SH), elbow (EL), and wrist (WR) joints.

sound. A crack heard during throwing would indicate something rather more serious; bones can break when throwing.

The type of throw shown in figure 9.3 is rather unlifelike, but the general proximal-to-distal sequence is a characteristic of more useful throwing. Baseball pitchers and outfield cricketers use many more body segments in throwing. The lower limbs are used first to rotate the pelvis about a vertical axis. As the upper body has considerable inertia, the shoulders lag behind the pelvis, putting the concentric muscles of the trunk into stretch. Such a situation is exemplified in discus throwing. These concentric muscles then contract and rotate the shoulders, leaving the upper limb behind and putting the shoulder muscles into stretch. So the sequence continues as described earlier, resulting in maximizing of work done. This is how the velocity of a thrown projectile is maximized. If we had unlimited anatomical range we could continue this motion until the currently active muscles reached their maximal intrinsic speed of shortening. At this stage no force would be developed and the velocity of the ball would not increase. This situation never occurs. Indeed we have to use muscles that are antagonist to the throw (figures 9.3*d* and 9.4) in order to avoid injury to ligaments, articular surfaces, and bones. Even so, this strategy is not always successful as indicated by the frequency of elbow injuries occurring in baseball pitchers.

Figure 9.5 represents a planar analysis of a soccer punt and shows hip (H) and knee (K) joint moments and thigh (T) and shank (S) segment angular velocities. The shank is known in this book as the more anatomically named leg. A number of important points illustrate the benefits of the proximal-to-distal sequence. Initially the hip flexor moment accompanies angular velocity of thigh flexion. Simultaneously there is a knee flexor moment associated with shank flexion that prepares knee extensors for their stretch–shortening cycle. By about 35 ms, the shank has maximal negative angular velocity, and knee extensor moment is in transition from flexor to extensor. By the time (about 72 ms) shank angular velocity reaches zero, the knee extensor moment is highly developed. As shank angular velocity increases, the knee extensor moment decreases according to the force–velocity relationship. Near the end of motion, shank extension stops increasing and a knee flexor moment is developed, undoubtedly to avoid injury from overextension of the knee joint. The danger of such an injury is evidenced by the large difference between thigh angular velocity (near zero) and shank angular velocity. Of course this difference represents knee joint angular velocity.

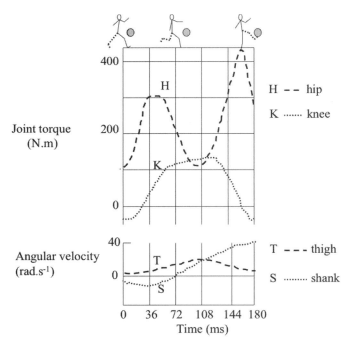

FIGURE 9.5 The proximal-to-distal sequence of segmental motion is seen in kicking.
Adapted from Putnam, C.A., 1991.

Striking

In striking we generally have a somewhat more difficult problem with control than in throwing, where the projectile begins in the hand. In many sports the object to be struck is moving, and in this respect striking requires the same perceptual qualities as catching. In golf the ball is stationary unless you wish to incur a penalty stroke. Even so, the ball is not connected to the club, so spatial perception is again required. We are not concerned here with perception but with mechanics. The major problem is that the length of the striking implement has an effect on the position of the striking point.

Figure 9.6 shows how the path of the head of a golf club is predictable if the axis of rotation is fixed in space. But if we have two or more segments involved, errors begin to creep in. In this case the axis of rotation of the hitting implement is not fixed, and contact with a ball depends on the relative angular

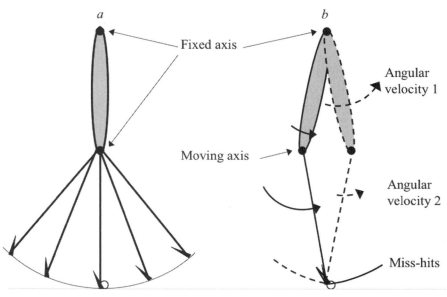

FIGURE 9.6 Relative segmental angular velocities and positions determine success in striking with a golf club.

velocities of the various segments from the starting position. For example, the angular velocity of the solid-bordered arm has been too small, and the club angular velocity has to be greater to allow contact. The converse is the case for the arm bordered by the broken line. Both of these conditions allow contact with the ball, but the contacts are mis-hits, as the club head is too high. However, we know from the throwing section that the use of multiple segments enhances the velocity of the end point of the most distal segment. Fixing the club, arms, and trunk as one segment and rotating about the longitudinal axis of the trunk are therefore good for putting a golf ball, provided that the pelvis is fixed in space. This technique is useless for hitting the ball a great distance; more segments must be used. The problem of accuracy versus velocity of contact is illustrated well in baseball. The batter rarely misses the ball when he "chokes" up on the handle of the bat in bunting. This makes the bat short, and there is little movement in segments of the body. When the batter is attempting to gain high speed of the ball, the norm is to miss completely because of the number of segmental angular velocities that require coordination. The more joints involved, the greater the problem becomes as exemplified by hammering a nail. In this case the positions and angular velocities of the shoulder, elbow, and wrist joints have to be coordinated.

The answer to the question of how many segments and how much angular motion to use is a compromise. Observation of golfers driving off the tee reveals two important characteristics. The proximal-to-distal sequence of segmental motion is always used, but the relative amounts of angular motion at the various joints vary considerably. The reasons for the latter do not seem to have been revealed. A biomechanical reason might be that the relative muscular strengths are different for different golfers. These properties affect the total amount of KE in the club head. If such differences exist, differences in relative timing of the proximal-to-distal sequence would be expected. Nonetheless, many budding golfers wish to emulate the swing of their favorite professional player. The plain fact is that they may not be biomechanically equipped to do so.

■ **Key Point**

Despite popular opinion, there is not a single ideal technique for throwing and striking that all individuals should aim to copy. The essential common factor is the proximal-to-distal sequence; the rest is under the influence of individual musculoskeletal differences.

Variations of Throwing and Striking

A limitation on the speed of the projectile is the need to release it at a certain position. Equation 9.1 shows how increasing the height of release *(h)* above the landing level adds to the horizontal range of the projectile. Unfortunately this equation refers to projectile motion under the influence of gravity only, while we perform in an air-resistant medium. Therefore the angle (θ) of projection requires modification if we are to attain the greatest range. If a compromise between *h* and θ is attempted, it is likely that the greatest amount of muscular work cannot be achieved in the proximal-to-distal sequence. This type of problem faces athletes, particularly in those sports in which the projectile has a small mass and travels a great distance. For example, aiming for maximal horizontal range in javelin throwing requires a value of θ much less than 45°. The reason is simply that the javelin is an aerodynamically shaped object and tends to remain airborne for a longer period than gravity would dictate. Again we may have to release the javelin before we have maximized velocity. This is a problem for the muscular control system. In choosing the correct motion in such a skill as throwing for distance, there are many factors to consider. These include the starting position and starting speed of the limb segments, the times of activation and deactivation of the various muscles, and the amount of activation throughout the throw. The amount of activation requires careful control because turning on all muscles maximally would destroy the proximal-to-distal sequence and reduce the KE transferred to the implement.

Many throwing actions involve rotation of the body about the longitudinal axis. The reason is that the velocity of the projectile can be maximized by the work done by the large rotators of the hip joints and trunk. Many short muscles produce rotation between adjacent vertebrae. These muscles can shorten by only a small amount, but their large number facilitates substantial work. More superficial trunk muscles span more than one intervertebral joint. They can produce considerable moment as they lie at some distance from the axis of intervertebral rotation, and they also can shorten by a substantial amount. Hip rotators can develop moment and angular displacement relative to feet, which are firmly planted on the ground. All of these actions can produce muscular work that can be transferred to the shoulder girdle and appear as KE. Provided that the shoulder musculature can resist the inertial effects of the arm, the KE generated in the shoulder girdle can be transferred to an arm holding a projectile. We now have an arm rotating about a longitudinal axis through the trunk and the projectile undergoing circular motion with an angular velocity ω and a radius of rotation *r*. As we know that $v = \omega \times r$ in circular motion, we conclude that the greater proximal work done by the trunk and hip musculature produces a velocity in the projectile that is greater than that produced by horizontal adduction of the shoulder joint alone.

The strong, proximally sited muscles provide virtually all of the work in discus and hammer throwing—so much so that a repeating Olympic champion of the past achieved his performance with a partly paralyzed left arm that could do little work. Discus and hammer throws, and some shot puts, show a throwing motion that is preceded by repeated whole-body rotation about its long axis. This allows the lower limbs to do work and build up rotational KE with each successive ground contact without severe anatomical restrictions. Similarly the javelin throw is preceded by a run-up, and the baseball pitcher uses leg muscles to give the body KE prior to the throw. The run-up in cricket bowling is particularly important because of the restrictions imposed on the muscles crossing the elbow in the straight-arm delivery. The final throw is then a means of adding further KE to the projectile.

In the 1950s, a group of javelin and discus throwers broke the javelin world records using a discuslike rotational throw. The technique was immediately banned as a dangerous modification of traditional javelin throws! Your author suspects that the problem was the need for a bucket of soapy water to allow the javelin to slide through the hands. This is true.

Enhancement of Throwing and Striking

While the velocity of the distal point of the distal segment in throwing and striking is clearly the result of the sequence of intersegmental coordination, the properties of the muscles and the dimensions and inertial characteristics of the segments are also implicated. Simulation of the planar type of throwing seen in figure 9.3 was performed by Edmondstone (1993) with muscle models driving the system (see figure 3.2 in chapter 3). Each muscle model was converted to a rotational model

Axial moments are common features of many throwing and striking skills, from discus throwing to hitting a baseball.

in which F, V, and L were converted to moment (M), angular velocity (ω), and angle (θ). Also incorporated were M–θ relationships of the parallel and series elasticity of the muscles. Single equivalent muscles represented a single moment generator at a joint. In total there were six rotational muscle models representing flexor and extensor moments at the shoulder, elbow, and wrist joints. A large search field of simulations was produced by activating the muscle models at different times. The best throw was chosen as the one that produced the best horizontal range of throw of a ball in the hand from the continually determined combinations of angle and velocity of the ball path. Once the optimal sequence of activation of the muscles that gave the best throw had been determined, new sets of search fields were obtained in the same manner when certain changes to the human model were made.

In general, improved performance resulted from increasing both the maximal isometric moment and the maximal shortening angular velocity of which the muscles were capable. A given percentage increase in these two parameters led to greater improvement in performance when applied to the shoulder muscles

■ **Key Point**

In most throwing and striking skills, the most proximal segments are the lower limbs.

followed by the elbow and wrist muscles. This improvement was due to the greater mechanical impulse obtained from the shoulder muscles. In some instances the improvement was seen in response to these changes when the timing of onset of muscular activation was changed. Increasing the series elastic stiffness of the elbow extensors also led to improvement of performance. Two points arise from this work. The first is that muscle training should stress enhancement of both maximal muscle force and maximal muscle shortening velocity. The second is that practice of the skill of the activity should accompany muscle strengthening, since the timing of onset of activation will need to change as the muscle properties change. This type of simulation has been described in some detail not only to demonstrate the sensitivity of performance to potential changes in the body, but also as a means of suggesting which characteristics of the performer should be focused upon in training.

Another aspect of striking with an implement is the events occurring during contact of bat on ball. It is not the purpose of this book to deal with such contact, which is largely an engineering problem. However, the actions that the person performs during contact are of biomechanical importance. All objects that strike and that are struck display components of elasticity and viscosity. Therefore they compress under the influence of the equal and opposite forces between implement and object struck. There is a tendency for the object to increase its velocity and for the implement to slow down as shown in figure 9.7.

The figure shows the results of a simulation in which a golf ball is struck by a golf driver. A spring-damper system is located between the club face and the ball to represent the combined effects of elasticity and viscosity of club and ball. At zero time, the club has a velocity of 44 m/s (VCL, 100 mph approximately) as shown by the black broken lines, and the ball is stationary. Three separate simulation conditions were zero force, 500 N, and 1000 N applied to the club head during contact. The major feature in all conditions is that the velocity and KE of the club head (VCL and KECL, respectively) decrease while the velocity and KE of the ball (VBA and KEBA, respectively) increase. The club head is clearly losing energy to the ball. During this process some of the decrease in KE of the implement serves to increase the KE of the object, and some is lost as heat and sound. The effect of increasing the force on the club head is to prolong the duration of club–ball contact. As the force increases, the club head loses less KE and velocity while the ball gains greater velocity and KE. The mutual force of contact multiplied by the distance traveled during contact (not shown) represents the work done on the object to be moved. Maximizing both of these variables will maximize the KE of the object. If muscles are relaxed at the instant of, or prior to, contact, the implement will lose more KE than if the muscles continue to contract. The popular term is to "hit through the ball" rather than "hit at the ball." Mechanically this means that every effort should be made to continue acceleration of the implement during contact.

Inherent in this argument is that the velocity of the implement at the first instant of contact is only loosely correlated with velocity of the hit object as it leaves the implement. For golfers in particular, it should be stressed that swing speed tells only part of the story. The values for velocity in this simulation are realistic and would result in a golf drive of about 300 yd (275 m) without air resistance. This will be significantly reduced in normal playing conditions. There is little doubt, however, that the spring and damping constants representing the combined effect of ball and club properties are inaccurate. The applied force is also guesswork and is chosen simply to illustrate the effects of continued application of force during contact. An extension of this simulation would be to match the club and ball properties to the muscular characteristics of the individual golfer so that optimal energy exchange between club and ball could be achieved. Despite the claims made by golf club manufacturers, the correct matching of the golf club and individual biomechanical characteristics is still in its infancy.

The major point revealed in the simulation is that one should continue to apply force to the striking implement during contact with the projectile if maximal velocity of projection is to be attained. Similar statements apply to kicking a soccer ball and hitting any object that has a reasonably high coefficient of restitution (see chapter 2, "Essential Mechanics and Mathematics," for coefficient of restitution).

In addition to striking implements, throwing implements can be used as extensions of the distal segment to enhance throwing velocity. People in some societies use a long piece of wood held in the hand with a depression in its widened distal part. The backward end of a spear is located in the depression, and a modified throwing action enables projection of the spear with great velocity. In the game of pelota (Spain) or jai alai (Mexico), a large curved woven basket is strapped to the throwing hand. During throwing, the ball rolls out of the basket distally, thus increasing its radius of rotation until it emerges from the basket. In this game and in spear throwing, the advantage is gained via an increase in the radius of rotation, which, for a given angular velocity, increases the linear tangential velocity ($v = \omega \times r$). A similar extension to the body that uses a different principle is the slingshot. In this case, the thrower builds up angular velocity over a period of time by repetitive circular motion rather than relying on a single throwing action.

Whether we use throwing or striking implements, or simply throw with the hand, we have been taught that a backswing is a good idea. But what is a backswing? In general it takes the form of moving the implement and parts of the body in a direction opposite to the intended direction of striking or throwing. This is usually performed immediately before the ball arrives and is the beginning of a continuous motion. In principle this is no different from the actions shown in figures 9.3 through 9.5. It is questionable whether a backswing has to be continuous from a biomechanical perspective. It is certainly true that this incorporates the benefits of prior muscle stretch. The result is that the force in the muscle has time to develop fully by the time the backswing velocity of a given segment reaches zero. But our research on human muscle has shown that only a small amount of stretch is necessary to incorporate the benefits of the stretch–shortening cycle

■ **Key Point**

Accelerating or "hitting through the ball" improves the transfer of energy between a striking implement and a struck object.

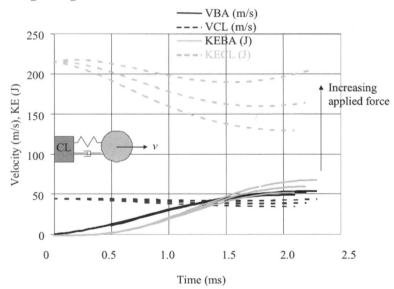

FIGURE 9.7 A model of club–ball interaction in striking in which continued application of force during contact produces the greatest ball velocity.

(Chapman et al., 1985). Large amounts of stretch do not offer significant improvement in distal segment end-point velocity. So biomechanically it is just as good to stand with the implement near to the final backswing position and to produce prior muscle stretch in the proximal-to-distal sequence by accelerating the more proximal segments first. The inertia of the distal segment will induce a "lagging behind," and arresting this lagging behind by distal muscles working eccentrically will provide the stretch–shortening cycle. This is not meant to argue against a continuous backswing–forward swing motion. Rather the point is simply that there is no biomechanical justification for large backward velocities. Whether a large looping backswing aids control is not of biomechanical concern.

Throwing and Striking Safety

The amount of deformation in a struck object depends upon its physical characteristics and the KE of the implement with which it is hit. The effect of such impacts upon the musculoskeletal system is rather difficult to appreciate from the numbers involved. Comparisons based on efforts we can make to deform a ball are of more use. When a soccer ball is kicked, photographs show that the ball almost completely envelops the foot. When the ball is headed, the player appears to be wearing a bulbous cap coming down to the ears. Attempts to produce deformation of a ball of this amount by pressing with the hands semistatically always lead to failure. In other words, the mutual force between player and ball is very high. Thankfully the duration of the force is small; otherwise the human tissues involved would break down. However, it is no surprise that most soccer players suffer from wearing of the articular cartilages in old age because of the number of times they have kicked a ball. Of course many other knee injuries have resulted from a specific trauma when someone is tackled.

Another biomechanical consequence of kicking is the development of the extensor muscles of the knee joint. These muscles must be strong to withstand repeated force that compresses a ball by the amount described. Long-time soccer players are recognizable by the protruding shape of the muscle vastus medialis. This muscle is located on the inner to front side of the thigh near the knee and is responsible for the greater part of the knee extensor moment in the last few degrees of knee extension.

Spinning a ball is good for the team and bad for the spinner. The good aspects are better discussed in books on baseball and cricket. The bad aspect of applying spin is that one's technique requires a great deal of rotation of the forearm about its long axis. This rotation does not involve a ball-and-socket type of rotation as we see in the hip and shoulder joints. Nor does it come from the wrist joint. It comes from the forearm, which comprises two bones, the radius and the ulna, that rotate in a coiling type of motion to provide axial rotation. When this is performed under low load with consequent small forces, there is little stress put upon the articular surfaces between the ulna and the upper arm bone (the humerus). But in order to spin the ball, we require large forces to give the ball large angular acceleration and subsequent angular velocity. The danger arises from not arresting this motion before the articular surfaces at the elbow or wrist (or both) are damaged. The tendency in trying to obtain as much angular velocity as possible is to delay use of the arresting forces until it is too late. Over a period of time the effects on the articular surfaces build up, and arthritis is a distinct outcome. These effects can be so drastic as to chip pieces of articular cartilage and even bone, which then tend to float about in the joint space and have to be removed surgically. Any person planning a career as a successful baseball pitcher must accept the inevitability of some form of elbow joint degeneration. Baseball is not the only culprit, as such injuries can occur in other throwing sports. There are also other types of elbow injury that come from throwing and are very sport specific but do not involve the articular surfaces. The advice "If it gets sore, don't do it" is good but unlikely to be followed by a person who wishes to achieve expert performance in any sport.

Impact between an implement and an object generates not only considerable equal and opposite forces, but also vibration. Mishitting a golf ball or a baseball can leave a stinging sensation that is due to excitation of pain receptors by high-frequency vibration. The solution to this problem is at variance with the need to maintain acceleration of the implement. If avoidance of vibration is the major aim, the solution is either to loosen the grip just prior to impact or to hit the "sweet spot" of the implement. Where on the implement the "sweet spot" resides is not of biomechanical concern to us.

■ **Key Point**

Eccentric contraction at the end of a throw is the major means of avoiding injury to ligaments and bone.

Practical Example 9.1

A discus thrower begins the final stage of the throw with the body rotating at an angular velocity of 10 rad/s, which is maintained until completion. The throw is completed by a horizontal adductor moment at the shoulder joint as seen in figure 9.8.

Calculate the magnitude of the linear velocity of the discus when the moment (M) reaches zero given that the relationship between moment and shoulder angle (θ) is

$$M = M_O(1 - K\theta).$$

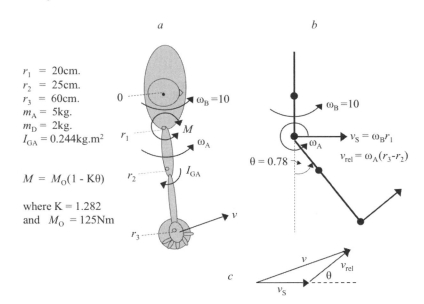

FIGURE 9.8 Kinematic and inertial values for the final stage of a discus thrown in the horizontal plane.

Solution

Since the motion is determined by moment, which is related to angle, we need to use the work–energy relationship for the solution.

$$\int_0^\theta M_O(1 - K\theta)d\theta = 0.5(I_{GA} + m_A r_2^2 + m_D r_3^2)(\omega_F^2 - \omega_I^2)0$$

$$[M_O\theta - M_O K\theta^2 \ / \ 2]_0^\theta = 0.5(0.244 + 5 \times 0.25^2 + 2 \times 0.6^2)(\omega_F^2 - 10^2)$$

M is zero when $K\theta = 1$, which is when $\theta = 1 \ / \ 1.282$ or 0.78 rad.

$$[(125 \times 0.78) - (125 \times 1.282 \times 0.78^2 \ / \ 2)] = 6.3825 \ \omega_F^2 - 638.25$$

$$\omega_F = 10.375 \text{ rad/s}$$

The velocity of the discus comprises two parts. One is the tangential velocity of the shoulder axis (v_S), and the other is the tangential velocity of the discus relative to the shoulder axis (v_{rel}). These velocity vectors are shown in figure 9.8. The resultant velocity is obtained by use of the cosine rule for the addition of two vectors, in this case:

$$v^2 = v_S^2 + v_{rel}^2 + 2 \ v_S \ v_{rel} \cos\theta$$

$$v_S = \omega_B r_1 = 10.0 \times 0.2 = 2.0$$

$$v_{rel} = \omega_A(r_3 - r_2) = 10.375(0.6 - 0.25) = 3.631$$

$$\theta = 0.78$$

$$v = (4 + 13.184 + 14.524 \times 0.711)^{1/2} = 5.245 \text{ m/s}$$

While the major purpose of this and all examples is to show the process of biomechanical calculations, the value for release velocity obtained here is very slow for discus throwing. Readers might gain some insight into the capabilities of top discus throwers by changing a number of the initial conditions and the moment-producing capabilities of the shoulder muscles.

Aim of Catching

A general definition of catching might be "controlled reduction of the speed of an airborne projectile." This definition includes catching a ball manually, but also bringing to rest a soccer ball by use of either the chest or foot. Catching usually finds its application in the sporting environment, but it is also an integral part of juggling and it occurs in the home for reasons of safety. The "control" aspect of catching is not the central issue of the biomechanics, but the biomechanical methods used to catch do help with control, as will be seen.

Mechanics of Catching

The problem in catching an object is to reduce its KE to zero while maintaining control over it. The following familiar equation relating change in KE to work done is

$$\int Fds + mv_i^2 \ / \ 2 = mv_f^2 \ / \ 2. \tag{Equation 9.2}$$

What we require is the final KE, $mv_f^2/2$, to equal zero. If the initial KE, $mv_i^2/2$, is considered positive, as it must be since v_i^2 will always be positive, then the work done must be considered negative and equal in magnitude to the initial KE. This indicates that either force F or displacement s must be negative. In this case force is opposite in direction to the initial velocity and displacement. This phenomenon is sometimes erroneously referred to as "negative work." The term is erroneous in the sense that work is the scalar product of the vectors F and s, and therefore, not having direction, it cannot be negative. In catching we are doing work to decrease the energy state of the system and to convert this energy into heat (a dissipative effect) or into potential energy in the form of strain energy in the object caught. From a purely mechanical perspective, only the two variables F and s can be varied to reduce the KE of the caught object to zero. From a biomechanical perspective it matters considerably how force and displacement are varied.

Biomechanics of Catching

The biomechanical problem is to satisfy the definition given previously using our particular human architecture. Our construction gives us a hand comprising multiple bones, tendons, ligaments, muscles, and other soft tissues such as fat and skin. It is tempting to state that we catch with our hand(s), but this is something we do at our peril. It is safer to employ some of the connected segments of the body for two reasons. The first is that the more segments we use, the greater the number of muscles that can contribute to energy reduction. This strategy reduces the stress on any individual muscle. The second reason is that more segments allow a greater range of movement over which force can be applied. In this way the peak force can be reduced in favor of increased displacement. In this case the peak pressure on the palm of the hand will be reduced.

As force and displacement are in opposite directions, reduction of a ball's KE in catching requires eccentric muscular contraction. Catching can be performed with muscle shortening, but this is to be discouraged for reasons that will become apparent. Simply, we are applying a pushing force against a projectile that is moving in the opposite direction to our push. The alternative is to keep our hand rigid and unmoving and let the KE of the ball be dissipated by force and displacement in the ligaments and articular cartilage, and sometimes in bone, which may fracture.

Accepting that we require limb movement to catch, why is catching easier than throwing? Reference to the relationship between force and velocity of muscle tells the story (figure 3.2c in chapter 3). We see that much greater forces can be produced when the muscle is lengthening than when shortening. So for any given velocity, the product of force and velocity is greater when the velocity is negative (muscle lengthening). This is exactly what we want in catching. As power is the product of force and velocity, a negative value of power means a rate of decrease of energy. Since the force is high, we need a smaller amount of muscle lengthening to dissipate the energy of the ball.

If catching is done with a glove as in baseball, only wrist motion is required to catch successfully as seen in figure 9.9. The aim in this figure is to represent the same area under the two force–displacement relationships and therefore the same work done in both types of catch. If the bare hands are used, more body

■ **Key Point**

Catching is an exercise in dissipating kinetic energy in which force at the hand acts in a direction opposite to the motion of the object to be caught.

■ **Key Point**

The more joints involved in catching, the greater the displacement of the force opposing the object; peak force can be reduced in this way.

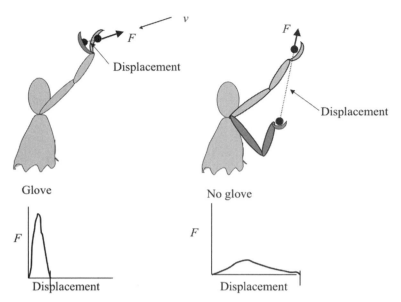

FIGURE 9.9 Catching with a baseball glove allows high force to be applied over a short displacement.

segments are needed to produce greater displacement because the high force results in a stinging sensation (or worse) in the hands. As only one player is allowed to catch with gloves in cricket, the nongloved players use a larger displacement strategy if there is time. However, broken fingers are not uncommon in this sport. This results from simply sticking out one's hand "in hope" due to the limited time available for detecting the flight of the ball.

The contrast between throwing and catching is that the throwing force has to be applied over a large displacement, since force production decreases as the velocity of the ball increases as described previously. In catching, minor movements are successful. I can catch anything that is thrown to me if sufficient time is available to judge the flight. For the same reason, I can land from any height to which I can jump up. The reasons lie in the biomechanical properties of muscle, particularly the increased force available as eccentric velocity increases.

From physics we know that it is possible to store energy and use it in some other form. This can even be done by the human body, although there is a very small time scale over which we can make use of some of that stored energy. If we reduce the KE of a ball to zero by bodily movement, eventually it dissipates to appear as energy in the form of heat. In some cases, such as the chest pass in basketball, there is evidence of use of the stretch–shortening cycle of muscle discussed previously. This action benefits from some stored elastic energy that is returned to the system. If stopping the ball is our only concern, we dissipate that stored energy.

The construction of the hand helps in our efforts to retain the ball during the catch. The momentum of the ball will tend to force the wrist into extension. Efforts to keep the fingers straight and in the same plane as the palm require considerable contraction of the finger extensor muscles. Accompanying this activity will be stretch of the finger flexors with some passive force in the flexor tendons. Therefore relaxation of the flexor and extensor muscles of the fingers will produce closure of the grip in response to the force tending to extend the wrist. You can easily demonstrate this mechanism by pushing on the palm of your right hand with the fingers of your left hand. You will observe automatic closure of the fingers of the right hand provided that their muscles are relaxed. In this sense the construction of the hand produces automatic catching. However, voluntary activity in finger flexors is required to keep the ball from rebounding from the palm of the hand.

Variations of Catching

Catching can be performed using many parts of the body, although some parts are better at dissipating the energy than others. Catching with a foot presents the

■ **Key Point**

Catching is an eccentric muscular activity in which the force-producing potential is greater than in the concentric mode (throwing).

■ **Key Point**

As the wrist is extended by the kinetic energy of the ball, the fingers automatically flex.

problem of being unable to grasp the ball. This skill is seen in soccer as a means of bringing the ball under control. In this case the mechanics underlying the skill are the same as for catching with the hands. The only requirement is to reduce the velocity of the ball to zero. Players do this by having the receiving foot high in the air and bringing it down at the beginning of ball contact at slightly less than the speed that gravity will produce. This action is depicted in figure 9.10*a*. Simply letting the ball hit a stationary foot would not suffice because the relatively high coefficient of restitution of the ball would lead to a rebound (see chapter 2, "Essential Mathematics and Mechanics," for coefficient of restitution).

The skill involved in this task is judging how much force can be applied to the ball. This in turn determines the speed changes in the downward motion of the foot. Initially the force applied by the foot against the ball will lead to compression of the ball. Strain energy will therefore be stored in the ball. At this stage a fixed foot would allow the ball to rebound, as the stored energy would be reconverted into KE. Consequently it is necessary to withdraw the foot in the direction of ball velocity. When ball velocity is near zero, the strain energy in the ball can be dissipated by rapid withdrawal of the foot, which leaves the ball with "nothing to push against," to use a common phrase. An alternative but similar technique is to withdraw the foot slowly, allowing high energy to be stored in the ball as the opposing force will be high (figure 9.10*b*). The foot is then removed faster than the downward velocity of the ball, so the stored energy in the ball simply dissipates. The whole process hinges upon never allowing the mechanical impulse applied by the foot to become large enough to impart an upward momentum to the ball.

Should the foot and lower limb be kept relatively rigid, the initial KE of the ball will be reduced and there will be significant strain energy in the ball at zero velocity (figure 9.10*c*). Maintaining the rigid lower limb will allow this strain energy to be returned and further work will be done, leading to a rebound. The strategy used in figure 9.10*b* allows the player to maintain control of the ball. In figure 9.10*c* the ball could be passed on in another direction. In fact the strategy shown in figure 9.10*c* is a means by which ball velocity can be enhanced in kicking. A ball that is kicked when it is initially moving toward a player has KE that can be stored as strain energy when its motion is arrested. Some of the stored energy is returned to the system and is added to the energy produced by the player's kicking muscles. This produces the force profile shown in figure 9.10*c*, which is similar to that seen in the countermovement jump (figure 7.8). Observation reveals that balls kicked

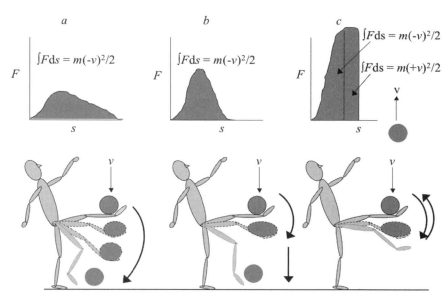

FIGURE 9.10 Catching with the foot requires a force profile that does not allow ball rebound.

© AP Photo/Andres Leighton

■ Key Point

All catching feats require any stored energy in the elastic structures of the object to be dissipated before the object can rebound.

■ Key Point

Throwing after a catch benefits from the stretch–shortening cycle of active muscle.

with maximal effort always travel faster when their initial velocity is in the direction opposite to their final direction.

A similar type of motion can be used to catch the ball on the chest, although this is pushing the definition of a catch to its limit. In this case the hip extensors help to withdraw the chest from the ball, and relaxation of the knee extensors allows the upper body to drop away from the ball. The ball then simply falls to the ground.

In water polo a relatively large ball is caught in one hand and frequently thrown in one continuous motion. Sometimes a number of fake throws intervene between the catch and the final throw. For the catch and throw to occur in a continuous motion, it is necessary to reduce the velocity of the ball to zero using muscular work and to propel it in the opposite direction using the benefits of the stretch–shortening cycle of muscle discussed previously. Fake throwing, known as a pumping action, requires the same muscular contribution as the catch and throw, but the duration of the shortening cycle is reduced and the forearm is axially rotated to bring the hand to the distal face of the ball. The ball is then pulled toward the player, and axial rotation in the opposite direction takes place to bring the hand to the proximal side of the ball. This action can be repeated numerous times, each representing a small catch–throw action. Basketball demonstrates the same type of maneuver, but using two hands means that axial rotation of the upper limbs is not required.

Finally, juggling is governed biomechanically by the catch–throw cycle but is more a problem of motor control than of biomechanics.

Enhancement of Catching

The chance of successful catching is increased by the time during which adjustments can be made. However, it appears that getting the initial conditions correct in terms of body kinematics and kinetics is the key. For example, once a ball rebounds from one's grasp, it is difficult to regain control. Fortunately a longer displacement over which force is applied to the ball will reduce the magnitude of the force. Concomitant with a longer displacement is a greater time of application of force. As we have discovered that longer displacement is the best biomechanical strategy for catching, it appears that further conscious increase in displacement (and therefore time) of the body will allow beneficial adjustment time.

Another way in which the chance of successful catching can be enhanced is to use a glove. A glove increases the surface area of the hand, resulting in three benefits. The first is that errors in the initial position of the glove in relation to ball position on contact can be minimized. The second is that force of contact will be partly dissipated by compression of the glove. In baseball the ball is caught in the webbing between

Cricketers who have to catch a very hard ball without the use of gloves will extend the contact time by reaching high and finishing in a crouched position.

the thumb and index finger and should never apply a direct force to the hand. Of course, the latter does occur because of errors of judgment. Finally, friction between the ball and glove helps to produce shearing forces that obviate lateral motion of the ball out of the glove.

Catching Safety

This section concerns the safety of the catcher's catching apparatus, usually the hands. We are not concerned with other parts of the body. For example, if a ball is directed toward your head, it is better to move your head than to attempt to capture the ball's flight path visually through open fingers.

Each tissue type that is deformed in catching has its own ultimate strength. This is the maximal stress below which the tissue will not be broken. Also toughness is individual to the type and size of each tissue. Toughness is the maximal energy that can be sustained before tissue fracture. In addition, the rate of change of stress affects the maximal force that can be sustained without fracture.

Fingers are the most likely parts of the body to suffer from unsuccessful catching. Injuries vary from bruising to lacerations to broken bones depending upon the direction of force application and its magnitude. People can avoid bruising and lacerations by using gloves. In sports in which gloves are disallowed, the alternative is to reduce force by increasing the displacement of the hands as discussed previously. For example, soft tissue can be split if force is applied to it when it overlies a bony prominence. This is attributable to the high pressure induced by the small area of contact. Phalanges and metacarpal bones are at risk if the force is so high as to exceed the ultimate stress. Such injuries can occur with force applied perpendicular to the long axis of the bones. A more dangerous situation occurs when perception is faulty. Closing the fingers too soon can lead to the ball's hitting the fingertips and producing a longitudinal compressive force. This type of force breaks bones and also is a major source of dislocation of the interphalangeal joints.

The message regarding safety in catching is to use gloves that do not require the force vector to pass directly through the hand. This allows all energy dissipation to be done by muscle flexion of the wrist and fingers rather than by compression of soft tissues overlying bone. In the absence of gloves, the catching technique should involve the greatest possible displacement of the point of contact with the object being caught. In addition, do not produce finger flexion too soon; let the force of contact close the fingers around the ball if it is small.

■ Key Point

One can enhance safety in catching by using gloves and increasing the amplitude of eccentric motion.

Practical Example 9.2

A person standing erect with arms by the side sees an object of mass 20 kg (44 lb) begin to fall after its supporting force is removed suddenly as shown in figure 9.11. It takes 0.15 s for the arms to be raised in a counterclockwise direction and subsequently catch the load with an initial clockwise angular velocity so that the downward velocity of the hands is equal to that of the load. Given the values shown, what is the constant counterclockwise shoulder moment that brings the system to rest when the arms are vertically below the shoulders?

Mass arms $= m_A = $ 8kg.
Mass load $= m_L = $ 20kg.
MI arms $= I_A = $ 0.44kg.m^2.
MI load $= I_L = $ 0.01kg.m^2
$l = $ 50.0cm.
$r = $ 33.5cm.
$M = $?
$s_1 = $?

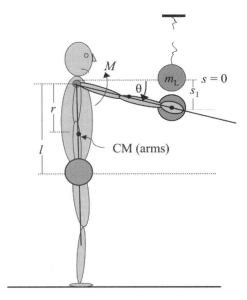

FIGURE 9.11 The mechanics of catching an object that falls unexpectedly.

Solution

The problem can be approached using the work–energy relationship. The first problem is to identify the downward velocity of the load in the elapsed time of 0.15 s. This will give the initial conditions for the beginning of the second phase of bringing the system angular velocity to zero.

For the first phase we know time, initial velocity, and acceleration, so

$$v_f = v_i + at \text{ or } v_f = gt \text{ or } v_f = 9.81 \times 0.15 = 1.47 \text{ m/s.}$$

The displacement of the load s_1 is required to identify the initial potential energy of the load and is obtained as follows:

$$v_f^2 = v_i^2 + 2gs_1, \text{ so } s_1 = (1.47^2 - 0) / (2 \times 9.81) = 0.11 \text{ m}$$

The value 1.47 m/s becomes the initial velocity v_i for the second phase. In order to allow us to deal with angular motion only, this v_i can be converted to angular velocity as follows. Using the general identity for the relationship between tangential and angular velocity of $v = \omega \times r$,

$$v_i \cos\theta = l\omega,$$

where $\theta = \sin^{-1}(s_1 / l) = \sin^{-1}(0.11 / 0.5) = 0.222$ rad.

$$\omega = 1.47\cos(0.222) / 0.5 = 0.287 \text{ rad/s}$$

In this phase the initial KE in the system plus the energy produced by gravity must be opposed by the energy produced by the shoulder moment.

About an axis through the shoulder joint,

Initial KE $= (I_A + m_A r^2)\, \omega^2 / 2 + (I_L + m_L l^2)\omega^2 / 2$

KE $= [(0.44 + 8 \times 0.335^2)2.87^2] / 2 + [(0.01 + 20 \times 0.5^2)2.87^2] / 2 = 12.7$ J

KE due to gravity $= (m_A r + m_L l)g \times \displaystyle\int_{\theta = \sin^{-1}(s_1/l)}^{\theta = \pi/2} \cos\theta d\theta = (m_A r + m_L l)g[\sin\theta \underset{\theta = \sin^{-1}(s_1/l)}{\overset{\pi/2}{]}}$

$= (m_A r + m_L l)g(1 - s_1 / l) = m_A g(r - r s_1 / l) + m_L g(l - s_1)$

$= [8\,(0.335 - 0.335 \times 0.11 / 0.5) + 20\,(0.5 - 0.11)] \times 9.81 = 97.02$ J

This expression will be recognized as the change in potential energy due to gravity as it involves mass, gravitational acceleration, and vertical displacement of each mass. The sum of these two expressions of energy must be equal to the energy produced by the shoulder moment M in order for the arm to come to rest at the vertical. The energy produced by the moment is

$$\int M d\theta = M\theta$$

since M is constant. The angle through which motion of the second phase occurs is

$$\pi / 2 - \sin^{-1}(s_1 / l) = (3.142 / 2) - 0.222 = 1.349 \text{ rad.}$$

So the energy produced by the shoulder moment is 1.349 M.

Finally, equating initial KE, KE due to gravity, and energy due to moment, we obtain

$$1.349\, M = 12.7 + 97.02 = 109.72;$$

$$M = 81.33 \text{ N.m.}$$

This value equates to producing a force of about 160 N or 16.51 kgf or 36.3 lbf at the hands perpendicular to the upper limb.

This analysis is only a rough approximation of the real situation. For example, the arms will be rotated counterclockwise initially as the load is falling, so the shoulder moment will need to be varied to rotate the arms clockwise to match the downward velocity of the load. In the second phase, the moment is unlikely to be constant due to the influence of both muscle length and velocity on force production. Yet the analysis does indicate that reduced perception due to age, poor vision, or being caught unaware will allow the speed of the load to increase. This will result in greater energy to dissipate and a smaller range of motion in which to do it before the load hits the knees. It also illustrates that poor muscular strength would be insufficient to arrest the load due to low moment-producing capabilities.

Summary

The velocity vector of a projectile at release determines projectile path. If distance thrown is the primary aim, the magnitude of velocity should be maximized since its value is squared in the equation determining the range of the projectile. As a purely mechanical process, the aim is to perform as much work as is possible and have it appear as KE of the projectile. This implicates the more proximal, larger muscles in the process. A proximal-to-distal sequence of activation of muscles is the solution. In this way the mechanical properties of muscles are used to their optimal effect.

A striking implement represents an added segment, making control more difficult. The problem is the timing of the angular velocities of linked segments when contact is made. Mis-hits are more likely in this case. The advantage of an added segment is generally increased linear velocity of the system end point (for a given segment's angular velocity) leading to increased KE imparted to the projectile.

One will maximize the transfer of energy from the implement to the projectile by attempting to maintain acceleration over the duration of impact. The extent to which this strategy is successful depends upon numerous factors, which include the interaction of the viscoelastic characteristics of both implement and projectile.

Safety in throwing is largely concerned with avoiding repetitive strain injuries. The insertions of muscle into bone are particularly subject to strain injury such as lateral epicondylitis. The elbow joint is also in danger from a combination of extension and outward rotation to the extent that bone chips can be produced.

Catching involves reduction or dissipation of the KE of an airborne projectile to zero by means of eccentric work. In other words, it is the application of muscular force while the muscle is being stretched. The greater the number of body segments and therefore muscles involved, the greater will be the displacement of the point of application of force. This increase in displacement will allow both peak force and average force to be minimized since work is the integral of force with respect to displacement. The skill of catching is to match the velocity of the point of application of force to the initial velocity of the projectile and to continue with negative displacement, which has an acceleration that avoids the occurrence of large peak forces. Such a strategy will avoid rebound of the projectile from the point of application of force and reduce the possibility of injury. Injury will also be reduced by use of a catching glove to enlarge the area of contact between the projectile and the body part in contact. A glove will also increase the linear displacement of the point of application of force for a given angular displacement of the wrist.

RECOMMENDED READINGS

Throwing, striking, and catching are fundamental to many sports. The references given in the body of this chapter have dealt with the basic underlying principles. Readers can find details on the application of these principles to specific sports in the scientific sport literature. The following are recommended.

Freivalds, A. (2004). *Biomechanics of the upper limbs: Mechanics, modeling and musculoskeletal injuries.* Boca Raton, FL: CRC Press.

Hay, J.G. (1978). *Biomechanics of sports techniques.* Englewood Cliffs, NJ: Prentice Hall.

McGinnis, P.M. (2005). *Biomechanics of sport and exercise.* Champaign, IL: Human Kinetics.

Vogel, S. (2001). *Prime mover.* New York: Norton.

CHAPTER 10

Climbing and Swinging

Climbing and swinging have in common the heavy use of upper body and upper limb musculature. Climbing is upward motion of the body using four limbs that pull and push against some object. In this activity our abilities pale in comparison with those of some of our nearest evolutionary cousins. A testament to this connection is that very young infants can grasp sufficiently well to support their weight. Older children do this at play quite naturally due to good grip strength. Swinging is a type of pendular motion indulged in by children and gymnasts specifically and accidentally by many other individuals. We can swing under, over, and around a bar. We swing for pleasure and for sport. Fortunately, we no longer have to swing for survival as do gibbons and other apes, even though we possess very similar musculoskeletal structure. We appear to retain our grip strength sufficiently well to support ourselves, but this ability decreases with age and inactivity. Occasional climbing and swinging are very dangerous, not because of an inherent incapacity to hold on, although that possibility exists, but due to a poor understanding of climbing and swinging techniques.

Aim of Climbing

The aim of climbing is to ascend objects and barriers using many muscles against opposing gravitational force. Adults climb for work, for sport, or more rarely to fix something on the roof. Our aim is sometimes confounded by weak musculature and sometimes by the lack of availability of places against which to apply force through the hands and feet. In contrast to the situation with some other activities, safety is a large component of climbing because the consequences of failing can be catastrophic.

Mechanics of Climbing

The basic mechanical aim in climbing is to increase the potential energy (PE) of the body by doing muscular work. The equations and discussion of this basic mechanical aim are covered in the section on mechanics of lifting and lowering in chapter 8. In climbing it is the center of mass (CM) of the body that is lifted by forces applied by muscles of the limbs. Figure 10.1 is a schematic representation of the mechanics of climbing.

In figure 10.1, all four limbs are exerting a downward force on their respective points of contact, resulting in upward forces on the body. In figure 10.1a, the left upper limb and right lower limb (broken outline) are capable of doing work, the former by bringing the shoulder toward the hand and the latter by extending the right lower limb. The same process of doing muscular work applies to the limbs in ladder climbing shown in figure 10.1b. The total work done is therefore the sum of the individual integrals of force with respect to displacement $\int Fds$ of each limb, since the forces are in parallel.

■ Key Point

Forces in parallel can be added to give a total force; those in series cannot.

Biomechanics of Climbing

Significant features of climbing are that the upper and lower limbs cannot produce the same force and that the same force cannot be applied throughout the full range of change in limb configuration. The former feature is clearly due to the considerably different muscle mass moving upper and lower limbs. The latter is due to the force–length relationship of muscle, which dictates through skeletal leverage that the joint torque will vary with joint angle. In each climbing task (figure 10.1, *a* & *b*), the left lower limb and right upper limb (solid outlines) have reached the limit of their range of movement. They cannot shorten or lengthen to do more muscular work. Consequently they require repositioning upward where they can be effective workers. The dark gray limb segments show

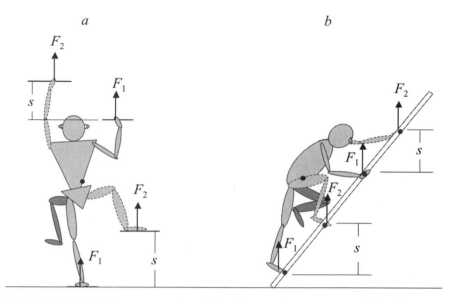

FIGURE 10.1 The segments with broken outlines are capable of doing work, during which the dark segments show repositioning of the left lower limb.

the lower limb in each task in the process of replacement. Only the lower limb repositioning is being shown to emphasize a safety feature. Repositioning one limb leaves three prior stable points of contact, which may not facilitate work but does enhance safety. If two limbs were repositioned simultaneously, only two points of contact would be available for stability.

In general, joint torques are maximal and vary by a minor amount in the middle of the range of joint movement. This fact certainly applies to elbow flexion, but during knee extension there is a "sticking point" that has to be overcome as discussed in the section on lifting and lowering in chapter 8. Such behavior argues for small increments of change in foot and hand position as we progress up some climbed object. For example, repositioning of the foot at a very high point in relation to the CM will result in severely flexed positions of knee and hip joints. Compensation for the compromised knee torque will require the arms to do greater work. Another compensating strategy might be to bounce or use the stretch–shortening cycle of muscle. Unfortunately the higher force and its rapid rate of change may result in overcoming limiting friction at the worst and provide difficulty in maintaining stability at the least.

Finger flexor strength in humans is large, yet it may prove to be the weakest link in the climbing chain of forces if handles are unavailable. Figure 10.2 shows four potential finger postures (*a-d*) that one might adopt when holding on to a right-angled corner. Arabic numerals represent the phalanges in this figure. In changing posture from *a* to *d* there is one less phalanx in contact with the horizontal supporting surface. The muscles producing force are not included in this figure because of their anatomical complexity. The general finger flexor torques are produced by the flexor digitorum profundus and sublimis with help from the lumbricals. These muscles have many different roles to play as well as synergistic actions. The extent to which each muscle can contribute to the total torque required will depend upon the axis about which the torque is necessary. In posture *a*, all finger flexors can make a contribution, while in posture *d* only muscles

■ Key Point

Extreme ranges of limb motion are subject to the combined effects of the force–length relationship of muscle and the mechanical advantage of each joint.

Fixing the lower end of the rope would enable the lower limbs to contribute to this rope climb.

© Arthur E. Chapman

that produce torque about the most distal phalangeal joint can contribute. From this point of view it would appear that the safest hand posture to adopt would be *a*. Although flexor muscles may not contribute useful torque, the muscle force helps to arrest distraction of each joint in a linear manner. Extensor muscles can also play a similar role, but should they reduce the net flexor torque their use should be avoided. For example, those extensor muscles that insert into all bones other than the last phalanx can play a supporting role. But use of any extensor of the most distal interphalangeal joint would be dangerous. The most dangerous situation is that in which the flexors are used for a prolonged period of time without the aid of the extensors, since fatigue will ensue quickly because of the smaller amount of muscle mass involved.

Variations of Climbing

Two examples of climbing illustrate different uses of the feet and the modifications we have developed for their use. Rope climbing does not benefit from the specific foot and hand positions afforded by a ladder. Furthermore, the rope is flexible, and intended regrasp positions can move away from the rope climber unless the rope is secured to the floor, often by a schoolmate in the gymnasium. Most accomplished gymnasts have sufficient upper body strength to perform this activity without recourse to the use of the lower limbs. Rope climbing by means of the upper limbs only is biomechanically simple but

■ Key Point

Prolonged isometric contraction of the finger flexors can induce rapid muscular fatigue, which compromises safety.

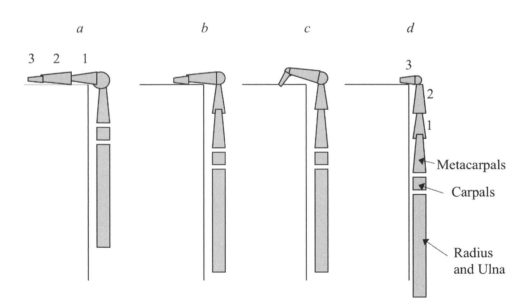

FIGURE 10.2 Different configurations of finger hold during climbing.

requires unusual strength of the elbow flexors and shoulder extensors. The latissimus dorsi muscle is one that is well developed and therefore an easily-seen attribute of gymnasts. Not surprisingly, this muscle is also well developed in arboreal primates.

For those humans who do not have such well-developed muscles, the technique is to support body weight with straight upper limbs and then regrasp the rope between the soles of the feet following unloaded knee and hip flexion. Knee and hip extension then follows, and when that is complete the hands are moved upward together in one movement. The problem here is that there are periods when only two limbs are able to do work and also to support the body precariously. It is also easy to lose sight of a place where the feet need to be repositioned, and the rope can swing away from the feet. People can overcome the latter problem by wrapping the rope around the leg, then sliding the leg through the loop of the rope for repositioning during knee and hip flexion. Rope climbing is also made easier if a partner can pull down on the rope to stabilize it. A rope ladder is easier to climb since it moves less freely than a single free rope.

In contrast, tree falling presents a separate set of problems. Obviously it is not possible to grasp the trunk of the tree with the hands or feet as in rope climbing. Tree fallers who take down a tree in pieces from the top have a particularly arduous task because they must find their way through the branches while being burdened with considerable mass such as chainsaws and handsaws and ropes hanging from their belts. These workers strap spikes onto the medial aspect of the leg. The spikes are driven into the tree trunk by extension of hip and knee joints, so it is not necessary to rely upon friction to provide upward force. Spikes also afford the opportunity to stand upright with the body being restrained from backward toppling by a belt that passes around the body and the tree trunk. In this manner workers can use both hands to reposition safety lines and to start and use a chainsaw. Most of this is straight mechanics rather than biomechanics.

Enhancement and Safety of Climbing

Climbing is a simple activity that is generally done slowly. This statement is not meant to suggest that climbing is without skill; it simply means that there is little complex integration of sensory information with complex muscular coordination as we might see in many gymnastics maneuvers. The skill in climbing has to do with planning strategy in terms of which actions are safe. In the sport of rock climbing, such planning is of paramount importance since misjudgment can leave the climber hanging with one point of support, or worse. So enhancement and safety go hand in hand in climbing.

In using muscles to do work, the sport climber must have high muscular strength and must train for it. In order that the force can be applied over a large displacement, the climber must also undertake joint mobility exercises. The emphasis on physical capabilities of occasional climbers lies more in knowing their limits since they are unlikely to train for an unknown occasion that requires climbing. Other climbers such as tree fallers, scaffolders, bridge workers, and even manual fruit pickers require initial training of upper body strength. If such activity is their daily occupation, the mere process of working should be sufficient to maintain the strength after the initial training.

■ **Key Point**

Muscular strength, particularly of the upper body, is essential for habitual climbers.

Body shape and proportion are of some biomechanical importance in climbing. Ideally any body mass should comprise muscle and then only muscle that can do climbing work. We are not constructed in this manner, however. But carrying excess fat is a burden that will have a direct correlation with the work required to raise the PE of the body since PE = *mgh*. Therefore climbers should aim for lean body mass as an enhancing factor in climbing and a safety factor in avoiding muscular fatigue.

■ **Key Point**

Gloves aid in reducing forces that can subject the skin to shear forces and burns.

Whenever it is possible, coverings such as gloves and shoes must be worn to enhance friction with the climbing surface and also to avoid friction burns if shearing of hands and feet with respect to the surface should take place. Most other safety considerations in climbing concern specialized equipment for various tasks. Such specialization is ergonomic safety rather than biomechanical safety.

Practical Example 10.1

Since most of the muscular factors related to applying force as in climbing have been covered heretofore, this example concerns a safety aspect of rock climbing.

Consider a climber who is joined to the rock face by a rope secured to the body. The climber (mass = m) falls and has an initial downward velocity of v_i when the rope becomes straight and begins to resist downward motion with a force F.

Part 1

Deduce an expression for the stretch *(s)* in the rope when the climber is at zero velocity for the first time.

Solution

Considering downward as positive we have the following equation of motion:

$$mg - F = ma \qquad \text{(Equation 10.1)}$$

as shown in figure 10.3.

Assuming that the rope has a linear force–length relationship of slope K (newtons per meter), which equals $\Delta F/\Delta s$, replacing F in the equation and replacing a with dv/dt gives

$$mg - Ks = mdv / dt.$$

Integrating this equation with respect to s to give work results in

$$\int mgds - \int Ksds = \int m(dv / dt)ds,$$

and since $ds / dt = v,$

$$mgs - Ks^2 / 2 = m\int vdv;$$
$$Ks^2 / 2 - mgs + mv^2 / 2 = 0.$$

The roots of this quadratic equation are

$$s = \{mg \pm [m^2g^2 - 4 \times (K / 2) \times mv^2 / 2]^{1/2}\} / (2K / 2) \text{ or } s = [mg \pm (m^2g^2 - Kmv^2)^{1/2}] / K.$$

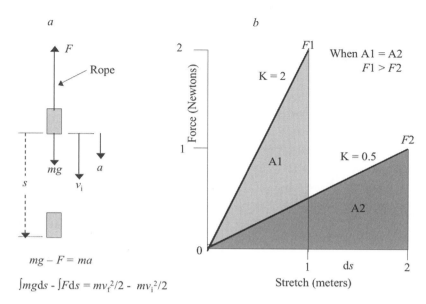

a

b

$mg - F = ma$

$\int mg\,ds - \int F\,ds = mv_f^2/2 - mv_i^2/2$

FIGURE 10.3 Arresting downward motion of a mass by means of ropes having different spring constants K = 2 N/m and K = 0.5 N/m.

Part 2

What are the implications of using a less stiff rope?

Solution

A less stiff characteristic of the rope means a smaller value of K such that there will be greater stretch (Δs) for a given force applied. The equation indicates that as K is reduced, the numerator will become smaller by a factor related to the square root of K. This will decrease the stretch in the rope. However, the denominator will get smaller, which will increase s. The final result is that as K is reduced, the stretch increases. The significance of this change in K can be seen in figure 10.3; here the work done by the rope is equal to the area bounded by the axes and the line of slope K = 2 and K = 0.5 in different cases, but the peak force is smaller for the K = 0.5 relationship. The climber will be subjected to a smaller force, which is desirable in order not to exceed the ultimate strength of any body tissues to which the force of the rope is applied. One way of reducing the stiffness of the rope is to decrease its cross-sectional area. Another strategy is to use a long rope so that there is a greater length of material to stretch. The elastic characteristics of the suspension used in bungee jumping take these strategies to the extreme. The disadvantage of this apparatus is the large mass of material to climb with. The final consideration is not to use a rope that is so compliant that the climber will hit the ground before the original kinetic energy (KE) is all stored as strain energy in the rope.

Practical Example 10.2

A worker of mass m_W = 80 kg (176 lb) is carrying a load of m_L = 30 kg (66 lb) and climbing a ladder that is inclined against a wall at 20° to the vertical. The worker is in contact with one step of the ladder, and the hands provide no significant propulsive force. What is the maximal acceleration a along the length of the ladder that the worker can produce safely if the step of the ladder is rated to fail at three times the worker's body weight?

Solution

There are two forces that contribute to the total force F_S on the step. One is the gravitational force acting vertically downward, and the other is the reaction to the force accelerating the worker in the direction of the ladder. The step will fail when the vector sum of these two forces exceeds the maximal step rating.

Resolving forces along the ladder,

$$F_S = (m_W + m_L) a + (m_W + m_L) g\cos(20°) = 110(a + g\cos20) = 110a + 1014 \text{ N}.$$

For the step not to break, this force must be less than $3 \times m_W \times 9.81 = 2354.4$ N. Therefore $110a + 1014 < 2354.4$. So

$$a < (2354.4 - 1014) / 110 = 12.19 \text{ m/s}^2.$$

The acceleration of about 12 m/s² is a large value for a human to produce, so the worker appears safe.

If the worker could produce a peak acceleration of 3 m/s², what would be the minimal safety rating of the step?

Referring to the first equation where $F_S = 110a + 1014$,

$$F_S = 110a + 1014 = 1344 \text{ N or } 1344 / (m_W g) = 1.7 \text{ times body weight}.$$

From these rough data it would appear that a failure force for each step of a ladder should be no less than two times body weight. When designing the failure rating for a ladder step, an ergonomist would need to know the peak force that a person could apply while on one foot and would develop a failure force for the maximum expected body mass and load carried, then double it for litigious workers.

Aim of Swinging

The aim of swinging by children appears to be the simple pleasure of the activity, perhaps unleashing some deeply buried motion pattern from our ancestral past. Children swing about an axis through the hands but appear to gain greater pleasure by using an apparatus aptly called a swing. Gymnasts swing to score points on such apparatus as the parallel, uneven, and high bars and also on the pommel horse, where the swing axis is unusual. Most gymnastics swinging has the aim of high amplitude of motion, but swinging on the rings loses points. In some cases the aim is to begin swinging from a stationary position with no contact with the earth other than through the object upon which the swinging takes place. Sometimes swinging begins from forces applied to the earth by the feet. In others cases the aim is to arrest the swinging motion, while hand-over-hand swinging achieves the aim of progression above the ground.

Mechanics of Swinging

The act of swinging is angular motion that relies heavily on gravity, although it is possible to swing by creating moments about the swinging axis by muscular contraction. So we are considering pendular motion. There are two types of pendulum; simple and compound. The simple pendulum comprises a point mass attached to an axis by a massless rod or string. This structure is unreal, although it can be approximated by a very small mass of high density attached to a very fine wire. Such a mechanism, which exists in the Science Museum in London, England, was used by early physicists to aid understanding of interplanetary motion, precession of the earth, and Coriolis acceleration, among other things. Alternatively a compound pendulum is a swinging body that has its mass distributed and not located at one point. Therefore it has a physical moment of inertia about an axis of rotation. Humans are a set of connected segments, each of which can be considered a compound pendulum. If the body swings as a single unit with no relative motion between segments, it can be considered to be a single compound pendulum.

For this single compound pendulum, shown in figure 10.4, the moment due to gravity produces angular acceleration according to the following equation:

$$-mgr\cos\theta = (I_G + mr^2)\alpha \qquad \text{(Equation 10.2)}$$

Provided that there are no dissipative forces, the PE decrease during swinging downward from a stationary handstand position will equal the KE gained at the bottom of the swing. The work done by gravity in decreasing the PE is the integral of moment with respect to angular displacement:

$$-mgr \int_{\theta = 90}^{\theta = 270} \cos\theta d\theta = -mgr[\sin\theta] = -mgr[\sin(270) - \sin(90)] = 2mgr$$

Two values of θ are shown in figure 10.4 in order to indicate the angle between the velocity vector and the vertical. It will be apparent that these two values are numerically identically as the person rotates.

This work leads to a gain in KE of $(mv^2 + I_G\omega^2)/2$ or $(mr^2 + I_G)\omega^2/2$. In fact there will be dissipation of energy due both to friction at the hands and to air resistance, which will obviate the possibility of continuing with this circular motion forever. The problem in the performance of grand circles on the horizontal bar is how to compensate for dissipative losses. With an inert body, this is impossible no matter how much the friction at the fulcrum is reduced. In human swinging we are fortunate that muscular work can replace frictional loss of energy and even lead to increase of the angular velocity of rotation. From the equation of motion of the compound pendulum as shown in figure 10.4, it is apparent that zero acceleration occurs when θ is 90° and 270° since the cosine of each of these is zero. Starting motion is therefore a problem from a position of 270°. Alternatively it is almost impossible to maintain a position of 90°, so gravity takes over to begin the swing.

Figure 10.5 shows a swinging motion of a single, solid compound pendulum and a double pendulum like a human, free to flex and extend at the hip joints. It is apparent that the equation shown in figure 10.4 does not apply to a double pendulum or any multisegmented body subject to swinging because the moment of inertia I_G and the value r change with changes in body configuration. What we see is greater displacement of the proximal segment in the early

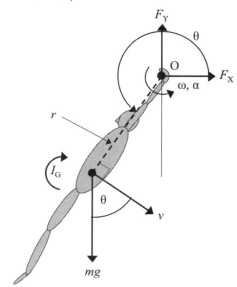

FIGURE 10.4 Mechanical definitions of a person swinging as a compound pendulum.

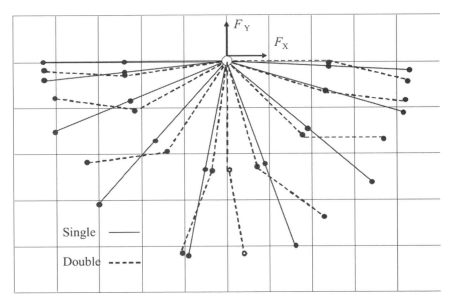

FIGURE 10.5 A multisegmented pendulum will not maintain a straight configuration when swinging.

stage of motion followed later by greater displacement of the distal segment. This motion is a natural property of a double pendulum and one that appears advantageous to the recovery phase of walking.

The components of force F_X and F_Y on the fulcrum will change continuously as the acceleration vector of the CM changes. These components are known as forces of constraint in that they fix the hands in relation to the earth. Whereas in rotation of a solid object these forces are predictable from the equation of motion, in the human they can be influenced by muscular contraction. The manner in which the human overcomes some of the problems described is revealed in the next section on biomechanics.

Biomechanics of Swinging

The previous section has shown that a double pendulum with no control of the intersegmental joint exhibits significant differences in angular velocity of each segment during a swing (see figure 10.5). It is apparent that the muscles must be used to maintain any fixed relationship between segments. Figure 10.6 shows the results of a computer simulation of a two-segment body representative of the upper and lower human body. The hip flexor moment required to maintain different piking angles in swings from a horizontal position against a small, constant dissipative frictional force at the hands is plotted with respect to time. While the moment varies throughout each swing, it is clear that the peak values are greater with greater hip angles (180° represents a straight posture). It is also apparent that the moment becomes negative or extensor at the later stage of the upswing. Even in a simple two-segment body, significant muscular involvement is required to maintain a fixed body configuration during swinging. In none of these simulations did the body CM rise above the horizontal before dropping back again. This is due to the small frictional moment acting against the motion. Even if the body were to begin from a handstand vertically above the bar, it would not return to the vertical on the upswing for the same reason. The problem to consider is how the gymnast solves the problem of completing a grand circle on the high bar. We know that the problem is one of frictional dissipation of energy, so the KE at the bottom of the swing is insufficient to raise the PE to its former starting value. Since the equation of motion includes values of r and I_G, there is a possibility that altering these values by piking on the upswing will reduce the negative value of α. Rearranging the equation of motion gives

$$\alpha = mgr\cos\theta\ /\ (I_G + mr^2).$$

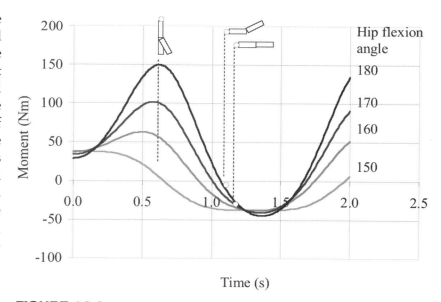

FIGURE 10.6 A greater hip flexor moment is required to maintain a greater hip flexion in a human swinging as a double pendulum.

The two terms in the denominator, I_G and mr^2, change as a function of the square of mass distribution. Therefore reducing r will have a greater effect on the denominator than on the numerator. Piking reduces r; therefore the greater the pike angle, the greater will be the value of α. Since α represents deceleration on the upward phase of the swing, the initial angular velocity at the bottom of the swing will be decelerated by a greater amount with piking. Simulation of a double pendulum as shown in figure 10.6 confirmed this hypothesis. The greater the hip joint angle (straighter body), the greater is the hip moment required to maintain the angle. Maintenance of a fixed posture indicates isometric contraction of the hip flexor or extensor muscles, or both. The consequent absence of muscle shortening means no muscular work done and therefore no possibility of compensating for frictional losses.

To investigate the applicability of using a hip flexor moment to change the hip joint angle throughout the swing, a set of simulations was performed in which the hip flexor moment was increased in magnitude incrementally between simulations. The motion began at a position of 91° because at 90° it would never occur. The results of these simulations are not shown because no reasonable solution was found in which hip musculature alone compensated for frictional loss and returned the model to a vertical position above the bar. Those solutions that worked required unreasonable anatomical ranges of movement between the segments. However, observation of gymnasts reveals that it is possible to swing through to finish in a position directly above the bar. Therefore we concluded that it was not possible to complete a grand circle using a hip flexor moment alone under the conditions of zero starting angular velocity and with the frictional moment used. Clearly the gymnast uses additional methods of compensating for the frictional energy loss. When the model was given initial angular velocity while passing through 90° above the bar, a grand circle was completed since the frictional energy loss was less than the initial rotational KE of the system.

When starting from a stationary hang below the bar (figure 10.7a), gymnasts solve the problem just described by flexing at the hips and extending at the

■ **Key Point**

An individual will not swing in a straight posture unless some muscular contribution is made, particularly at the hip joints.

■ Key Point

Swinging motion from rest can be induced by correctly timed muscular contraction.

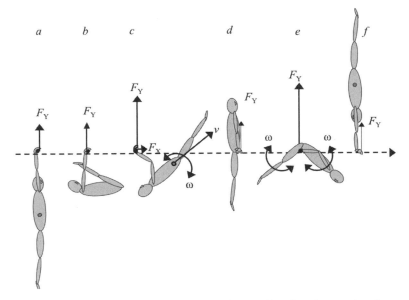

FIGURE 10.7 Movements required by muscular contraction in moving from a stationary hang to a handstand above the high bar and their effects upon force components at the hands.

shoulders, which brings them into the position shown in figure 10.7*b*. Then rapid extension of the hips is performed in concert with a vigorous pull by the arms (figure 10.7*c*). In this way the vertical force on the bar exceeds gravitational force, and the CM is raised. Simultaneously the body rotates, and the gymnast finishes with the final position shown (figure 10.7*d*). The next maneuver is a rapid piking (figure 10.7*e*) followed by hip extension and rotation, which results in the final position shown (figure 10.7*f*). Small pushes with the arms can then allow rotation to begin with a significant angular velocity, and a grand circle can be completed.

The actions described and simulated were done on the assumption that the bar is rigid. There is, however, a large amount of flexure in the bar. The bar therefore acts as a mechanism in which energy can be stored as strain energy at a suitable time and returned to the gymnast when conditions for its use are more favorable. Typically the energy is released to add height and velocity to the gymnast prior to release from the bar in order to increase dramatically the appearance of the dismount.

An increase in vertical force similar to that described for starting from rest is seen on a child's swing. Backward rotation occurs with simultaneous pulling on the flexible vertical suspensions. Translational and rotation KE are induced in this way, and the swinging motion can begin. With each successive upward phase, the arm pulling can be used to add more energy and increase the PE to which the swing is raised. A more effective method is to stand on the seat and use the jumping muscles to add greater energy with successive swings.

Strain energy stores in the horizontal bar can be returned to increase the kinetic energy of the gymnast.

The results of simulation in which hip flexor and extensor moments were used to generate a swinging motion are shown in figures 10.8 and 10.9. In figure 10.8 the initial hip flexor moment gives the system (mostly the lower limbs) counterclockwise angular motion QB with little angular displacement of the torso and arm segment QA. The subsequent hip extensor moment returns the body to a straight posture but with a net counterclockwise angular displacement and angular velocity at time 1.7 s. The kinematics at this point are sufficient to oppose gravity to allow the body to reach a horizontal position at about 2.2 s. The body then experiences clockwise rotation induced by the moment due to gravity. This shows how muscular work can begin a swinging motion from rest. Figure 10.9 shows the large magnitude of the simultaneous forces at the hands in the same simulation (F_X and F_Y to the right and upward, respectively).

Changes in these forces of constraint are generally opposite in sign after about 1.7 s. This is characteristic of pendular motion, but it is modified somewhat in this case because the internal moment is needed to maintain a straight body. Of course these forces, in concert with that of gravity, when divided by mass give the acceleration of the CM in the horizontal and vertical directions.

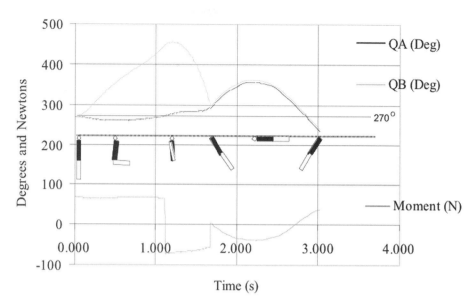

FIGURE 10.8 Hip flexor moment required to generate a swinging motion from rest.

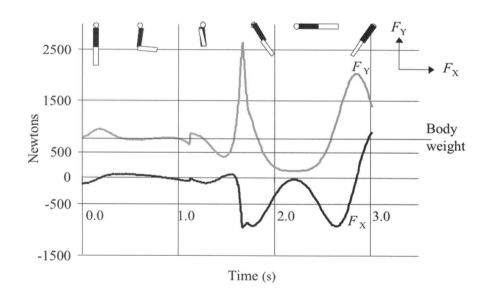

FIGURE 10.9 Muscular contraction produces components of force on the bar which in turn lead to acceleration of the CM of the person.

Variations of Swinging

Swinging is used either as a precursor to or during widely different maneuvers. In this section we will consider three variations. The first is grasp-regrasp maneuvers of the bar following a grand circle on the high bar during which the gymnast is airborne. The second is horizontal progression using swinging, and the third is abseiling (or rappelling) down a mountainside.

■ Key Point

Muscular work can modify constraint forces at the hands, resulting in addition of energy to the system.

The grasp-regrasp is a dangerous maneuver, particularly when an airborne somersault intervenes. The problem to solve is where during the rotation this maneuver is best performed. As discussed previously, an object undergoing angular motion about a fixed axis experiences a radial acceleration a_r in a direction toward the axis according to the relationship $a_r = v^2 / r$. The force producing a is ma, and it is transmitted through the gymnast's hands. During a grand circle the gymnast feels this force, and at most positions in the motion its sudden loss would leave the gymnast following a translational path at a tangent to the arc of motion at release. However, there is a point where the force of gravity, in acting vertically downward, can replace the force at the hands. If the value of a required is equal to g as the gymnast passes vertically over the bar, there need be no grasp by the hands. Should v be small, a will be small, and there will be a need to push on the bar to act against the greater value g. A required value of a greater than g will necessitate a pull. Nonetheless, the smallest values of force at the hands will occur with the CM vertically above the bar, and this is therefore where the grasp-regrasp should be done. This is not a perfect solution to the problem, since loss of the grasp will result in parabolic motion of the CM rather than circular. Yet this region of motion remains the best mechanical solution.

Traveling from hand to hand on a horizontally oriented ladder is relatively simple if we can hang on. All that is required is an initial angular velocity that brings the CM outside a vertical line through the fulcrum in the direction of progression. At this stage, one hand releases its grip and grasps the next rung of the ladder; the remaining hand lets go as the CM is now outside the vertical through the fulcrum in the opposite direction to progression. Gravity takes over and the next swing begins. Progression will be arrested unless some muscular work is done to overcome the friction in the system. Our arboreal cousins are much more adept than we are at this means of progression due to the large latissimus dorsi muscles they possess. In fact, these muscles are so strong that they can project the body upward with significant velocity to give airborne motion that is arrested only when the performer catches the next branch with one hand.

Abseiling (rappelling) is a rapid means of letting oneself down a vertical or near-vertical wall. The performer pushes the body away from the wall and lets the body downward along a rope that runs through a body harness. Should the wall or rock face not be planar, there may be obstructions or indentations the abseiler wishes to avoid. The person does this by pushing off to the left or right and creating rotation of the system in two planes simultaneously. In one sense the task becomes easier as the length of the human pendulum increases. This pendulum can be considered to be a simple pendulum if its length is great in comparison with the stature of the person. Since the time of swing of a simple pendulum is equal to $2\pi\sqrt{(l/g)}$, the longer the pendulum the greater the period of the swing. This allows the person more time to make decisions as to how to contact the rock face and where to place the feet.

■ Key Point

Release and regrasp maneuvers on the high bar should be performed when the gymnast is above the bar, when the constraint force at the hands is least.

Enhancement and Safety of Swinging

Enhancement of swinging activities is almost entirely achieved through increase in muscular strength. This applies particularly to the shoulder girdle and shoulder joint musculature, as well as that of the trunk and hips. The only other factor of significance, though it is not of a biomechanical nature, is a means of increasing the immense courage needed to perform most of the gymnastics swinging activities. Gymnasts use leather pads to offset the effects of the considerable frictional shearing forces on the palms of the hands. These pads are usually chalked so that no perspiration is present to reduce friction and induce slippage. Other safety aids are a harness on pulleys under the control of an experienced coach and landing pits of compliant material to dissipate the KE on landing.

■ **Key Point**

High-bar swinging maneuvers cannot be performed successfully without enhanced upper body strength.

Practical Example 10.3

A gymnast is performing a series of grand circles on the horizontal bar and wishes to perform a maneuver between releasing and regrasping the bar. What is the maximal angular velocity of the gymnast at the position of release that will ensure that a regrasp can occur given the values shown in figure 10.10?

Solution

Regrasping can occur only if the velocity of the CM is less than that which would project it farther horizontally than shown in figure 10.10. This means that the vertical and horizontal components of velocity of the CM are equal in magnitude in their respective directions and in the two positions shown. Fortunately only gravitational force is involved in the airborne phase, and that gives a constant downward acceleration. Initially it is necessary to establish airborne time so that the horizontal displacement can be obtained. Defining downward as positive and using the equation of uniform motion $v_f = v_i + at$, we obtain the following :

Vertically, $v\sin\theta = -v\sin\theta + gt$; so $t = (2v\sin\theta) / g$.

The horizontal displacement of the CM must be less than or equal to $2r\sin\theta$. The horizontal displacement obtained from the horizontal component of velocity and time is $v\cos\theta t$. So,

$v\cos\theta t \leq 2r\sin\theta$.

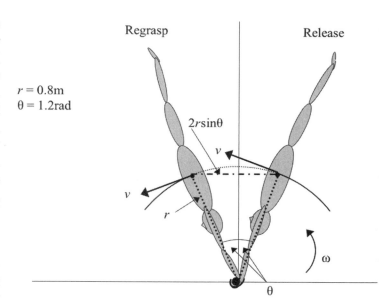

FIGURE 10.10 The velocity vector is crucial for allowing a release and regrasp maneuver on the high bar.

$r = 0.8\text{m}$
$\theta = 1.2\text{rad}$

Substituting for t from the prior equation gives

$v\cos\theta \times (2v\sin\theta) / g \leq 2r\sin\theta$ or $v^2 \leq rg / \cos\theta$ or $v \leq (rg / \cos\theta)^{1/2}$.

Finally we obtain the value for angular velocity ω from the identity $v = r\omega$:

$r\omega \leq (rg / \cos\theta)^{1/2}$ or $\omega \leq (g / r\cos\theta)^{1/2}$;

$\omega \leq \{9.81 / [0.8\cos(0.35)]\}^{1/2}$ or $\omega \leq 3.61$ rad/s

An angular velocity of less than this value would allow regrasping, but not in the same posture and also with perhaps less time to perform the aerial maneuver. An angular velocity of a greater value spells disaster.

Summary

Climbing involves increasing the PE of the body by means of concentric muscular work. Within this aim is the secondary requirement of safety. While all four limbs can be used simultaneously in one phase of the upward motion, anatomical ranges of limb movement require removal and subsequent replacement of the feet and hands on the climbing surface. A general guide to safety is to retain three points of contact with the climbing surface while one limb is repositioned. Safety is also enhanced by ensuring a good handhold. Human beings generally are able to support their body weight by means of finger flexor activity. Searching for a good grip on the climbing surface is of paramount importance in comparison with a good foothold because the latter is dependent to a large extent upon friction. Rarely do hands slip off the rungs or runners of a ladder, but feet do. In this context, gloves can increase safety by enhancing frictional force and reducing friction burns should sliding occur. There are many other pieces of apparatus that can aid safety and that are specific to the climbing task. For example, safety ropes and spikes are used by tree fallers.

Muscular strength training is essential for sport climbers and should form part of the training program for those whose occupation involves climbing. Emphasis should be on increasing strength throughout the full range of limb motion. In this way a greater vertical displacement will be possible with all four limbs secured to the climbing surface. The result will be fewer changes of point of application of force on the climbing surface per vertical distance climbed.

Swinging is a modified form of pendular motion in which the person is effectively attached to the earth by the hands gripping the swinging apparatus. Through this arrangement, internal muscular forces can find their external expression and achieve swinging motion from rest. As the human body is segmented, it behaves differently from a rigid pendulum such that muscle moments are necessary to maintain a straight body posture. The swinging person is generally playing with gravity, which produces an advantageous moment on the downswing and opposes motion on the upswing. At the bottom of the downswing, the gripping hands have to produce force against the bar to equal body weight plus the centripetal force producing circular motion (mv^2/r or $mr\omega^2$). In performing a grand circle on a horizontal bar, a gymnast will perform regrasp maneuvers at the top of the swing since the gravitational force is aiding circular motion so that grip force is least or even absent.

Success in swinging activities relies almost entirely upon muscular strength. Safety in swinging activities relies upon a compliant energy-dissipating landing surface and skill and attention of supporters, who should be placed appropriately to catch the performer.

RECOMMENDED READINGS

Climbing and swinging are largely confined to recreation and sport. Readers are encouraged to search for books on sports and recreational pursuits in which these activities are common.

Airborne Maneuvers

Most people do not undertake airborne maneuvers voluntarily because for them flying through the air is usually the result of an accident. Yet some of the most astonishing and dramatic feats of which humans are capable are to be seen in airborne maneuvers. Such feats are largely restricted to gymnastics and diving competitions, in which points are awarded for successful execution and style, and in the circus, where employment is related to the number of "oohs and ahs" generated in the audience.

Few people spend much time off the ground in the course of their daily activities; others do so voluntarily or involuntarily. The main focus of this chapter is the biomechanics of voluntary airborne maneuvers that can lead to success in a competitive environment. Involuntary airborne motion usually comes unexpectedly, but there are certain maneuvers that can allow us to avoid injury in such circumstances. Unless teaching of at least some simple airborne maneuvers is included in a physical education program, we are doing an injustice to our next generation.

The mathematics required to perform a thorough analysis of three-dimensional motion are beyond the scope of this book. In some cases the results of performing bodily actions are given without the underlying equations of motion. Those students who wish to gain a further understanding should pay particular attention to the references.

Aim of Airborne Maneuvers

The problem is to perform rotations while airborne, either about one principal axis or about more than one simultaneously. Further aims are to change both angular velocity and the axis about which rotation occurs. These changes are possible because of the segmented nature of the human body.

Mechanics of Airborne Maneuvers

Problems in airborne motion depend upon the complexity of the aim. However, there are two major mechanical principles that cannot be contravened. The first is that the angular momentum at takeoff is fixed. Airborne motion occurs under the influence of one external force, namely gravity. Since gravity acts on all parts of the body simultaneously, it cannot induce rotation about the center of mass (CM) where none exists after the body has left the ground. Equations 11.1a and b represent the translational and rotational impulse–momentum relationships respectively, where the subscripts TO and F represent takeoff and final times.

■ **Key Point**

Mechanical impulses produced by muscles internal to the human system are equal and opposite while the body is airborne, so the total local angular momentum remains constant.

Momentum

$$\int F dt + m v_{TO} = m v_F$$

(Equation 11.1a)

Angular momentum

$$\int M dt + I \omega_{TO} = I \omega_F$$

(Equation 11.1b)

The second mechanical principle is that any muscular force F and joint moment M created by F will be internal to the system, so the integrals will equal zero.

They obey the action–reaction law, and therefore there will be no change in total body angular momentum.

The problem facing a person who wishes to change motion patterns of the body while airborne is how to work within the constraints of these two principles. A single solid object will continue with constant angular momentum and angular velocity about the initial axis of rotation through the CM. However, a segmented body can change its moment of inertia such that while angular momentum remains constant, decreases in moment of inertia lead to increases in angular velocity and vice versa (figure 11.1).

Note that angular momentum (H_{CM}) is represented as a vector coincident

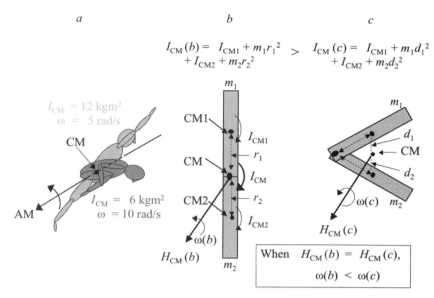

FIGURE 11.1 Body moment of inertia depends upon configuration of segments.

with the axis of rotation. This primary axis is known as the axis of momentum (AM) as shown in figure 11.1a. The symbol AM is used here strictly to denote axis of momentum and to avoid confusion. Angular momentum is symbolized by the traditional H, H_{CM} being the angular momentum about an axis through the CM. Because angular momentum is the product of moment of inertia and angular velocity, any change of moment of inertia will change the angular velocity of rotation. In this manner the angular velocity can be controlled within limits as seen in figure 11.1a. Angular momentum is represented by an arrow of given length and direction, perpendicular to the plane of the page in which motion takes place. Angular momentum is a vector quantity resulting from the product of a scalar (moment of inertia) and a vector (angular velocity). In figures 11.1b and c, the parallel axis theorem is used to represent the moment of inertia of a two-segment body in two configurations. Since the values of r in figures 11.1b are greater than those of d in figure 11.1c, $I_{CM}(b)$ is greater than $I_{CM}(c)$. Therefore, for equal values of angular momentum $H_{CM}(b) = H_{CM}(c)$, angular velocity $\omega(b)$ is less than $\omega(c)$.

■ Key Point

While the angular momentum vector remains constant in airborne motion, the angular velocity vector can change with change in the orientation of body segments.

Planar Motion

The segmented nature of the body also facilitates rotation of different segments and can appear to give the body angular velocity where none existed at takeoff. In a similar manner, angular momentum can be redistributed among segments so that certain segments remain at zero angular velocity. Figure 11.2 shows simulation of a segmented body in which all particles are rotating instantaneously at the same angular velocity of 1 rad/s. The orientations of the drawings are those occurring at the times shown at the bottom of the plots. At zero time, a moment is created that rotates the wheel in relation to the remainder of the body. The wheel (W) and remainder (R) of the system show, respectively, increasing and decreasing clockwise angular velocity when the moment is being applied. When R reaches zero angular velocity at 0.215 s it has a small angular displacement,

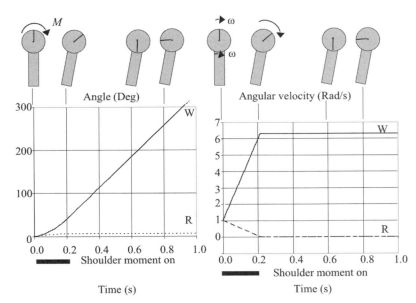

FIGURE 11.2 An internal moment redistributes angular momentum among segments.

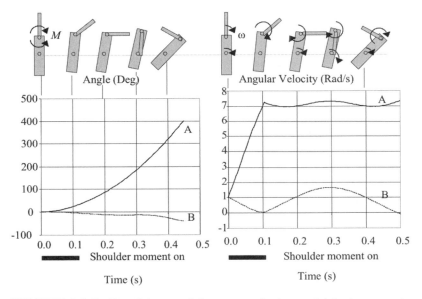

FIGURE 11.3 Parallel use of the upper limbs redistributes angular momentum among segments in a nonuniform manner.

while W has a much greater angular displacement. At this time (0.215 s) the moment is turned off, and the kinematics at that instant continue unchanged. R maintains its zero angular velocity and undergoes no further angular displacement, while W maintains its current angular velocity and subsequent increasing angular displacement. The fact is that the initial whole-body angular momentum finally appears as angular momentum in W only.

In the following section on biomechanics we shall see how this strategy of redistribution of angular momentum is used by athletes to their advantage. In this simulation, the wheel has a constant moment of inertia relative to the system CM because of its regular distribution of mass relative to the system moment of inertia. Therefore the profiles of angle and angular velocity change in a uniform manner. These changes should be compared with those shown in figure 11.3. In the latter, the wheel has been replaced by an arm segment which is in fact both arms rotating while remaining parallel to each other. In this case the moment of inertia relative to the CM changes as the arm segment rotates. The equation for I_{CM} given in figure 11.1 shows how the angular momentum about an axis through the CM changes with the position of the extended segment.

The mass of the arm is at its greatest distance from the system CM in the initial position and at its least distance when it is beside the trunk. The initial conditions for figure 11.3 are shown, with the body (B) and the arm (A) having identical angular velocities of 1 rad/s as in figure 11.2. The moment is present again for a period of 0.215 s in this simulation, and the arm segment is located artificially at a midpoint between the shoulders. As before, B and A, respectively, lose and gain angular velocity due to the applied moment until the moment is stopped. At this point the moment of inertia of A is decreased relative to the system CM so that its angular momentum is decreased correspondingly. As a consequence B increases its angular momentum and gains angular velocity, which peaks when the moment of inertia of the arm is least (i.e., when it is parallel to B at 0.3 s). When A continues to rotate to its position of greatest moment of inertia, B loses angular momentum and ceases to rotate. Continued rotation of A would produce the observed changes in angular velocities repeatedly. The exact same motion would be seen if a person were to perform this motion with both arms acting as the single arm segment seen here. The major point is that the angular velocity of each segment is not constant but oscillates about some mean value. Also it is apparent that a body that has a single rotating segment will share its angular momentum with other segments as the segmental moments of inertia change relative to the CM. If both arms were used at an angle of 180° to each other, this new system would behave like the wheel, and the nonuniform change in angular velocities would become uniform.

So far we have observed the effects of an applied moment that lasts for a small period of time. In figure 11.4, the arm extensor moment is continuous and leads to counterclockwise rotation of the body. We observe a continual increase in the angular velocity of A, while B has an initial decrease in angular velocity. Yet the changes are not uniform because of the changing moments of inertia of each segment relative to the CM. Of significance is that B shows counterclockwise angular velocity in response to the counterclockwise internal moment applied to it. Finally, figure 11.5 shows redistribution of segmental angular momentum in the absence of an internal moment and only rotation of the arm segment initially. Here we see angular momentum of segments increasing and decreasing as their moments of inertia relative to the system CM change. While both A and B segments change their moments of inertia, the moment of inertia of A has much greater changes than that of B because the CM of B is very close in position to the system CM at all times.

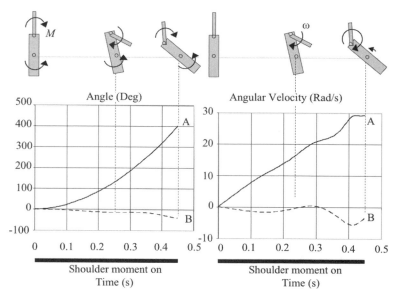

FIGURE 11.4 Redistribution of angular momentum is continuous when the internal moment is continuous.

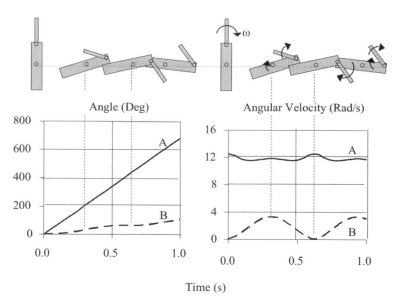

FIGURE 11.5 Variation of the arms' moment of inertia relative to the CM gives variation in the redistribution of angular momentum among segments.

In summary, production of internal moments will induce intersegmental redistribution of angular momentum, but the total angular momentum of the system will not change as a result of the moment. Intersegmental changes in angular momentum will appear oscillatory in nature (e.g., using one arm) unless segments are changing their angular momenta in a manner that does not change their combined moments of inertia relative to the system CM (e.g., using two arms, 180° out of phase).

■ **Key Point**

Angular momentum of the body can be partitioned among segments by muscular contraction.

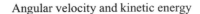

Angular velocity and kinetic energy

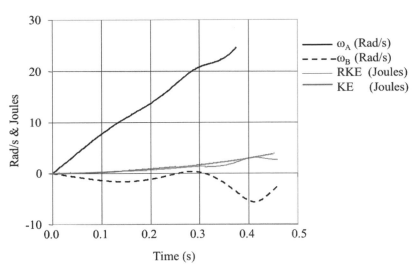

FIGURE 11.6 Internal muscular work changes the KE so a net angular displacement of the body is achieved.

■ **Key Point**

Whereas internal angular momenta cancel due to their vectorial nature, work, being scalar, is done at both ends of the muscle; a net angular displacement of the body can therefore occur without any external force.

So far only angular momentum has been considered in this analysis of intersegmental airborne motion. However, observation of figure 11.4 shows a net displacement of the body due to an internal moment even though no segments had initial angular momentum. The fact is that internal moments can do work since they are due to muscles that produce force at each end in the same direction as movement of each end. When the moment is integrated with respect to the angular displacement of each segment to which it is applied, a value for work is obtained. Therefore kinetic energy (KE) is produced, which leads to a net change in angular position when the moment stops. Figure 11.6 shows changes in total KE and rotational kinetic energy (RKE) along with angular velocities (ω_A and ω_B) previously shown in figure 11.4 (initially zero angular velocity and continual moment).

A net gain in energy is seen due to the integral of moment with respect to time. It should be noted that the total energy (KE) is largely due to RKE, the remainder being due to the sum of the values $mv^2/2$ of the CM of each segment. In this case muscular work while the body is airborne can lead to a net angular displacement even though no angular momentum was present at zero time. In contrast, figure 11.5 shows that a net angular displacement can be achieved in the absence of moment provided that there is initial angular momentum.

Three-Dimensional Motion

The concept of angular momentum can be applied to two principal axes simultaneously. In this case neither principal axis coincides with the AM. The AM lies somewhere between these two axes as seen in figure 11.7. Note that the AM lies in the plane formed by the individual angular momentum vectors H_T (transverse) and H_L (longitudinal). These two vectors are added by the normal rules of vector addition using the right-hand rule as described in chapter 2. The cylinder shown in figure 11.7 has a third or sagittal axis perpendicular to the original two, although its moment of inertia is identical to that about the transverse axis. The human body has different moments of inertia about all three principal axes, so the AM would not lie in any of the three planes formed by any two principle axes since the third would be excluded. If the cylinder were to represent a person, we would have a combination of somersaulting and twisting. These motions would take place symmetrically about the AM, which would have a fixed orientation in space, even though the CM would be accelerating due to gravity. The analysis of three-dimensional motion of an airborne body requires mathematical techniques that are largely beyond the scope of this book. However, numerous publications treat this area of study in depth from the point of view of engineering mechanics (Beer et al., 2007). Some of the material presented here is intended to provide an introduction to further study for those students who wish to pursue this area of biomechanics in greater depth.

Principal axes of the body are mutually perpendicular and meet at the CM. In the case of a sphere of uniform density, we are unable to identify where the three principal axes are located, because the moments of inertia about all axes passing through the CM are equal. For symmetrical bodies, which the human body approximates in the standard anatomical posture, there are three principal axes. The axes and their respective moments of inertia are depicted in figure 2.7c in chapter 2. Principal axes are located with respect to the body, and they are necessary in order for us to appreciate the angular velocity and angular momentum of the airborne system. Figure 11.8 shows the principal axes along with a representation of the angular velocity and angular momentum vectors. The principal axes of the body (X', Y', Z') are embedded in the body itself and centered at the CM (point O). This provides a "body" reference frame that can be located in translation and rotation with respect to the fixed "global" frame of reference (X, Y, Z) embedded in the earth. The mathematics relating these two frames of reference are not dealt with here.

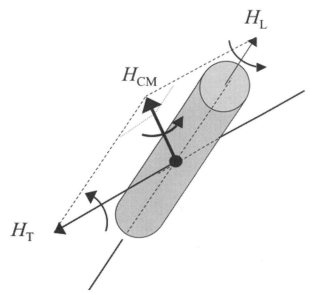

FIGURE 11.7 Angular momentum about principal axes summed vectorially to yield a whole-body axis of momentum of given magnitude and direction.

Note that the resultant angular velocity ω_O and angular momentum H_O vectors are neither in the same direction nor of the same magnitude. For example, the angular velocity about the longitudinal axis $\omega_{Z'}$ is relatively large, but the angular momentum about the same axis $\omega_{Z'}$ is relatively small. The reason is that the angular momentum is the product of moment of inertia and angular velocity; in this case the moment of inertia $I_{Z'}$ is very small in comparison with $I_{X'}$ and $I_{Y'}$. When both translation and rotation of the body frame occur, only an external force eccentric to the CM can change the angular momentum vector. In human airborne motion this eccentric force is absent, so the AM can move under the influence of gravity but cannot change either its magnitude or its direction relative to the global reference frame; on the other hand, the angular velocity vector can. A small complication is that it is the angular velocities about the principal axes that we see when observing multidimensional airborne motion, not the angular momentum. Therefore we need to be able to relate angular momentum to angular velocity if we are to understand three-dimensional angular motion.

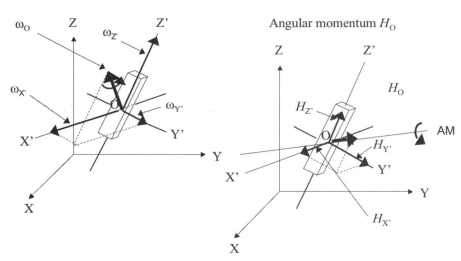

FIGURE 11.8 Angular velocity and angular momentum summed vectorially do not necessarily coincide.

Figure 11.8 shows general motion where the AM is not coincident with any of the three principal axes of the body. Such will be the case in three-dimensional activities like diving somersaults with twist. In a plane somersault dive, the AM will be coincident with the transverse principal axis, as will the angular velocity vector. Since the human body can change configuration of segments, it is possible to reorient the principal axes relative to the AM. The result of such maneuvers is to vary the angular velocities of somersaults and twists.

Figure 11.8 shows how the separate angular momenta can be summed according to the head-to-tail operation of vector addition to give the final AM. We see a global reference system (X, Y, Z) fixed relative to the earth and a local reference system coincident with the principal axes of the body (X', Y', Z'). Thick arrows indicate angular momentum about each principal axis ($H_{X'}, H_{Y'}, H_{Z'}$). By placing them head to tail, along X' and then parallel to Y' and parallel to Z', we find the resultant H_O, which is the angular momentum about the CM (O) of the body in both magnitude and orientation. Figure 11.8 shows that the final angular momentum vector H_O and its coincident AM are not parallel to any of the principal axes. Unfortunately the complexity of three-dimensional motion is significantly greater than that of two-dimensional motion. The graphical process described here can be characterized mathematically; this is rather tedious yet necessary if the motion is to be analyzed numerically. The mathematics become even more complex when we are dealing with three-dimensional motion of segmented bodies. Yet we must understand such complexity if we are to understand how H can be redistributed among the three axes in human airborne motion.

Biomechanics of Airborne Maneuvers

The major biomechanical principles of airborne maneuvers to be discussed are translational airborne motion, two-dimensional rotation, the task of creating a net angular displacement in two dimensions while one is airborne, and the task of inducing multiaxial motion.

Translational Airborne Motion

While airborne motion is generally of a rotational nature, there are some activities that are restricted to pure translational motion or curvilinear motion. The basketball shot in which the player jumps vertically is an example of overall linear motion, although some segments undergo rotation in this action. The problem in this case is to produce appropriate velocity of the basketball. In using the term velocity we imply both magnitude and direction, since velocity is a vector quantity. Any activity requiring such precision will benefit from having the organs that sense spatial orientation as stationary as possible. Since these organs are located in the inner ear and the eye sockets, it is beneficial to keep the head as still as possible. More specifically, it is better to avoid acceleration of the head. Unfortunately, gravity gives us whole-body acceleration, but there are ways in which the acceleration of the head can be minimized. The strategy is to ensure that the ball is projected when the body is near to the apex of its flight.

■ Key Point

Simultaneous rotation about two or three principal axes leads to symmetrical rotation of the body about the axis of momentum passing through the CM.

So, how near to the apex should we be, and why? Since we can reorient the position of our CM by repositioning our limbs, upward acceleration of the ball and arm (or arms), which occurs in late upward body motion, has a reaction that accelerates the remainder of the body in the opposite downward direction. Should the resulting downward velocity match the upward velocity of the CM, the remainder of the body, including the sense organs, will be stationary. The timing of events is crucial to success because adoption of this strategy after one reaches the apex of the flight will produce a worsening effect. This worsening effect can be offset somewhat if the knee joints are in a flexed position at the beginning of downward motion. With the body falling, extension of the knee joints will offset the downward motion by tending to raise the remainder of the body. It is important to remember that the CM has a predictable path in this action, and the skill is to time segmental motion so that the head experiences a different acceleration profile from that of the CM. We shall see later, in the section on variations, how such a strategy is used to good visual effect by ballet dancers.

■ **Key Point**
Redistribution of momentum can be used to minimize the acceleration of certain body parts, particularly the head.

Two-Dimensional Rotation

In this section the case considered is motion in two dimensions or planar motion about a fixed body axis. In the anatomical posture the human body can loosely be considered as a rectangular prism that has a different moment of inertia about each principal axis. The greatest moment of inertia is that related to the sagittal axis since the mass of the body has its greatest distribution away from that axis. This is followed in descending order by moments of inertia about the transverse and longitudinal axes. Labeling axes fixed in the body as X, Y, and Z corresponding to sagittal, transverse, and longitudinal axes, we can state their relative moments of inertia as $I_X > I_Y > I_Z$.

Consider the case in which a diver leaves the board with angular momentum about axis Y only, as in performing a forward somersault. The diver is required to perform a single somersault and enter the water headfirst, in fact hands first. The single somersault for a diver who takes off from the feet is actually an angular displacement of one and a half revolutions or $360° + 180° = 540°$. The problem then is to rotate through $540°$ in the time (t) during which the diver is airborne. Since gravity is the only external force acting while the body is airborne, the airborne time is a function of both the height (h_{TO}) of the CM above the water level and the vertical component of initial velocity v_i of the CM at takeoff. Since acceleration is constant (g), with the downward direction positive, and we know h_{TO} and v_i, we can use equations of uniformly accelerated motion containing these values and t as follows:

$$v_f = v_i + at \qquad \text{(Equation 11.2)}$$

$$v_f^2 = v_i^2 + 2as \qquad \text{(Equation 11.3)}$$

From Equation 11.2 we obtain

$$t = (v_f - v_i) / a.$$

From Equation 11.3 we obtain

$$v_f = (v_i^2 + 2as)^{1/2}.$$

Combining these equations and substituting gives

$$t = [(v_i^2 + 2gh_{TO})^{1/2} - v_i] / g,$$

which is the time available for rotation of the diver. Note that we would also obtain this equation by using Equation 2.2 ($s_f = s_i + v_i t + at^2 / 2$), which is of quadratic form. Should the diver maintain the initial posture at takeoff, the angular velocity (ω) required will be

$$\omega = 540 / \{[(v_i^2 + 2gh_{TO})^{1/2} - v_i] / g\}°/s.$$

■ Key Point

For a given angular momentum, the angular displacement in planar rotation can be modified by changes in both moment of inertia and time while the body is airborne.

If the angular velocity is insufficient, there are three possible solutions. One is to increase the angular velocity at takeoff. A second is to increase v_i, which, being negative in sign, will increase the denominator and decrease the required angular velocity. The remaining solution is to use the concept that moment of inertia about a principal axis through the CM (I_{CM}) multiplied by angular velocity is angular momentum (H_{CM}). In the absence of external moment, H_{CM} will remain constant; but decreasing I_{CM} will lead to an increase in ω, and the required angular displacement will be achieved. This strategy is adopted to perform double, triple, and even quadruple somersaults in diving.

The two major means of reducing I_{CM} are to pike the body, which is a very stylish way of touching one's toes with the forehead touching the patellae. The other is to bunch into a fully tucked posture in which all body segments are brought as close to the CM as possible. The latter gives the lowest value for I_{CM}. The postural changes described cannot be performed at random because dives are prescribed as "piked" or "tucked." All dives begin in the extended posture, and the rate of tucking determines the rate at which the angular velocity increases; the reverse is the case when one is extending the posture. Tucking too quickly may give too great an angular velocity and an "overshoot" on the dive. The consequence of untucking too quickly will be an undershoot. A further biomechanical problem is the muscular force necessary to tuck or pike quickly. Circular motion of the body requires a force toward the CM equal to $r\omega^2$, where r is the distance from the CM to the mass of the segment in question. The resulting centripetal force $mr\omega^2$ exists even if there is no tucking. During tucking or piking, additional muscular force is necessary to move the segments toward the CM. This is particularly so in piking, in which the trunk and hip flexors are subject to considerable loading.

It takes skill to judge how fast to pike or tuck and how fast to open out again.

Piking of the body that has initial angular momentum can lead to an apparent illusion. Figure 11.9 shows two superimposed dives that have equal velocity (*v*) and angular momentum (H_{CM}) at takeoff, one of the two involving a piking action. Note that H_{CM} is coincident with the axis of rotation, which is perpendicular to the page and indicates clockwise rotation according to the right-hand rule. The unfilled figures indicate stages of the dive in the extended posture. The pike dive is represented by the darker figures, the middle one of which is in the darkest shade only for the purpose of clarity. What appears to happen is that the orientation of the lower limbs is fixed with no angular velocity on the upward stage of the dive, and the remainder of the body alone rotates about the hip joint axis. The remainder of the body has similar fixation on the downward motion, with lower limbs now appearing to rotate about the hip joint axis.

What is happening is that the counterclockwise angular velocity of the lower limbs due to piking equals the initial clockwise angular velocity of the body. Therefore the lower limbs have no angular velocity in the upward stage of the dive. The reverse is the case for the remainder of the body in the downward phase. Although divers are able to perform a large number of wonderful tricks, the ability to fix the orientation of a body segment is not one of them. Doing this would require an external moment, and there is not one in the current example. What we are seeing is the constant total angular momentum of the system appearing in the upper body initially and the lower limbs finally due to the internal muscular forces inducing piking. This example illustrates that segments of an airborne body can rotate in relation to another segment that is fixed in orientation only if there is initial angular momentum in the system. The apparent fixed angular orientation of the lower limbs initially and of the remainder of the body subsequently are due to the similarity of angular velocity of piking and initial total body angular velocity. The illusion is that the diver can fix each of these segments at will. Calculations of internal moments and forces based on this assumption would yield incorrect results. The system has to be analyzed as a whole and not in what appear to be convenient parts.

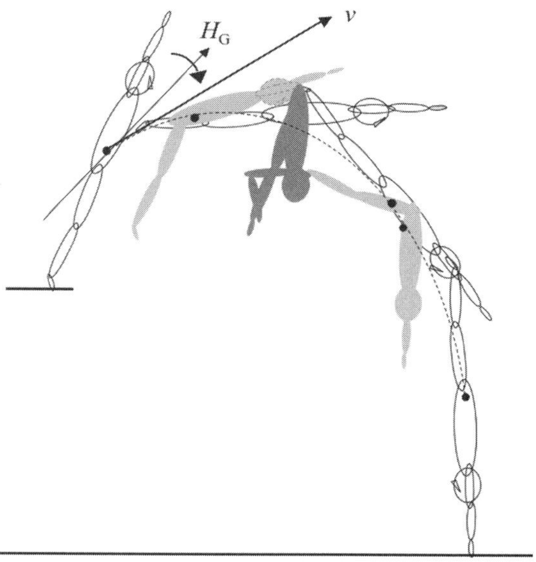

FIGURE 11.9 Changing body configuration can give the illusion of being able to fix a segment in space.

Creating a Net Angular Displacement in Two Dimensions While Airborne

The ability to avoid toppling by use of arm motion has been discussed in chapter 4. One can achieve a similar rotational effect while airborne; in this case the problem is to create a net angular displacement of the whole body. Consider the situation in which a body is in freefall either in curvilinear or pure translational motion with zero angular momentum. The flight path of the CM cannot be altered by intersegmental motion, as the only external force available is gravity acting through the CM. Also there can be no external moment, only competing internal muscular moments. However, muscular contraction represents work done because each end of the muscle is moving in the same direction as its force vector. Therefore KE of some form must be produced, whether translational or rotational. Such effects were examined in the previous section on mechanics as depicted in figures 11.2 through 11.6, where the arms are used. A typical example of this effect

■ Key Point

When a piking action is performed in concert with angular velocity in a forward dive, the angular momentum appears initially only in the upper body, followed subsequently by the lower limbs.

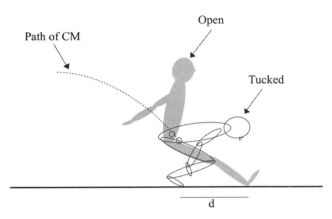

FIGURE 11.10 Improving a long jump by using internal moments to reorient the body.

is seen in long jumping. The measure of success in the long jump is the horizontal distance between the edge of the takeoff board closest to the sandpit and the most rearward mark in the sand. The body configuration on landing is therefore of prime importance in terms of the horizontal distance between the feet and the horizontal position of the CM on landing. Figure 11.10 shows two postures on landing that yield different foot positions in the landing pit. In the tucked and open positions, the feet land vertically below and in front of the CM, respectively.

The distance gained by the open position is the value "d" shown. The advantage seen is produced as follows. Assuming that the kinematics at takeoff are identical, the path of the CM in the two cases is identical. However, there is a smaller angular displacement for the open posture during flight, for two possible reasons. One is that in the open position the body is maintained in an elongated posture. The large moment of inertia will lead to a smaller angular velocity about the transverse axis for any given angular momentum at takeoff. Tucking will have the opposite effect. Another reason is as described in the section on mechanics with reference to figure 11.3. In this case the angular momentum is localized in the rotating upper limbs, and the remainder of the body will show zero angular velocity. The lower limbs can also be used in a cycling fashion to add to this observed localization of angular momentum. The total effect is known as "running in the air" and also as the "hitch-kick." The lower limbs cannot produce the same angular motion as the arms due to restriction of motion of the hip joint. Yet they can be brought backward in a straight posture and brought forward with the knee joints flexed. This results in a clockwise rotation of the CM of the lower limbs in much the same manner as the rotating arms but with less amplitude of motion.

In long jumping it is very difficult to take off without clockwise angular momentum (clockwise with respect to figure 11.10), so any clockwise motion of the mass of the limbs will serve to arrest, and even reverse, the rotation of the remainder of the body. Additionally the number of rotations of the limbs will determine the effect we are discussing. Although troublesome while airborne, the clockwise angular momentum at takeoff provides a final advantage. If the jumper progresses from an open to a tucked position during landing, the moment of inertia will decrease and the angular velocity will increase as a consequence. This maneuver will allow less time for gravity to produce a moment tending to rotate the jumper backward. It will also decrease the lever arm between the foot and the CM, which will reduce magnitude of the counterclockwise moment due to gravity. While success in long jumping is primarily based upon the height, magnitude, and direction of the velocity vector of the CM at takeoff, the airborne maneuvers described can have a significant additional effect on success.

Inducing Multiaxial Motion

Multiaxial motion can result from angular impulses produced against the ground prior to takeoff, changes in the orientation of principal axes while one is airborne, or both in combination. Figure 11.11 represents two views of a diver leaving

■ **Key Point**

The ability to create angular displacement while airborne can aid in landings that are advantageous depending upon the activity.

the board, with angular velocities $\omega_{Y'}$ and $\omega_{Z'}$ about the transverse (Y') and longitudinal (Z') axes. It is assumed that the diver maintains the elongated takeoff posture. It is also assumed for simplicity that the body is a cylinder of uniform density. Therefore the moments of inertia $I_{Y'}$ and $I_{X'}$ are identical while $I_{Z'}$ is considerably smaller. The products $I_{Y'}\omega_{Y'}$ and $I_{Z'}\omega_{Z'}$ equal the angular momentum about each principal axis $H_{Y'}$ and $H_{Z'}$, respectively. Therefore H_{CM} lies in the plane defined by the Y' and Z' axes. What follows is an examination of the subsequent airborne motion of the diver. The motion is complex and depends upon the relative angular momenta and angular velocities about each principal axis. However, it must be stressed again that the orientation and magnitude of H_{CM} cannot change until an external moment is applied by the water on the diver at entry. In one sense H_{CM} represents an axis about which the motion of the body occurs symmetrically.

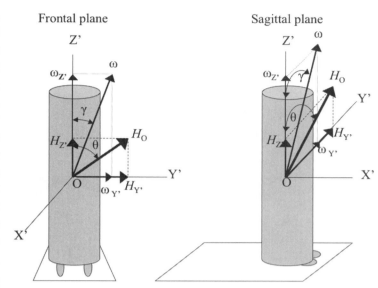

FIGURE 11.11 Two views of the calculation of the body's angular velocity ω and angular momentum H_{CM} vectors.

Angles γ and θ represent the orientation of the Z' axis to the angular velocity vector ω and to H_{CM}, respectively, and their significance is as follows.

$$H_{Z'} = H_{CM}\cos\theta \quad H_{Y'} = H_{CM}\sin\theta \quad H_{X'} = 0 \qquad \text{(Equation 11.4)}$$

$$H_{Z'} = I_{Z'}\omega_{Z'} \quad H_{Y'} = I_{Y'}\omega_{Y'} \quad H_{X'} = I_{X'}\omega_{X'} \qquad \text{(Equation 11.5)}$$

$$\omega_{Z'} = H_{CM}\cos\theta / I_{Z'} \quad \omega_{Y'} = H_{CM}\sin\theta / I_{Y'} \quad \omega_{X'} = 0 \qquad \text{(Equation 11.6)}$$

Since the value of $\omega_{X'}$ is zero, there is no rotation about the X' axis and the value of υ remains constant. The body therefore undergoes what is known as precession about H_{CM}. Precession can be best described diagrammatically as shown in figure 11.12 for two cases of different initial angular momenta and angular velocities. Figure 11.12a has the same kinematics and kinetics as shown in the frontal plane view of figure 11.11a. Shown in figure 11.12 (a and b) are two cones, a body cone whose axis of symmetry is H_{CM} and a space cone whose axis is $\omega_{Z'}$. They touch at a tangent to each other along the line ω. The surface of the body cone rotates around the surface of the space cone, which has a fixed orientation in space since its axis of symmetry is H_{CM}. It can be seen that H_{CM} is greater in figure 11.12b than in figure 11.12a. This greater H_{CM} is due to a greater initial value of $H_{Y'}$ as $H_{Z'}$ is unchanged (refer to figures 11.11 and 11.12). As a consequence H_{CM} is directed at a smaller angle to the horizontal Y' axis, and the half angle of the space cone $(\theta - \gamma)$ is greater. This results in a more vertical orientation of the body in figure 11.12b than in figure 11.12a after half a revolution of the body cone around the surface of the space cone. In the cases given of the initial orientation and angular momentum about the longitudinal and transverse axes, H_{CM} can never be horizontal. Therefore the body will never achieve a vertical orientation when upside down. However, if the longitudinal axis Z' was inclined initially at an angle γ clockwise, the body would be vertical when upside down.

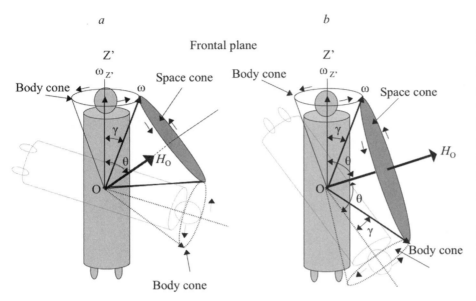

FIGURE 11.12 The body cone rolls around the surface of the space cone, and the two always touch at a tangent where surfaces meet.

This discussion illustrates how a diver cannot take off with a purely vertical orientation of the longitudinal axis (Z') and contact the water headfirst with a vertical orientation of Z' if angular momentum is created about two axes at the instant of takeoff. A strategy enabling this to occur is dealt with later in this section.

The adoption of a variety of postures will affect the relative magnitudes of the moments of inertia about principal axes. Such changes will affect the nature of precession observed. Rearrangement of Equation 11.6 results in Equation 11.7 as follows:

$$\omega_{Y'} / \omega_{Z'} = (H_{CM} \sin\theta / I_{Y'}) / (H_{CM} \cos\theta / I_{Z'}) = (I_{Y'} / I_{Z'})\tan\theta, \qquad \text{(Equation 11.7)}$$

and noting from figure 11.11 that $\omega_{Y'} / \omega_{Z'}$ is equal to $\tan\gamma$, we obtain

$$\tan\gamma = (I_{Y'} / I_{Z'})\tan\theta.$$

The type of precession obtained is dependent upon the ratio $(I_{Y'}/I_{Z'})$. When $I_{Y'} \leq I_{Z'}$, $\gamma \leq \theta$; and we obtain the type of precession shown in figure 11.12. The space and body cones are at a tangent externally, and the spin and precession are in the same direction, known as *direct precession*. When $I_{Z'} > I_{Y'}$, $\gamma > \theta$; and the body cone lies inside the space cone and the vector $\omega_{Z'}$ is directed negatively along the Z' axis. The space cone and the body cone have opposite senses known as *retrograde precession*. This can be approximated by the piked posture so that the nature of the airborne motion is influenced by postural changes and when they occur during the precession motion. One can appreciate the precession effect by throwing a cylindrical form in the air with different relative amounts of angular momentum about the transverse and longitudinal axes. Students should try this with a pencil and a can of food and note the different effects.

In the absence of angular momentum after the body has left the ground, there is no possibility of producing continuous motion of one part of the body unless another part is moving as seen in figure 11.13. However, we have seen previously that creating angular motion about more than one principal axis induces precession, which can lead to difficulties in entering the water with a vertical longitudinal axis. Another method of producing airborne motion with multiaxial rotation is to leave the ground with angular motion about one principal axis and subsequently to reorient parts of the body as shown in figure 11.14. In this figure the combination of initial rotation about the transverse axis and the subsequent tilt of this axis away from the AM induces a twist angular velocity.

This strategy is seen diagrammatically in figure 11.15. Initially H_{CM} coincides with the transverse axis during pure rotation about the transverse axis. The

■ **Key Point**

The presence of angular momentum about one axis and the creation of angular momentum about a second axis induce rotation about a third axis.

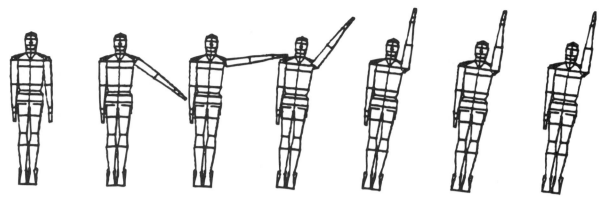

FIGURE 11.13 Abduction of the upper limb leads to body tilt in the absence of initial angular momentum.
Reprinted, by permission, from M.R. Yeadon, 1984, *The mechanics of twisting somersaults* (Leeds, UK: Loughborough University of Technology), 285.

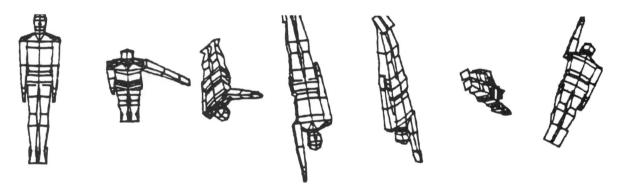

FIGURE 11.14 Angular momentum about the transverse axis plus production of angular momentum in one arm about a sagittal axis gives rotation (spin) about the longitudinal axis.
Reprinted, by permission, from M.R. Yeadon, 1984, *The mechanics of twisting somersaults* (Leeds, UK: Loughborough University of Technology), 285.

coincident orientation of these axes is removed by the tilt reaction to adduction of the left arm. Since H_{CM} cannot change, some of the angular momentum ($H_{Y'}$) is lost and is gained ($H_{Z'}$) vectorially about the longitudinal axis Z′ to represent an angular velocity called twist. It should be noted that this effect will begin as soon as the arm begins to adduct and that the final motion will also depend upon the rate of abduction. To enter the water vertically without twist, one must perform an opposite motion with the arm in order to locate the transverse axis coincident with the AM. Whether the body enters the water vertically is a matter of chance unless this latter strategy is timed to occur with the body in a specific orientation. This is a complicated effect, and readers who wish to pursue this area further must read the comprehensive treatment by Yeadon (1984).

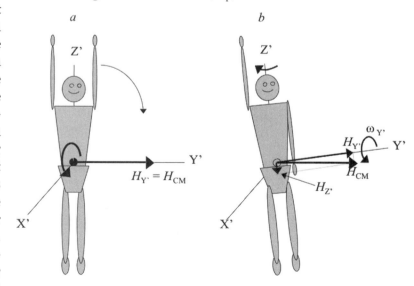

FIGURE 11.15 Upper limb adduction displaces the transverse axis Y′ away from H_{CM} and induces a spin angular momentum $H_{Z'}$.

Variations of Airborne Maneuvers

Expert ballet dancers appear to be able to "hang in the air" during a *grand jeté*. This term is loosely translated as "large throw," and it is a running jump with the lower and upper limbs raised to the horizontal. The appearance of hanging in the air is due to the path of motion of the head. Were the body to remain rigid following takeoff, we would expect the head and all other parts of the body to follow a parabolic path under the influence of gravity. In fact what we see is an initial movement of the head upward as the limbs are being raised. The expected upward motion of the CM is within the body due to the upward motion of the limbs. The remainder of the body does not accelerate upward as would be expected. The culmination of this movement is a relatively flat trajectory of the head, which gives the appearance of "hanging" for a period of time at a relatively constant horizontal level. As the limbs lower to the vertical, the reverse effect is seen, with the downward acceleration of the CM due to the lower limb motion. This effect of "hanging in the air," which can be quite dramatically achieved by good ballet dancers, is due simply to our watching one part of the body since we cannot see the CM. Plainly, nothing can hang in the air, however useful we might think this could be.

Bar clearance in the high jump occurs with the back of the body facing the bar. The jumper literally goes over backward. All that is required is that the parts of the body be above the bar when they get to it. This is achieved by body rotation in a backward somersaulting type of action with the body extended and arched backward. After the hips have cleared the bar, the hip joints flex to lift the lower limbs upward and away from the bar. These actions result in what may seem to be a rather odd phenomenon. That is, the body can clear the bar while the CM actually passes underneath it. Of course this means that all parts of the body not directly over the bar must be as low as possible on either side of the bar. In one respect gymnasts could be better high jumpers than track and field athletes since they can adopt a semipiked posture during a somersaulting motion. They could roll around the bar with the feet leading in a backward somersault. If you are not convinced by this explanation, think of a rotating horseshoe. Gymnasts can elevate their centers of mass to great heights when required to perform multiple somersaults. Unfortunately, they do so by taking off on both feet, which is not allowed in high jumping.

A further effect in high jumping is precession of the body. The takeoff occurs with multiaxial rotation, and appropriate values of angular momenta and angular velocities lead to precession, which aids bar clearance.

When one is clearing hurdles in a sprint, the aim is to spend as little time in the air as possible. This requires the CM to be raised to the lowest safe level that will allow clearance of the hurdle. Initially the hurdler drives a leading lower limb forward by hip flexion, with the knee joint as flexed as possible. By the time the heel of the leading limb reaches the hurdle, the limb is straight and close to horizontal depending upon the stature of the athlete. At this stage the trunk is leaning well forward to keep the CM as low as possible. This posture is maintained for a period of time when the CM is moving forward. The problem now is how to negotiate the hurdle with the trailing limb. The trailing limb cannot be brought forward through hip flexion alone in a vertical plane, since the thigh would strike the hurdle. In fact it is brought forward in a horizontal plane, with hip flexion superimposed upon hip abduction. This motion represents angular momentum about a vertical axis through the hip joint. The reaction to this is an opposite angular momentum induced in the remainder of the body. To keep all of the remainder of the body from rotating, which would make landing and continued running impossible, the upper limb on the same side of the body does the job by sweeping backward in a horizontal plane. In this way the trunk can be retained in a forward-facing posture for landing and continued running. One should also appreciate that if the upper and lower limb actions cancel each other out in terms of angular momentum, the whole body motion will not be multiaxial so the problems of precession will be avoided.

Consider another example, that of keeping a hula hoop rotating around the waist, as most children are capable of doing. During the motion the body goes from hip flexion through lateral spinal flexion, hip extension, opposite lateral flexion, and back to hip flexion. The action will not work if the body is maintained in an erect posture. The feet remain stationary, and all parts of the body are rotating in one direction as seen from a view along a vertical axis. There will be an axial moment created at the feet for the movement to start and a further smaller moment to overcome joint friction and other energy-dissipating sources. Now imagine performing the same action on a frictionless turntable that is free to rotate in a horizontal plane. As the body motion continues, the turntable rotates in the opposite direction since the moment at the feet is internal to the system in the horizontal plane. In this way it is possible to produce angular displacement of the whole body so that it ends up facing in the opposite direction. Throughout the movement, the net angular momentum in the system is zero since no external moment is applied.

This action is the basis for a trampoline maneuver in which the athlete starts with a seat bounce, rises in the air, rotates through 180° about a longitudinal axis, and ends up with a seat bounce but facing in the opposite direction. This is also the basis for the righting reflex of a cat that falls initially from a supine orientation, although other explanations of more complexity have been given (e.g., Batterman, 1974; Lanouie, 1940; Dyson, 1973). The whole maneuver can be performed without any initial angular momentum, as observed in expert divers who hang from the board and after release are told to turn right or left. It is also the basis for some maneuvers involving multiaxial rotation as described by Yeadon (1984).

Enhancement of Airborne Maneuvers

The rate at which the moment of inertia about any principal axis can be changed is determined by

Rotation of the upper and lower limbs about their transverse axes can offset forward angular momentum gained at takeoff in the long jump.

moments produced by muscles. Therefore the facility that an airborne athlete possesses for complex multiaxial rotation will depend upon muscular strength. Strength training is a necessary accompaniment to technique training in all airborne maneuvers. It is rare to see a diver, trampolinist, or gymnast who does not have well-developed upper body musculature. These athletes require training of the shoulder musculature particularly, since the upper limbs are used to redistribute angular momentum among principal axes. In uniaxial rotation exhibited by long jumpers, the hip, knee, and trunk flexors and extensors must be trained to allow the hitch-kick technique to be used effectively. Fortunately these muscles increase their strength as a result of training for other parts of the event.

Safety in Airborne Maneuvers

Safety is a primary concern in airborne maneuvers since they all end with landing on solid ground, water, or sand. A particular example is seen in the circus, where the astonishing release and regrasp maneuvers of performers are undertaken with safety nets. These nets are highly compliant and undergo large displacement when the performers land on them as is usual at the end of the performance. Clearly the large displacement requires only a small peak force to do the work of dissipating the KE of falling. Nets of this type are not allowed in trampoline and gymnastics competitions. However, the inherent danger involved in some gymnastics maneuvers is recognized, and a coach is allowed to stand near to the athlete during competition when particularly difficult and dangerous parts of a gymnastics routine are being performed.

The major need for safety is in training when new maneuvers are being learned. In this case landing should occur on surfaces that have a large amount of compliant material such as foam rubber. In this way the athlete can concentrate upon the technique unencumbered by the constant threat of injury on landing. Trampolinists take another approach by practicing on diving boards so that water can serve as the landing surface.

Unfortunately there is no landing surface that can guarantee perfect safety. An alternative is to provide a harness, which is located around the athlete's waist. The harness is attached to a rigging of pulleys and ropes, which gives the supporter or coach a large mechanical advantage in arresting downward motion of the performer. These harnesses have a circular ring that can slide inside an outer ring so that twisting movement can be performed. A drawback in the use of harnesses is that they cannot be located perfectly symmetrically about the CM and therefore can induce unwanted external forces, as well as change the position of the CM of the system by virtue of their mass. Obviously the person who controls the harness requires training to avoid inducing excessive external forces on the performer.

It is clear that all future physical educators require training in how to provide a safe environment for children who are learning airborne skills. Part of such training should be how to teach the nonperforming children to support the performer. If this is done correctly, many children will not only experience the thrill of learning how to do clever tricks with their bodies but also be better able to appreciate the expertise of top-class performers as a result of their own attempts. The alternative is perfect safety, with children asked to do nothing that represents a physical challenge. If you will permit some authorial license, I suggest that we are not doing our children any service by denying them the opportunity to perform physically simply because the odd accident happens.

Practical Example 11.1

A diver in an extended posture with arms above the head leaves a platform that is 5 m (5.4 yd) above the level of the water with a vertical component of velocity of the CM of 2.5 m/s. The distance r between the CM and the feet is 1.2 m (1.3 yd), and the distance between the feet and hands is 2.2 m (2.4 yd).

Part 1

If the posture is unchanged, what must be the angular velocity ω at takeoff for a dive of 2.5 somersaults to be completed?

Solution

The angular displacement while the diver is airborne is 2.5 somersaults, which is $2.5 \times 2\pi$ radians. This value divided by the time of flight t will give the required angular velocity. The CM begins and ends at heights of 5 m plus 1.2 m and 1.0 m above the water, respectively. The total distance the CM falls is therefore 5.2 m. Since gravity is the only force affecting the flight of the CM, we can use one of the equations of uniform motion, which includes displacement s, acceleration g, and the required time. Taking downward as positive gives

$$s = v_i t + at^2 / 2 \text{ so } 5.2 = -2.5t + 9.81t^2 / 2 \text{ or } 4.905t^2 - 2.5t - 5.2 = 0.$$

The solution of this quadratic equation is

$$t = \{2.5 \pm [(-2.5)^2 - 4 \times 4.905 \times (-5.2)]^{1/2} \}/ (2 \times 4.905) = 1.316 \text{ s or } -0.8 \text{ s}.$$

Only the time of 1.316 s makes sense, so the angular velocity ω required is

$$\omega = 2.5 \times 2 \times 3.142 / 1.316 = 11.94 \text{ rad/s}.$$

This value converts to $11.94/2\pi$ or 1.9 revolutions per second.

It will be noticed that this calculation is independent of any mass or moment of inertia. Therefore the analysis applies to divers of all shapes and sizes, providing that the position of the CM is known relative to the feet and hands.

Part 2

The diver has an initial moment of inertia of 8 kg.m² in the extended posture and the ability to decrease this value to 2 kg.m² instantaneously, to spend time in the tuck, and to extend instantaneously. How much time should be spent in the tucked posture for a dive of 3.5 somersaults to be completed?

Solution

Since the initial and final kinematics are identical to those in Part 1, the diver spends the same time $t = 1.316$ s airborne. We also know that the angular momentum H_{CM} is unchanged irrespective of the body posture, since no external moment is applied, and is equal to $8 \times 11.94 = 95.52$ kg.m^2/s. Therefore the angular momentum in the extended posture $I_{CME}\omega_E$ is equal to that in the tucked posture $I_{CMT}\omega_T$. We also know that the sum of the times in the extended and tucked postures $(t_E + t_T)$ equals the total time t. So the angular displacement required (3.5 revolutions) is related to the time in each posture as follows:

$$\omega_T t_T + \omega_E t_E = 3.5 \times 2\pi = 22.0 \text{ rad.}$$

$$t_T + t_E = t \text{ or } t_E = t - t_T.$$

Since $\omega_T = H_{CM} / I_{CMT}$ and $\omega_E = H_{CM} / I_{CME}$,

$$(H_{CM}/I_{CMT}) t_T + (H_{CM}/I_{CME})(t - t_T) = 22.0 \text{ rad} = t_T[(H_{CM}/I_{CMT}) - (H_{CM}/I_{CME})] + t (H_{CM}/I_{CME}).$$

So $t_T = [22.0 - t (H_{CM}/I_{CME})] / [(H_{CM}/I_{CMT}) - (H_{CM}/I_{CME})].$

$$t_T = [22.0 - 1.316(95.52 / 8)] / [(95.52 / 2) - (95.52 / 8)]$$

$$t_T = 0.176 \text{ s}$$

It is unlikely that a diver could tuck and extend rapidly enough to spend only 0.176 s in the tucked posture, so the values given here are unrealistic. During tucking the angular velocity will increase at a rate dependent on the rate of change of the moment of inertia. The obverse is true for extending from the tucked posture. This illustrates that the diver has control over the total angular displacement and therefore the body position at entry to the water.

Practical Example 11.2

The diver shown in figure 11.15 achieves an angular displacement of the transverse axis (Y') of $\theta = 10°$ about the sagittal axis (X'), by clockwise angular motion of the left upper limb. The initial angular momentum is $H_{CM} = 100$ kg.m^2/s, resulting from the product of moment of inertia of $I_{CMT} = 10$ kg.m^2 and an angular velocity of 10 rad/s about the transverse axis. Calculate the final angular velocities about the transverse ω_T and longitudinal ω_L axes if the moment of inertia about the longitudinal axis (Z') is $I_{CML} = 2$ kg.m^2.

Solution

The total angular momentum and its vector direction will not change, but it will be redistributed between the longitudinal and transverse axes. The angular momentum about the transverse axis will reduce to

$$H_{CMT} = H_{CM}\cos10 = 100 \times 0.985 = 98.5 \text{ kg.m}^2\text{/s.}$$

Since $H_{CMT} = 98.5 = I_{CMT}\omega_T = 10 \times \omega_T$, $\omega_T = 98.5 / 10 = 9.85$ rad/s.

$$H_{CML} = H_{CM}\sin10 = 100 \times 0.174 = 17.4 \text{ kg.m}^2\text{/s.}$$

Since $H_{CML} = 17.4 = I_{CML}\omega_L = 2 \times \omega_T$, $\omega_T = 17.4 / 2 = 8.7$ rad/s.

As a check on the calculations, the square root of the sum of squares of the individual angular momenta should equal the original angular momentum as follows:

$$(H_{CMT}^2 + H_{CML}^2)^{1/2} = (98.5^2 + 17.4^2)^{1/2} = 100 \text{ kg.m}^2\text{/s} = H_{CM}$$

The angular velocity of spin about the longitudinal axis is large despite its small gain in angular momentum. This indicates the sensitive nature of redistributing angular momentum among the principal axes of the body, and therefore the need to time changes in body configuration very carefully.

Summary

An airborne human body cannot change its angular momentum by means of muscular contraction because the muscular forces are internal to the system. As angular momentum is the product of moment of inertia and angular velocity, it is possible to vary the angular velocity by changing the configuration of body segments. This allows a diver to control the number of rotations so that entry into the water can occur with a vertical orientation of the body.

The segmented nature of the body allows angular momentum to be redistributed among segments. A body that has zero angular momentum at takeoff can demonstrate angular momentum in one part while another part shows angular momentum of equal and opposite magnitude and direction. This allows certain segments to experience increased angular velocity at the expense of others that lose it, often completely. Such an effect can be used to reposition the body favorably prior to landing on the feet.

In contrast to the case of unchanging total angular momentum, internal muscular moments acting through a joint or segment angular displacement do work that appears as a change in KE of the system. This work is responsible for a net change in angular displacement of the whole body.

Angular momentum and angular velocity are described relative to three principal axes of the extended human body. Angular velocities about the principal axes are added vectorially to give total body angular velocity, which has both magnitude and direction. Angular momenta are added similarly. In general, the angular momentum and angular velocity vectors will not coincide because of the differing moments of inertia about the principal axes.

During airborne motion, changes in the relative positions of segments can be used to make any part of the body appear to deviate from a path that would be dictated by gravity. This allows the head to be maintained relatively stationary for a small period of time, with the advantage that the spatial sense organs are subject to relatively small acceleration. In addition this technique can provide pleasing visual effects in dance, gymnastics, and diving.

Multiaxial rotation can be produced through the application of moments against the ground about more than one principal axis during takeoff. Alternatively, angular momentum about only one axis can be redistributed among the other two principal axes, leading to multiaxial rotation. When a body that has angular momentum about one axis has that principal axis tilted about another axis by limb motion, angular momentum is produced about the third axis. This process can be reversed by limb segment motion, but it needs to be timed correctly to be effective in diving.

Multiaxial rotation while the body is airborne is mathematically and mechanically complex and difficult for performers to control. An in-depth analysis of this type of activity is beyond the scope of this book, and readers are referred to Yeadon, 1984, for a comprehensive treatment of this material.

RECOMMENDED READINGS

Airborne maneuvers are largely confined to diving, gymnastics, trampolining, and the jumps in track and field athletics. The first of the following books addresses the conceptual nature of three-dimensional motion of rigid bodies, and the second deals with the numerical analysis.

Beer, F., Johnston, E.R., Eisenberg, E., Clauser, W., Mazurek, D., and Cornwell, P. (2007). *Vector mechanics for engineers.* New York: McGraw-Hill Ryerson.

Fogiel, M. (1986). *The mechanics problem solver.* New York: Research and Education Association.

Hay, J.G. (1978). *Biomechanics of sports techniques.* Englewood Cliffs, NJ: Prentice Hall.

McGinnis, P.M. (2005). *Biomechanics of sport and exercise.* Champaign, IL: Human Kinetics.

Appendix A

Mathematical and Mechanical Symbols

For ease of reference and simplicity, this list is in alphabetical order of English language script followed by Greek language script, and it ends with some mathematical operators. In general, symbols representing vector quantities appear in *italics*. The appropriate combinations of mass, length, and time appear under Dimensions.

Symbol	Meaning	Units	Dimensions
ENGLISH			
A	Area	Meters squared	l^2
a	Acceleration	Meters per second squared	l / t^2
CM	Center of mass	Unitless	Dimensionless
CR	Coefficient of restitution	Unitless	Dimensionless
cm	Centimeter	Meters $\times$ 100^{-1}	l
d	Distance	Meters	l
E	Energy	Joules	Dimensionless
F	Force	Newtons	$m \times l / t^2$
g	Gravitational acceleration	9.81 Meters per second^2	l / t^2
g	Gram	Kilogram $\times$ 1000^{-1}	M
H	Angular momentum	Kilogram $\times$ Meters squared/second	$m \times l^2 / t$
I	Moment of inertia	Kilogram $\times$ Meters squared	$m \times l^2$
k	Radius of gyration	Meters squared	l^2
l	Length	Meters	l
m	Mass	Kilograms	m
M	Moment of force	Newton $\times$ Meters	$m \times l^2 / t^2$
N	Normal force	Newtons	N
p	Pressure	Pascals	$m / l \times t^2$
r	Length	Meters	l
rad.	Angle	Radians	Dimensionless
s	Displacement	Meters	l

Symbol	Meaning	Units	Dimensions
s	Time	Seconds	T
t	Time	Seconds	t
v	Velocity	Meters per second	l/t
W	Work	Newton × Meters	$m \times l^2/t^2$
W	Weight	Newtons	$m \times l/t^2$
Y	Young's Modulus	Stress /strain	$m \times l \times t^2$
GREEK			
α	Angular acceleration	Radians per second^2	T^{-2}
β	Angular displacement	Radians	Dimensionless
ε	Strain	Change in l/original l	Dimensionless
θ	Angular displacement	Radians	Dimensionless
μ	Coefficient of friction	F/N	Dimensionless
π	Circumference / diameter	3.14159	Dimensionless
σ	Stress	Newton × Meter^{-2}	$m/l \times t^2$
ω	Angular velocity	Radians per second	T^{-1}
MATHEMATICAL			
Δ	Change of	Any variable	NA
e	Natural number	2.71828	NA
Σ	Summation	Any variable	NA
∫	Integral sign	Any variable	NA

Mechanical Formulae

The most fundamental forms of equations of motion are summarized here and presented with their equation numbers found in the main body of the book. Equations are categorized under major mechanical and mathematical headings. See appendix A for the meaning of mechanical symbols.

Translational Kinematics

For conditions of constant acceleration:

$$v_f = v_i + at \qquad \text{(Equation 2.1)}$$

$$s_f = s_i + v_i t + at^2 / 2 \qquad \text{(Equation 2.2)}$$

$$v_f^2 = v_i^2 + 2as \qquad \text{(Equation 2.3)}$$

Horizontal range R of an airborne projectile with initial velocity v, angle of projection to the horizontal θ, and height above the ground h.

$$R = [v^2 \sin\theta\cos\theta + v\cos\theta\sqrt{(v^2\sin^2\theta + 2gh)}] / g \qquad \text{(Equation 9.1)}$$

Translational Kinetics

Force

The most fundamental equation in mechanics, which states that force accelerates mass and that the ratio of force to mass determines the magnitude of acceleration.

$$F = ma$$

When force is due to gravitational attraction

$$F = mg \qquad \text{(Equation 3.1)}$$

Impulse–Momentum

Product of constant force and time (impulse Ft) results in a change in momentum (mv) of a mass.

$$Ft = mv_f - mv_i \qquad \text{(Equation 2.12)}$$

The integral of variable force with respect to time determines the final momentum of a mass.

$$\int F dt = mv$$

Specific bounds of integration determine the change in momentum from initial (t_i) to final (t_f) times.

$$\int_{t_i}^{t_f} F dt + mv_i = mv_f \qquad \text{(Equation 2.13)}$$

Work–Energy

The product of a constant force and displacement of the point of application of the force in the direction of the force (work Fs) results in a change in kinetic energy $(mv^2/2)$ of a mass.

$$Fs = mv^2 / 2 \qquad \text{(Equation 2.21)}$$

The integral of variable force with respect to displacement determines the change in kinetic energy of a mass.

$$\int F ds = mv_f^2 / 2 - mv_i^2 / 2 \qquad \text{(Equation 2.20)}$$

In a vertical direction with gravity the only force, a loss in kinetic energy $((mv^2)/2)$ leads to an equal gain in potential energy (mgh) and vice versa.

$$mv^2 = 2mgh \text{ or } (mv^2) / 2 = mgh \qquad \text{(Equation 2.24)}$$

In a vertical direction, work done $(\int F ds)$ leads to an equal gain in potential energy (mgh).

$$\int F ds = mgh \qquad \text{(Equation 8.10)}$$

The sum of the work done by forces 1 to n is equivalent to the work done by a single force acting at the CM.

$$\int_{i=1}^{i=n} F_{CM} ds_{CM} = \Sigma \int_{i=1}^{i=n} F_i ds_i \qquad \text{(Equation 7.3)}$$

Power

Average power is the work done divided by the time of the work. Instantaneous power is the rate of change of work (or energy) with respect to time and is the vector product of force and velocity of the point of application of the force.

$$W / t = \text{Power} = Fs / t = Fv \qquad \text{(Equation 2.26a)}$$

Another expression for the rate of change of energy with respect to time.

$$dE / dt = \text{Power} \qquad \text{(Equation 6.5)}$$

Angular Kinematics

For conditions of constant angular acceleration:

$\omega_f = \omega_i + \alpha t$

$\theta_f = \theta_i + \omega t + \alpha t^2 / 2$

$\omega_f^2 = \omega_i^2 + 2\alpha s$

Angular Kinetics

Moment of Force

Product of force and perpendicular distance creates a moment M which produces angular acceleration α of a moment of inertia I.

$Fd = M = I\alpha$ (Equations 2.15 and 2.16)

Angular Impulse–Angular Momentum

Integral of moment with respect to time (angular impulse, $\int Mdt$) results in a change in angular momentum ($I\omega$) of a moment of inertia.

$\int Mdt + I\omega_i = I\omega_f$ (Equation 2.17)

Work–Energy

The product of constant moment of force and angular displacement changes the rotational kinetic energy ($I\omega^2/2$).

$M\theta = I\omega^2 / 2$ (Equation 2.22)

The integral of variable moment of force with respect to angular displacement changes the rotational kinetic energy.

$\int Md\theta + I\omega_i^2 / 2 = I\omega_f^2 / 2$ (Equation 2.23)

Total muscular work is the sum of work done by muscles 1 to n.

$$\text{Muscle work} = \sum \left(\int_{i=1}^{i-n} (M_i d\theta_i) \right)$$ (Equation 7.4)

Power

Instantaneous power in rotation is the product of moment of force and angular velocity.

$\text{Rotational power} = M\omega$ (Equation 2.26b)

Circular Motion

The instantaneous tangential velocity at the circumference of a circle of radius r is the product of the radius and the angular velocity.

$$v = \omega \times r \qquad \text{(Equation 2.8)}$$

The instantaneous radial acceleration in circular motion is the product of the radius and the square of angular velocity.

$$a = v^2 / r = \omega^2 r \qquad \text{(Equation 2.9)}$$

The instantaneous centripetal force in circular motion is the product of the radius, the square of angular velocity, and the circling mass.

$$ma = mv^2 / r = m\omega^2 r \qquad \text{(Equation 2.10)}$$

Properties of Tissues

Normal stress is equal to the ratio of force and perpendicular area.

$$\sigma = F / A$$

Normal strain is the ratio of change in length and initial length.

$$\varepsilon = \Delta l / l$$

Young's modulus of elasticity is equal to the ratio of stress and strain.

$$Y = \Delta\sigma / \Delta\varepsilon$$

Toughness is the integral of stress with respect to strain prior to rupture when a tissue is stressed.

$$\text{Toughness} = \text{Energy to fail} = \int F ds$$

Pressure is force per unit area.

$$p = F / A \qquad \text{(Equation 5.3)}$$

Some Techniques of Calculus

$$\text{If } y = x^n, dy / dx = nx^{(n-1)} \qquad \text{(Equation 2.4)}$$
$$\text{If } y = x^n, \int y dx = \int x^n dx = x^{(n+1)} / (n+1) + C \qquad \text{(Equation 2.5)}$$
$$d(r\sin\theta) / dt = r\cos\theta \, (d\theta/dt) = r\omega\cos\theta \qquad \text{(Equation 2.6a)}$$
$$d(r\cos\theta) / dt = -r\sin\theta \, (d\theta/dt) = -r\omega\sin\theta \qquad \text{(Equation 2.6b)}$$
$$d(xy) / dt = x dy / dt + y dx / dt \qquad \text{(Equation 2.7)}$$

Other Equations and Expressions

The moment of inertia of an extended body about any axis A is equal to the moment of inertia about a parallel axis through the CM (G) plus the product of mass and square of the distance *(r)* between the two axes.

$$I_A = I_G + mr^2 \qquad \text{(Equation 2.14)}$$

The coefficient of restitution of a rebounding falling body is the ratio of the velocity immediately following rebound v_R and the velocity immediately prior to impact v_D. This can be expressed as the square root of the ratio of rebound height to height of drop.

$$CR = v_R / v_D = (h_R/h_D)^{0.5} \qquad \text{(Equation 2.25)}$$

The coefficient of limiting friction is the ratio of the force *(F)* parallel to the surface of potential sliding and the perpendicular force *(N)*.

$$F / N > \mu \qquad \text{(Equation 5.1)}$$

The roots of a quadratic equation of the form:

$$ax^2 + bx + c = 0, \text{ are}$$

$$x = [-b \pm \sqrt{(b^2 - 4ac)}]/2a$$

Appendix C

Problems

This section is categorized by mechanical concept rather than by fundamental human movement as is done in the chapters of the text. The questions are generally in the order of increasing complexity. In many problems the data are approximations of real data, and in some cases such data have not been discovered experimentally. The major aims are to indicate data that would need to be collected to solve the problems and to provide solutions by which readers can test their grasp of biomechanical analysis. No diagrams are given for these questions, so readers should pay attention to the diagrams accompanying the worked examples in the text of each chapter.

Since no diagrams are provided, the first action required to solve the problems is to draw the mechanical scenarios from the descriptions given in the questions. Some geometrical and mathematical techniques are not covered in this book and must be found from other sources.

Kinematics (K)

K1. Calculate the distance traveled by a person who walks for 1 h at a speed of 1.5 m/s.

K2. Calculate the absolute angular displacement of the upper limb of a discus thrower who performs three full body rotations and rotates the upper limb in the same direction through 2.0 rad relative to the body.

K3. Calculate the average speed of running of a person who has a step length of 2.5 m with duration of 0.5 s.

K4. Calculate the average speed of walking of a person who has a step length of 0.9 m with a frequency of 1.25 steps per second.

K5. Calculate the translational displacement of a football that undergoes displacements of 30 m parallel to the sideline and 15 m parallel to the goal line.

K6. A person walks from point A to B along a path that is five blocks in a northerly direction and three blocks to the east. Proportionally how much shorter would the journey be if a tunnel connected points A and B in a straight line?

K7. The long axis of the upper limb (length = 50 cm) of a manual operator lies at an angle of 1.0 rad to the x-axis in an x-y plane that has its origin at the shoulder joint axis. Calculate the position of the distal end in the x and y directions.

K8. A lever is free to rotate about one end, and a manual operator rotates it through an angle of 1.5 rad by applying force at 20 cm from the axis. How far does the hand move along its path of motion?

K9. A person climbs a ladder that has a distance between each step of 25 cm according to the following table:

step	0	1	2	3	4	5	6	7	8	9	10
time	0	1.6	2.7	3.4	4.0	4.5	5.0	5.6	6.3	7.4	9.0

Use numerical differentiation to determine the greatest climbing speed, the step at which this occurs, and the step at which acceleration is zero.

K10. The relationship between displacement *(s)* and time *(t)* for the starting phase of a sprint may be described as follows: $s = 2t^2 - 0.11t^3$. Calculate the speed at $t = 6.0$ s, the acceleration at 2.0 s, and the time at which the acceleration is greatest.

K11. A baseball that has a velocity of 40.0 m/s is decelerated by a catcher applying an opposing deceleration *(a)* according to the following equation: $a = 6000(1 - 200t / 3)$. Calculate both the times when the velocity of the ball is zero and the displacements during those times.

K12. A cart at rest experiences acceleration of 0.2 m/s² from $t = 0$ s to $t = 0.5$ s and of 1.0 m/s² from $t = 0.5$ s to $t = 1.0$ s. Calculate both velocity and displacement at $t = 1.0$ s. Repeat the calculation with the order of accelerations reversed.

K13. Two high jumpers have different kinematics at takeoff. Jumpers A and B have their CM at 1.2 m and 1.26 m above the ground with vertical velocities of 3.5 m/s and 3.33 m/s, respectively. Which jumper raises his CM to the greatest height?

K14. A gymnast wishes to perform two full somersaults with her CM at the same height at takeoff and landing. If the average angular velocity of rotation is 10.0 rad/s, calculate the magnitude of the required vertical velocity vector at takeoff.

K15. A long jumper takes off with the CM at 0.2 m and 1.2 m in horizontal and vertical directions, respectively, in relation to the front edge of the takeoff board. Takeoff velocity is 8.0 m/s at an angle of 0.7 rad to the horizontal. On landing, the CM is 0.5 m above the ground and 0.8 m behind the heels. How far is the jump?

K16. The following table shows the output from a transducer of angular velocity (ω) during elbow flexion.

ω (rad/s)	0	0.49	0.92	1.27	1.49	1.57	1.49	1.27	0.92	0.49	0
t (s)	0	0.2	0.4	0.6	0.8	1.0	1.2	1.4	1.6	1.8	2.0

Use numerical differentiation and integration over periods of 0.4 s and 0.8 s to calculate angular acceleration (α) and angular displacement (θ). What are the percentage differences between these values at 1.6 s and the correct values of 1.81 rad and 1.99 rad/s²?

K17. A book lies at displacements of 0.05 m, 0.0 m, and –0.25 m in the sagittal, transverse, and longitudinal directions from the shoulder, respectively. The right hand is used to place the book on a shelf that is 0.3 m, 0.2 m, and 0.1 m in the same direction from the shoulder. Calculate the translational displacement of the book between initial and final positions.

K18. A control panel lies in a vertical plane facing an operator who must have access to all controls. Calculate the radius of the circle within which reachable controls should be located on the panel for the operator, whose upper limb length is 50 cm, if the shoulder joint axis is fixed at a distance of 30 cm from the panel.

K19. An object is projected by elbow extension with the upper arm fixed horizontally. Elbow joint angular velocity and time are related by $\omega = 4t^{1/2}$. The distance between the elbow axis and the object is 45 cm, and the elbow angle is 2.1 rad at $t = 0$ (full extension is zero radians). Calculate the magnitude and direction of the linear velocity of the object when the elbow angle reaches 1.0 rad.

K20. Two individuals are twirling identical objects manually at identical constant angular velocities of 4π rad/s in horizontal planes by means of a string attached to each object. Calculate the acceleration of each object for strings of length 0.5 m and 1.0 m.

K21. A gymnast rotates in a vertical plane about a horizontal bar gripped by the hands with a distance between the CM and the bar of 1.2 m. Calculate the angular velocity when the gymnast passes through the vertical position above the bar such that no gripping force is required.

K22. A hurdler is landing with the lead lower limb straight and vertical and a distance between the hip joint and the toe of 0.8 m. If the horizontal component of velocity of the hip joint is 8.0 m/s, calculate the angular velocity of the lower limb such that the toe is traveling at zero angular velocity relative to the ground.

K23. In the recovery phase of walking, the lower limb is undergoing motion in the vertical plane only. Instantaneously the thigh is aligned vertically below the hip joint; the included angle between the thigh and leg is 2.8 rad; and the thigh and leg have angular velocities of 0.3 rad/s and 0.5 rad/s, respectively, and lengths of 30 cm and 40 cm, respectively. If the respective horizontal and vertical components of velocity of the hip joint axis are 3 m/s and 0.1 m/s, calculate the instantaneous velocity of the distal point of the leg.

K24. A body segment of length $l = 50$ cm, rotating in a fixed x-y reference plane about an axis through one end, has its angle θ with respect to the x-axis related to time by $\theta = 3t^{1/2}$. Calculate the components of translational velocity and acceleration of the distal end in the x and y directions when $\theta = 1.0$ rad.

K25. An object is accelerated in a horizontal plane by elbow flexion starting from a straight upper limb aligned along an axis X, the upper arm stationary, and the forearm of length *l*. Derive expressions for the position *x*, velocity v_x, and acceleration a_x of the object in the X direction in relation to angular position θ, angular velocity ω, and angular acceleration α.

Forces and Moments (FM)

FM1. Calculate the force applied to the top of the cranium when a stationary person balances a bucket of water of mass 10 kg on the head.

FM2. Two workers push a cart of mass 40 kg by applying equal forces of 20 N at angles of 30° on either side of the intended direction of motion. What is each worker's force perpendicular to this direction, and what is the acceleration of the cart in the intended direction?

FM3. The strongest man in the world challenges five other men to a tug-of-war competition. If each man can apply a force to the rope of 3000 N, what force must the strongest man produce?

FM4. A rope supporting a mass of 50 kg vertically is wound around a horizontal cylinder of radius 5 cm. What perpendicular force must be applied by a manual operator at the end of a lever of length 0.5 m attached to the cylinder in order to hold the load stationary?

FM5. What perpendicular force must be applied by the manual operator in Problem FM4 to accelerate the load at 0.15 m/s²?

FM6. What perpendicular force must be applied by the manual operator in Problem FM5 if there is an opposing frictional force at a tangent to the surface of the cylinder of 2.5 N?

FM7. Calculate the acceleration of a boat of mass 50 kg when an oarsman of mass 100 kg applies a force of 500 N with each oar while the water produces a resisting force on the boat of 50 N.

FM8. Calculate the angular acceleration of a discus about an axis through its CM ($I_G = 0.2$ kg.m²) to which is applied a couple of 1.0 N.m.

FM9. A person supports a load $m_L = 10$ kg in the hand at a distance of $r_L = 45.0$ cm from the elbow joint axis with the upper limb lying in a vertical plane, the elbow joint at 90°, and the forearm horizontal. The mass of the forearm is $m_A = 2.5$ kg with its CM at a distance of $r_A = 22$ cm from the elbow axis. Two muscles, biceps brachii (BB) and brachioradialis (BR), maintain this posture. The insertions of BB and BR are located at horizontal distances of $r_{BB} = 5$ cm and $r_{BR} = 30$ cm, respectively, on the forearm. The angles of pull of each muscle are 80° (BB) and 5° (BR) to the long axis of the forearm. If BB produces a force of 500 N, what must be the force produced by BR?

FM10. What are the magnitudes and directions of the compressional forces produced at the elbow joint by the muscles in Problem FM9? What are the magnitude and direction of their resultant relative to the vertical direction?

FM11. Lower limb recovery in running is occurring in a vertical plane with respective instantaneous horizontal velocity and acceleration of the transverse knee joint axis of $v = 8.0$ m/s and $a = 0.0$ m/s² from left to right. The distance between the knee joint and the leg CM is $r = 15.0$ cm. The angle of the leg is $\theta = 4.7$ rad from the right horizontal through the knee joint axis, and the absolute angular velocity and angular acceleration of the leg are $\omega = 1.5$ rad/s and $\alpha = -0.5$ rad/s², respectively (positive is counterclockwise). If the mass of the leg is $m = 7$ kg and its moment of inertia is $I_G = 0.1$ kg.m², calculate the magnitude and direction of both the joint force *(F)* and moment *(M)* vectors. Is the muscular contraction producing the moment M concentric, isometric, or eccentric?

FM12. A person maintains a vertically oriented stationary stance while both arms are experiencing angular velocity of $\omega = 5.0$ rad/s and angular acceleration $\alpha = 4.0$ rad/s² in the sagittal plane when passing upward through a horizontal orientation. Calculate the force and moment vectors at the ankle joint given the following: shoulder to CM of the arms $r_A = 0.3$ m, shoulder to ankle distance $L_B = 1.5$ m, total mass of the arms $m_A = 8.0$ kg, mass of the remainder of the body $m_B = 70$ kg, and moment of inertia of the arms $I_{GA} = 0.17$ kg.m².

FM13. An airborne individual is vertically oriented, with both upper limbs instantaneously vertically oriented below the shoulder joints having an angular acceleration of $\alpha_A = 2.0$ rad/s² in the sagittal plane. Calculate the angular acceleration of the remainder of the body α_B given the following values: shoulder to CM of the arms $r_A = 0.3$ m, shoulder to CM of the remainder of the body $L_B = 0.5$ m, total mass of the arms $m_A = 8.0$ kg, mass of the remainder of the body $m_B = 70$ kg, moment of inertia of the body $I_{GB} = 15$ kg.m², moment of inertia of the arms $I_{GA} = 0.17$ kg.m².

FM14. A diver is leaning forward at an angle $v = 1.25$ to the horizontal axis with a distance $r = 1.1$ m between the CM and the center of pressure at the feet. If the force vector at the feet is of magnitude $F = 800$ N and inclined at an angle of $\theta_F = 1.4$ rad to the horizontal, calculate the instantaneous translational and rotational acceleration of the body of mass $m_B = 70$ kg and moment of inertia about the CM $I_G = 10$ kg.m².

FM15. Calculate the force vector (F_B) on a horizontal bar if a gymnast of mass $m = 80$ kg is performing a grand circle in a vertical plane with an angular velocity $\omega = 3.5$ rad/s when passing directly under the bar. The distance from the bar to the CM of the gymnast is 1.3 m.

FM16. Calculate F_B if the gymnast in Problem FM15 produces an instantaneous hip flexor moment of 500 N.m given the following: masses of upper body $m_U = 50$ kg and lower limbs $m_L = 30$ kg; moments of inertia of upper body $I_{GU} = 4.0$ kg.m² and lower limbs $I_{GL} = 3.0$ kg.m²; and distance from the hands to CM of upper body $r_U = 0.75$ m, from the hands to hip joint axis $L_U = 1.15$ m, and from hip joint to CM of lower limbs $r_L = 0.3$ m.

FM17. A frontal plane view of a person with the knee condition known as genu varus shows the hip joint vertically above the ankle joint with the medial femoral condyle located 5 cm lateral to the vertical line through the hip joint during the midstance phase of walking. If the only contact between the femur and the tibia is between their medial condyles, and the vertical force on the hip joint is found to be 700 N, calculate the moment required to maintain the angle of genu varus constant.

FM18. If the moment in Problem FM17 is maintained by tension in the iliotibial band plus other muscles whose vertical component of force at the knee joint occurs at a lateral distance of 6 cm from the medial condyle, calculate the contact force between medial condyles.

Impulse–Momentum (IM)

IM1. An operator can apply a force of 25 N to a trolley of mass 10 kg that is intended to run on rollers at a speed of 2.0 m/s. For how long should the force be applied?

IM2. A soccer ball of mass 0.25 kg is kicked along a horizontal path with a momentum of 20 kg.m/s. The goalkeeper then applies a vertical force of 100 N for a period of 0.01 s in a vertical direction. What are the resultant direction and magnitude of the velocity vector of the ball?

IM3. Calculate the final velocity of the body of $m_{BODY} = 80$ kg and the ball of mass $m_{BALL} = 2$ kg when a person standing on a frictionless surface projects the ball horizontally with a force of 100 N for a period of 0.15 s.

IM4. Calculate the change in velocity of the boat if the forces indicated in Problem FM7 are applied for 0.25 s.

IM5. A spherical golf ball ($m = 46$ g, $I_G = 8505.4$ g.mm², $r = 21.5$ mm) is struck with a force of 200 N for a brief period of 0.01 s at a tangent to its surface. Calculate its resulting translational and angular velocities.

IM6. A gymnast of mass $m = 60$ kg lands, having fallen for 2.0 m. Calculate the mechanical impulse required to arrest body motion completely.

IM7. An individual of mass m who is stationary in a crouched posture applies a vertical force to the ground according to the following relationship: $F = 2mg(1 - t)$. Calculate the position of the CM above its starting position and its velocity if the force is applied for 0.5 s.

IM8. The quadriceps muscles produce a constant extensor moment of 20 N.m when the knee joint is flexing in a horizontal plane at an angular velocity of $\omega = 10$ rad/s. If the thigh remains stationary, calculate both the time taken for ω to reach zero and the angular displacement θ during this time given the following: mass of leg and foot $m = 4.5$ kg, distance from the knee joint axis to the CM of leg and foot $r = 0.25$ m, and moment of inertia of the leg and foot about a transverse axis through the CM of $I_G = 0.15$ kg.m².

IM9. A person who is standing erect in the anatomical position maintains a rigid configuration of the body while applying a flexor moment about each shoulder joint axis. Calculate the angular velocity required of the arms immediately prior to arrested of their motion when horizontal, which will result in raising the CM of the whole body 1 cm off the ground. Shoulder to CM of the upper limb $r_U = 0.3$ m; mass of each upper limb $m_U = 4.0$ kg;

mass of the remainder of the body $m_R = 60$ kg; moment of inertia of the upper limb $I_{GU} = 0.2$ kg.m^2.

IM10. Calculate the net angular impulse at each shoulder required to achieve the result obtained in Problem IM9.

IM11. An ice skater is rotating about her longitudinal axis with the upper limbs held vertically below the shoulders and with an initial angular velocity ω = 9.0 rad/s, which she reduces by extending her arms laterally. Calculate the final angular velocity when the upper limbs reach the horizontal given the following: moment of inertia of the body minus arms about the longitudinal axis $I_{GB} = 3.0$ kg.m^2, moment of inertia of each upper limb about a transverse axis $I_{GU} = 0.13$ kg.m^2, distance from longitudinal axis to shoulder joint $r_{SH} = 0.22$ m, distance from shoulder joint to upper limb CM $r_U = 0.3$ m, mass of each upper limb $m_U = 3.0$ kg. Assume that the moment of inertia of the upper limb about its longitudinal axis is insignificant.

IM12. A gymnast is rotating in a straight body configuration ($I_G = 12$ kg.m^2) in a vertical sagittal plane about a horizontal bar gripped by the hands. The grip is released when the longitudinal axis of the body is 1.0 rad in front of a vertical line through the bar. The aim is to perform one full somersault plus the angular displacement required to land with the feet vertically below the CM of the body. Calculate the angular velocity required at release if the position of the CM undergoes a net downward displacement of 1.5 m from release to landing and the distance from the hands to the body CM is $r = 1.3$ m.

IM13. If the gymnast described in Problem IM12 spends 0.25 s in a tucked posture ($I_G = 5$ kg.m^2) while airborne and has the same initial angular velocity at release, what will be the total angular displacement?

IM14. An airborne individual is instantaneously stationary and inclined facing forward at an angle of 0.5 rad to the vertical. The aim is to land with the CM vertically above the feet, having undergone a vertical displacement of 2.0 m. To perform this motion the arms are rotated about a transverse axis through the shoulders in a vertical sagittal plane such that their combined CM is located on the shoulder joint axis. Calculate the required constant angular velocity of the arms (ω_A) given the following: moment of inertia of arms about the shoulder joint axis $I_A = 1.0$ kg.m^2, moment of inertia of remainder of the body about the whole-body CM $I_R = 11.0$ kg.m^2, total mass of arms $m_A = 5$ kg, distance from shoulder joint axis to the whole-body CM $r_B = 0.5$ m, and distance from shoulder joint axis to arm CM $r_A = 0.18$ m.

IM15. A diver in a straight body configuration has an angular velocity and moment of inertia about the transverse axis of $\omega_T = 0.5$ rad/s and $I_{GT} = 12$ kg.m^2, respectively. The upper limbs are rotated very rapidly about axes through the shoulder joints parallel to the sagittal axis. This action produces angular displacement of the longitudinal axis of the body about the sagittal axis of $\theta_S = 0.2$ rad. Calculate the angular velocity of spin of the body about the longitudinal axis ω_L if the moment of inertia about this axis is $I_{GL} = 2.0$ kg.m^2. Calculate the final value of ω_T.

Work–Energy (WE)

WE1. Calculate the work done by a person who lifts a load of mass $m_L = 10$ kg through a vertical height of 1.5 m while the CM of the body $m_{CM} = 80$ kg is lifted through 1.0 m. What is the total work done if the person lifts a load of 5 kg on two occasions sequentially?

WE2. Calculate the work done by a person of mass $m = 75$ kg who reduces his velocity to zero after having fallen through a vertical displacement of 2.0 m prior to ground contact.

WE3. A person who is landing on one lower limb dissipates energy with a maximal force of 2000 N perpendicular to the surface of the semilunar cartilages of the knee joint, which have linear elastic behavior with a modulus of elasticity of $Y = 15 \times 10^5$ N/m. Calculate the maximal deformation of the cartilage.

WE4. Elbow flexion is performed from a straight upper limb posture with a load of 10 kg held in the hand and with the humerus fixed vertically below the shoulder joint. The elbow flexor moment is $M = 30(1 + \sin\theta)$ N.m where θ is the angular displacement from the starting position. Calculate the total work done by the elbow flexor moment as the load passes through a vertical displacement of 60.0 cm given the following: moment of inertia of forearm-hand segment about its CM $I_A = 0.1$ kg.m^2, moment of inertia of the load about its CM $I_L = 0.005$ kg.m^2, mass of the forearm-hand $m_A = 3.0$ kg, distance from elbow joint axis to the forearm-hand CM $r_A = 20.0$ cm and to the load $r_L = 45.0$ cm. How much of this work appears as KE and how much as PE? What is the angular velocity of elbow flexion at this instant?

WE5. A hospital worker pushes a cart of mass $m = 40$ kg up a slope of 10° by applying a force F that is related to the distance s as follows: $F = 600e^{-s}$. The coefficient of friction of the wheels is a constant value of $\mu = 0.2$. Calculate the velocity v when a displacement of $s = 4$ m is attained and determine how much energy is lost to friction. The cart is stationary at $s = 0$.

WE6. A person who is airborne rotates each arm through one revolution in a sagittal plane about a transverse axis through the shoulders with a total constant moment of $M = 5$ N.m in order to produce an angular displacement of the whole body. Calculate the resulting angular velocity ω_B and angular displacement θ_B of the body given the following: total moment of inertia of arms about the shoulder joint axis $I_A = 1.0$ kg.m^2, moment of inertia of remainder of the body about the whole-body CM $I_R = 11.0$ kg.m^2, total mass of arms $m_A = 5$ kg, and distance from shoulder joint axis to whole-body CM $r_B = 0.5$ m.

WE7. The relationship between maximal knee extensor moment M and included knee joint angle θ is $M = 250 - 70\sin\theta$ ($\theta = \pi / 2$ represents full extension). The body mass $m_B = 80$ kg, and thigh and leg lengths are, respectively, $L_T = 46$ cm and $L_L = 40$ cm. Calculate, from an energetic perspective, the maximal load m_L that can be lifted in a "squat lift" to full knee extension from an initial knee joint angle of 0.5 rad. Can this aim be achieved? Assume negligible velocity of lifting, and confine the combined CM of body and load to a vertical line through the hip and ankle joints.

WE8. The lift described in Problem WE7 ends with complete lower limb extension. Calculate the metabolic cost of knee extension if the concentric muscular efficiency is 0.3. What would be the cost in lowering the load if eccentric muscular efficiency was 1.6 times that of concentric?

WE9. A cyclist exerting maximal effort requires an average moment at the axle of $M = 400$ N.m to maintain a constant speed against friction of the tires and air resistance. The relationship between the moment and angular velocity of the crank is a modified version of the translational force–velocity relationship for muscle as follows: $(M + a)(\omega + b) = (M_O + a)b$, where the maximal moment under isometric conditions $M_O = 1700$ N.m, $a = 430$, and $b = 7.5$. Calculate the angular velocity of pedaling and the power output.

WE10. In Problem WE9 the cyclist can produce the same road speed by changing gears. If the angular velocity of pedaling in the prior problem is either halved or doubled, calculate the average moment in each case and determine whether both changes are possible.

Appendix D

Answers

These answers include only the solutions and the important elements that are necessary to obtain the solutions, and not the full development as shown in the examples within each chapter.

Kinematics (K)

K1. Distance = 3600 s × 1.5 m/s = 5400 m or 5.4 km.

K2. Angular displacement = 3 rev + 2 rad = $(3 \times 2\pi + 2)$ rad = 20.85 rad.

K3. Average speed = 2.5 m / 0.5 s = 5.0 m/s.

K4. Average speed = 0.9 m × 1.25 steps/s = 1.125 m/s.

K5. Distance = $(30^2 + 15^2)^{1/2}$ = 33.5 m.

K6. Proportionally shorter by $(8 - 5.83) / 8 = 0.27$.

K7. X = 0.5cos(1.0 rad) = 0.27 m. Y = 0.5sin(1.0 rad) = 0.421 m.

K8. Distance s = 1.5 rad × 0.2 m = 0.3 m.

K9. Greatest speed = 0.5 m/s at step 5 and acceleration = zero.

K10. $v = 4t - 0.33t^2$; $a = 4 - 0.66t$. At 6 s, v = 12.12 m/s and a = 0.03 m/s². a max at t = 0 s.

K11. From Equation 2.5,

$$v_t = -6000\int_0^t (1 - 200t / 3)dt + 40 = (6000 \times 200)t^2 / 6 - 6000t + 40.$$

From chapter 7 and Appendix B, the quadratic solution when v_t is zero is:

$t = \{6000 + [6000^2 - 4 \times 40(6000 \times 200) / 6]^{1/2}\} / [2(6000 \times 200) / 6].$

t = 0.1 s and 0.2 s.

Similarly,

$$s = \int_0^t vdt, \text{ where } s_0 = 0; st = 40t - 6000t^2 / 2 + 6000 \times 200t^3 / 18.$$

s at 0.01 s = 16.6 cm.

s at 0.02 s = 13.3 cm.

K12. From Equations 2.1 through 2.3 on uniformly accelerated motion:

v at 1.0 s = $(0.2 \times 0.5) + (1.0 \times 0.5) = 0.6$ m/s.

s at 1.0 s = $(0.5 \times 0.2 \times 0.5^2) + [(0.2 \times 0.5)0.5 + (0.5 \times 1.0)0.5^2] = 0.2$ m.

Order of accelerations reversed:

v at 1.0 s = $(1.0 \times 0.5) + (0.2 \times 0.5) = 0.6$ m/s.

s at 1.0 s = $(0.5 \times 1.0 \times 0.5^2) + [(1.0 \times 0.5)0.5 + (0.5 \times 0.2)0.5^2] = 0.4$ m.

K13. From Equations 2.1 through 2.3 on uniformly accelerated motion:

Jumper A: $s = 1.2 + 3.5^2 / (2 \times 9.81) = 1.824$ m.

Jumper B: $s = 1.26 + 3.33^2 / (2 \times 9.81) = 1.825$ m.

K14. $\theta = 2 \times 2\pi$ rad at 10 rad/s gives $t = 1.257$ s airborne. From Equation 2.1 with positive upward:

$-v = v - gt; v = 9.81 \times 1.257 / 2 = 6.16$ m/s.

K15. From Equations 2.1 through 2.3 on uniformly accelerated motion, time airborne is obtained from:

$(1.2 - 0.5) = -8.0\sin(0.7)t + gt^2 / 2; t = 1.172$ s or -0.122 s.

The latter would be the time to move upward from 0.5 m to 1.2 m; therefore $t = 1.172$ s. Horizontal displacement while airborne is:

$8\cos(0.7) \times 1.172 = 7.17$ m. Therefore
total horizontal displacement = $7.17 + 0.2 + 0.8 = 8.17$ m.

K16. Over 0.4 s, at 1.6 s $\alpha = -1.95$ rad/s, 2.0% different; $\theta = 1.744$ rad, 3.6% different.

Over 0.8 s, at 1.6 s $\alpha = -1.86$ rad/s, 6.5% different; $\theta = 1.560$ rad, 13.8% different.

As Δt increases, error increases.

K17. Displacement = $[(0.3 - 0.05)^2 + (0.2 - 0)^2 + (0.1 + 0.25)^2]^{1/2} = 0.474$ m.

K18. $r = (50^2 - 30^2)^{1/2} = 40$ cm.

K19. $\theta = 2.1 - \int\omega \, dt = 2.1 - \int 4t^{1/2} \, dt = 2.1 - 4(2 / 3)t^{3/2}$.

$\theta = 1.0$ rad when $t^{3/2} = 3.3 / 8$ or $t = 0.554$ s.

ω at 0.554 s = $4 \times 0.554^{1/2} = 2.977$ rad/s.

$v = r \times \omega = 45 \times 2.977 = 133.965$ cm/s at a downward angle of $(\pi / 2 - 1)$ rad from the horizontal.

K20. Acceleration is radial and equal to $\alpha = r\omega^2$.

For string = 0.5 m, $\alpha = 0.5 \times (4\pi)^2 = 91.52$ m/s².

For string = 1.0 m, $\alpha = 1.0 \times (4\pi)^2 = 183.04$ m/s².

K21. No grip indicates no possible radial acceleration. Radial acceleration can be produced only by gravity. Therefore $\alpha = r\omega^2 = g$, $\omega = (9.81 / 1.2)^{1/2} = 2.86$ m/s².

K22. From $v = r \times \omega$; $\omega = 8.0 / 0.8 = 10$ rad/s.

K23. Let X and Y be positive horizontally and vertically upward, respectively. Summing hip velocity, knee velocity relative to hip, and distal point velocity relative to knee in X and Y directions:

$v_X = 3.0 + 0.3 \times 0.3 + 0.4 \times 0.5\cos(\pi - 2.8) = 3.278$ m/s.

$v_Y = 0.1 + 0 - 0.4 \times 0.5\sin(\pi - 2.8) = 0.033$ m/s.

Instantaneous velocity $v = (3.278^2 + 0.033^2)^{1/2} = 3.278$ m/s

at $\theta = \tan^{-1}(0.033 / 3.278) = 0.01$ rad or $0.58°$ above the horizontal.

K24. Tangential velocity $v = r\omega = rd\theta / dt = r3t^{-1/2} / 2$. When $\theta = 1$ rad, $3t^{1/2} = 1$, so $t = 0.1111$ s.

Therefore $v = 0.5 \times 3 \times 3.0 / 2 = 2.25$ m/s.

$v_X = 2.25\sin(1 \text{ rad}) = 1.893$ m/s.

$v_Y = 2.25\cos(1 \text{ rad}) = 1.216$ m/s.

K25. $x = l\cos\theta$.

$v_X = \mathrm{d}(l\cos\theta) / \mathrm{d}t$. Using Equations 2.6a and 2.6b:

$v_X = \mathrm{d}(l\cos\theta) / \mathrm{d}t = [\mathrm{d}(l\cos\theta) / \mathrm{d}t]\mathrm{d}\theta / \mathrm{d}\theta = \mathrm{d}(l\cos\theta) / \mathrm{d}\theta \times \mathrm{d}\theta / \mathrm{d}t = -l\omega\sin\theta$.

$a_X = \mathrm{d}[-l\omega\sin\theta] / \mathrm{d}t = -l\, \mathrm{d}[\omega\sin\theta] / \mathrm{d}t$.

Both ω and θ vary with respect to time, so the following identity must be used:

$\mathrm{d}(xy) / \mathrm{d}t = x\mathrm{d}y / \mathrm{d}t + y\mathrm{d}x / \mathrm{d}t$. (from Equation 2.7)

Substituting ω for x and $\sin\theta$ for y we obtain the following:

$a_X = -l\, (\omega\mathrm{d}\sin\theta / \mathrm{d}t + \sin\theta\mathrm{d}\omega / \mathrm{d}t) = -l\, (\omega^2\cos\theta + \alpha\sin\theta)$.

This expression is often seen as $-l\, (\dot{\theta}\cos\theta + \ddot{\theta}\sin\theta)$. A similar process can be used for Y and Z directions.

Forces and Moments (FM)

FM1. $F = mg = 10 \times 9.81 = 98.1$ N.

FM2. $F = ma$, so $a = F / m$. Therefore $a = (2 \times 20 \times \cos30) / 40 = 0.866$ m/s^2.

FM3. Since each man is in contact with the ground, their forces are in parallel and can be added.

$F = 5 \times 3000 = 15000$ N or 1529 kgf or 3364 lbf or 1.5 tonsf.

Good luck! If only one person touched the ground, forces of the remainder would be ineffectual.

FM4. Moment due to load $= 50 \times 9.81 \times 0.05 = 24.525$ N.m. This must be opposed by a moment of $F \times 0.5$ N.m. Therefore

$F = 24.525 / 0.5 = 49.05$ N.

FM5. Moment due to load = $[(50 \times (9.81 + 0.15)] \times 0.05 = 24.9$ N.m. Therefore

$F = 24.9 / 0.5 = 49.8$ N.

FM6. Moment due to load = $[(50 \times 9.81) + 2.5] \times 0.05 = 24.65$ N.m. Therefore

$F = 24.65 / 0.5 = 49.3$ N.

FM7. $F = ma$; therefore

$a = F / m$: $a = (2 \times 500 - 50) / (100 + 50) = 6.33$ m/s^2.

FM8. $M = I_G\alpha$; therefore

$\alpha = M / I_G$: $\alpha = 1.0 / 0.2 = 5$ rad/s^2.

FM9. Moment due to load plus forearm = $(m_L r_L + m_A r_A)g$. Opposing moment due to the muscles to maintain stationary conditions is $(r_{BB}F_{BB} \sin80 + r_{BR}F_{BR} \sin5)$. Therefore

$F_{BR} = [(m_L r_L + m_A r_A)g - (r_{BB}F_{BB} \sin80)] / r_{BR}\sin5$.

$F_{BR} = 953.1$ N.

Other muscles will contribute to reduce the forces in these two muscles.

FM10. Let R be the joint reaction force. Resolving forces horizontally and vertically:

$R_X = F_{BB}\cos80 + F_{BR}\cos5 = 1036.3$ N.

$R_Y = F_{BB}\sin80 + F_{BR}\sin5 - (m_L + m_L)g = 452.9$ N.

$R = 1130.9$ N at $\theta = \tan^{-1}(1036.3 / 452.9) = 66.4°$ to the vertical, upward.

Articular cartilage is strong.

FM11. Joint force F is obtained from ma of the CM of the leg. Using the expressions developed in Answer K25 with appropriate modifications for directions, component accelerations of the CM of the leg are:

$a_X = -r(\omega^2\cos\theta + \alpha\sin\theta) = -0.15[1.5^2\cos4.7 + (-0.5)\sin4.7] = -0.071$ m/s^2, and

$a_Y = -r(\omega^2\sin\theta - \alpha\cos\theta) = -0.15[1.5^2\sin4.7 - (-0.5)\cos4.7] = 0.338$ m/s^2.

Component forces at the shoulder joint are:

$F_X = 7 \times (-0.071) = -0.497$ N; $F_Y = 7 \times (0.338 + 9.81) = 71.036$ N.

$F = (-.497^2 + 71.036^2)^{1/2} = 71.038$ N at $\theta = \tan^{-1}(71.036 / 0.497) = 1.578$ rad from the right horizontal.

Calculating moments about the knee joint axis eliminates the knee joint force and

$M - mgr\cos\theta = (I_G + mr^2) \alpha$ (N.B. parallel axis theorem).

$M = \{[0.1 + 7 \times (0.15)^2] - 0.5 + 7 \times 9.81 \times 0.15\cos(4.7)\} = -0.2564$ N.m.

The moment is clockwise; therefore a flexor moment at the knee, which combined with a counterclockwise angular velocity indicating knee extension, represents eccentric muscular contraction of the knee flexors. Note that translational velocity of the knee joint does not appear in the calculations. A finite translational acceleration of the knee joint would modify the equations.

FM12. Ankle kinetics requires calculation of shoulder kinetics as internal inputs. Using the techniques of Answer FM11 with a right horizontal representing the X-axis, Y vertically upward, and positive angular motion counter-clockwise, kinematics of the CM of the arms are:

$$a_X = -r(\omega^2\cos\theta + \alpha\sin\theta) = -0.3(5^2\cos0 + 4\sin0) = -7.4 \text{ m/s}^2.$$

$$a_Y = -r(\omega^2\sin\theta - \alpha\cos\theta) = -0.3(5^2\sin0 - 4\cos0) = 1.2 \text{ m/s}^2.$$

Component forces and the moment at the shoulder joint are:

$$F_X = 8 \times (-7.4) = -59.2 \text{ N}; F_Y = 8 \times (1.2 + 9.81) = 88.08 \text{ N}.$$

$$M = (I_G + mr^2)\alpha + mgr\cos\theta = [(0.17 + 8 \times 0.3^2)4 + 8 \times 9.81 \times 0.3\cos0] = 27.104 \text{ N.m.}$$

Equal and opposite values of these forces and this moment represent external inputs at the shoulders. Therefore it is now possible to calculate the force F_A and moment M_A at the ankle, which maintain $v_{XB}, v_{YB}, a_{XB}, a_{YB}, \omega_B,$ and α_B of the remainder of the body equal to zero.

$$F_{XA} - F_X = m_B a_{XB}; F_{XA} = -59.2 \text{ N}.$$

$$F_{YA} - F_Y - m_B g = m_B a_{YB}; F_{YA} = 88.08 + 70 \times 9.81 = 774.78 \text{ N}.$$

Note that the extra 88.08 N represents an increase in vertical force above body weight, which provides an upward acceleration to the whole-body CM.

Only F_X and M at the shoulders create external moments if the reference axis is through the ankle, so:

$$M_A - M - F_X L_B = 0 \text{ since } \alpha_B \text{ is zero. } M_A = 27.105 + (59.2 \times 1.5) = 115.905 \text{ N.m.}$$

The moment is plantarflexor at the ankle, and the moment created by the force at the shoulders dominates the total moment because of its large moment arm.

FM13. If moments are taken about the transverse axis through the shoulder joints, the equal and opposite forces at the shoulder joints need not be calculated. The moment producing angular acceleration of the arms is:

$$M_A = (I_{GA} + m_A r_A^2)\alpha_A = (0.17 + 8 \times 0.3^2)2 = 1.78 \text{ N.m.}$$

The acceleration of the remainder of the body is:

$$-M_A = (I_{GB} + m_B L_B^2)\alpha_B; \alpha_B = -1.78 / (15 + 70 \times 0.5^2) = -0.548 \text{ rad/s}^2.$$

Note that angular velocity plays no part in this problem.

FM14. Total moment counterclockwise about a transverse axis through the body CM is:

$$M = F\cos\theta_F r\sin\theta - F\sin\theta_F r\cos\theta = Fr(\cos\theta_F\sin\theta - \sin\theta_F\cos\theta) = Fr\sin(\theta - \theta_F).$$

$$M = 800 \times 1.1 \times \sin(1.25 - 1.4) = -131.5 \text{ N.m and therefore clockwise.}$$

$$\alpha = -131.5 / I_G = -131.5 / 10.0 = -13.15 \text{ rad/s}^2 \text{ (therefore clockwise).}$$

$$a_X = F\cos\theta_F / m_B = 800 \times 0.17 / 70 = 1.94 \text{ m/s}^2.$$

$$a_Y = (F\sin\theta_F - m_B g) / m_B = (800 \times 0.985 - 70 \times 9.81) / 70 = 1.45 \text{ m/s}^2.$$

$$a = 2.42 \text{ m/s}^2 \text{ at an angle of } \tan^{-1}(1.45 / 1.94) \text{ to the horizontal} = 0.64 \text{ rad.}$$

FM15. F_B is the sum of the centripetal force (which in this case is vertically upward) required to produce circular motion plus that to support the body against gravity.

$$F_B = 80 \times 1.3 \times 3.5^2 + 80 \times 9.81 = 2058.8 \text{ N.}$$

FM16. Since the hip flexor moment is internal to the system, there will be no change in the force at the hands. Also there cannot be an instantaneous change in force in this system. See Problem IM15 for the effect of a prolonged moment.

FM17. The moment is simply the vertical force multiplied by the perpendicular distance from its line of action to the contact point between medial condyles.

$$M = 700 \times 0.05 = 35.0 \text{ N.m.}$$

FM18. Compression C is caused by force due to gravity plus combined iliotibial band and muscular force F. Summing moments gives:

$$M = F \times 0.06; F = 35 / 0.06 = 583.33 \text{ N}; C = 700 + 583.33 = 1283.33 \text{ N.}$$

Impulse–Momentum (IM)

IM1. $Ft = mv; t = mv / F = 10 \times 2 / 25 = 0.8 \text{ s.}$

IM2. Initial horizontal momentum = 20. Vertical momentum = $500 \times 0.02 = 10$.

Resultant momentum $L = (20^2 + 10^2)^{1/2} = 22.36 \text{ kg.m/s.}$

Velocity of ball $v = L / m = 22.36 / 0.25 = 89.44 \text{ m/s}$ at an angle of $\tan^{-1}(0.5)$ to the horizontal = 26.6°.

IM3. With no friction the total momentum is zero continuously; therefore

$$m_{BALL} \times v_{BALL} = -m_{BODY} \times v_{BODY} = 100 \times 0.15 = 15 \text{ N.s or kg.m/s.}$$

$$v_{BALL} = 15 / 2 = 7.5 \text{ m/s}; v_{BODY} = 15 / 80 = -0.1875 \text{ m/s.}$$

IM4. Impulse = $[(2 \times 500) - 50]0.25 = 237.5 \text{ N.s}; \Delta v = 237.5 / (100 + 50) = 1.583 \text{ m/s.}$

IM5. Translation $v = Ft / m = 200 \times 0.01 / 46 = 0.0435 \text{ N.s/g} = 43.5 \text{ m/s.}$

Rotational $\omega = Frt / I_G = (200 \times 21.5 \times 0.01) / 8505.4 = 5055.6 \text{ rad/s} = 804 \text{ rev/s.}$

N.B. Conversions between measurement units are required. The constants for a golf ball are correct, so such an unlikely angular velocity must be the result of mistaken assumptions concerning club–ball contact.

IM6. Velocity on landing $v = (2gs)^{1/2} = 6.264$ m/s. Impulse $= mv = 375.84$ N.s.

IM7. $F - mg = ma$; therefore

$$a = [2mg(1 - t) - mg] / m = g(1 - 2t).$$

$$v = g \int_0^{0.5} (1 - 2t)dt = g[t - 2t^2 / 2]_0^{0.5} = g[0.5 - 0.25] = 2.45 \text{ m/s}.$$

$$s = g \int_0^{0.5} ([(t - t^2)]dt = g[t^2 / 2 - t^3 / 3]_0^{0.5} = g[0.125 - 0.0417] = 0.82 \text{ m}.$$

IM8. Angular impulse = change in angular momentum. With a negative moment:

$$\int -Mdt = (\omega_F - \omega_I)(I_G + mr^2) \text{ or } \omega_F = [\int -Mdt / (I_G + mr^2)] + \omega_I.$$

$$\omega_F = 0 \text{ when } [\int_0^t Mdt / (I_G + mr^2)] = \omega_I = [Mt / (I_G + mr^2)]. \text{ Therefore:}$$

$$t = (I_G + mr^2) \omega_I / M = (0.15 + 4.5 \times 0.25^2)10 / 20 = 0.216 \text{ s}.$$

$$\theta = \int \omega dt = \int\{[\int -Mdt / (I_G + mr^2)] + \omega_I\}dt = \int\{[-Mt / (I_G + mr^2)] + \omega_I\}dt.$$

$\theta = \{[-Mt^2 / 2 (I_G + mr^2)] + \omega_I t\}$ and by substitution and from $t = 0$ to 0.216:

$$\theta = \{[-20 \times 0.216^2 / 2 (0.15 + 4.5 \times 0.25^2)] + 10 \times 0.216\} = 1.078 \text{ rad}.$$

IM9. Momentum of arms prior to stopping relative to the body must equal total body momentum immediately after stopping; therefore vertically:

$$m_U v_U = (m_U + m_R)v_{R+U}, \text{ but } v_U = r_U \omega_U, \text{ therefore } \omega_U = (m_U + m_R)v_{R+U} / m_U r_U.$$

To raise whole-body CM by 0.01 m, $v_{R+U} = (2g \times 0.01)^{1/2}$; therefore

$$\omega_U = 12.55 \text{ rad/s}.$$

IM10. Net angular impulse (AI) = net angular momentum (H_{SH}).

$$AI = H_{SH} = (I_G + mr^2) \omega_A = (0.2 + 4 \times 0.3^2)12.55 = 7.028 \text{ N.m.s}.$$

IM11. Angular momentum H remains constant $= (I_{GB} + 2m_U r_{SH}^2) \times \omega_I = 29.61$ kg.m²/s.

Upper limbs horizontal: $H = [I_{GB} + 2I_{GU} + 2m_U(r_{SH} + r_U)^2] \times \omega_F = 4.88 \times \omega_F$.

$$\omega_F = 29.61 / 4.88 = 6.07 \text{ rad/s}.$$

IM12. This problem is a little complicated mathematically since we require a relationship between initial angular velocity and angular displacement, which in turn needs time while the body is airborne.

Net angular displacement $= (4\pi - 1)$ rad, and angular velocity $\omega = (4\pi - 1)$ / t. However, both ω and t are unknown. We know that initial vertical velocity at $t = 0$ is $v_{VI} = r\omega\sin(1)$ and that displacement of the CM $s = -1.5$ m, so:

$v_{VF} = r\omega\sin(1) - gt$; $v_{VF}^2 = r^2\omega^2\sin^2(1) + 2gs$. Therefore:

$r^2\omega^2\sin^2(1) + g^2t^2 - 2r\omega\sin(1)gt = r^2\omega^2\sin^2(1) + 2gs$. Therefore:

$2r\omega\sin(1)gt = g^2t^2 - 2gs$. But $t = (4\pi - 1)$ / ω, so we obtain:

$2r\sin(1)(4\pi - 1) = [g(4\pi-1)^2 / \omega^2] - 2s$, which results in:

$\omega = \{g(4\pi - 1)^2 / [2r\sin(1)(4\pi - 1) + 2s]\}^{1/2}$.

$\omega = 7.664$ rad/s.

IM13. Local angular momentum H_G about an axis through the CM is both known and conserved while the body is airborne. Initial angular velocity is known, from which airborne time can be calculated.

$H_G = I_G \times \omega = 12 \times 7.664 = 91.97$ kg.m^2/s.

$t = \Delta\theta$ / $\omega = (4\pi - 1)$ / $7.664 = 1.509$ s.

0.25 s is spent in the tucked posture with $I_{GT} = 5$. Therefore from $H_G = I_G \times \omega$:

$\omega_T = 91.97$ / $5 = 18.39$ and $\omega_T t_T = \Delta\theta_T = 18.39 \times 0.25 = 4.599$ rad.

The remaining time in the extended position is $t_E = 1.509 - 0.25 = 1.259$ s. Therefore

$\omega_E t_E = \Delta\theta_E = 7.664 \times 1.259 = 9.649$ rad.

Total angular displacement $= 4.599 + 9.649 = 14.25$ rad $= 2.27$ somersaults.

IM14. Angular momentum of the arms H will produce equal and opposite angular momentum of the remainder of the body during the period of t s of the fall of s m. $s = gt^2$ / 2; therefore

$t = (2s$ / $g)^{1/2} = 0.639$ s.

Angular displacement of the remainder $\Delta\theta = \omega t$; therefore

$\omega = 0.5$ / $0.639 = 0.783$.

H of remainder $= (I_R + m_A r_B^2)$ $\omega = 9.592$ kg.m^2/s.

$H_A = (I_A + m_A r_A^2)$ ω_A; $\omega_A = 9.592$ / $(1.0 + 5 \times 0.18^2) = 8.255$ rad/s.

IM15. See figures 11.8 and 11.15. Initial angular momentum H_I is conserved and shared vectorially between transverse and longitudinal axes.

$H_I = I_{GT}$ $\omega_T = 12 \times 0.5 = 6$ kg.m^2/s.

$H_L = H_I \sin\theta_S = 6 \times 0.197 = 1.192 = I_{GL}$ ω_L; therefore $\omega_L = 0.596$ rad/s.

$H_T = H_I \cos\theta_S = 6 \times 0.980 = 5.880 = I_{GT}$ ω_T; therefore $\omega_T = 0.490$ rad/s.

Work–Energy (WE)

WE1. W = mgh = $(10 \times 1.5 + 80 \times 1.0)9.81$ = 2256.3 J. For successive lifts:

W = 2 $(5 \times 1.5 + 80 \times 1.0)9.81$ = 3041.1 J.

WE2. Work = $\Delta KE = mv^2 / 2$. $v^2 = 2gh$. W = mgh = 1471.5 J.

WE3. $Y = \Delta F / \Delta s$. Therefore $\Delta s = 2000 / (15 \times 10^5)$ = 1.3 mm.

WE4. When load is raised 60 cm, $\theta = 90 + \sin^{-1}[(0.6 - 0.45) / 0.45)]$ = 1.911 rad.

$$W = \int_0^{\theta = 1.911} M d\theta = 30 \int (1 + \sin\theta)d\theta = 30[\theta - \cos\theta 0 \;]_0^{\theta = 1.911} =$$

$$30[(1.911 - 0) - (-0.333 - 1)] = 97.34 \; N.m.$$

PE = $m_A g \Delta h_A + m_L g \Delta h_L$ where $\Delta h_A = 0.45 + r_A \sin(109.47 - 90) = 0.517$ m, and $\Delta h_L = 0.6$ m.

Therefore PE = 15.21 + 58.86 = 74.07 J. Since W = KE + PE:

KE = 97.34 – 74.07 = 23.27 J.

Angular velocity ω can be obtained from the general relationship $\Sigma \int M d\theta = I\omega^2 / 2$.

$97.34 - \int (m_A g r_A + m_L g r_L)\sin\theta d\theta = (I_A + I_L + m_A r_A^2 + m_L r_L^2) \omega^2 / 2$.

$30.615 = 1.0125 \omega^2$.

Therefore $\omega = 5.5$ rad/s.

WE5. Work done by the force is $\int F ds = \int (600e^{-s} - F_F)ds$. This integral is

$$\text{Total work} = [-600(e^{-s}) - F_F s \;]_0^{s=4} = -600(e^{-4} - 1) - 4F_F = 589.01 - 4F_F.$$

$F_F = \mu N = 0.2 \times (40 \times 9.81 \times \cos 10) = 77.29$. Energy lost to friction = 309.15 J.

Work to move load = 589.01 – 309.15 = 279.86 J.

Work to raise PE = mgh = $40 \times 9.81 \times 4\sin 10$ = 272.56 J.

Work to produce KE = 279.86 – 272.56 = 7.3 J. = $mv^2 / 2$.

$v = 0.6$ m/s.

In this case great energy savings can result from decreasing both slope and friction of wheels.

WE6. Work done by shoulder moment = $M \times 2\pi$ = 31.42 J.

From RKE = $I\omega^2 / 2$, $31.42 = (I_R + m_A r_B 2) \omega_B^2 / 2$. $\omega_B = 2.26$ rad/s.

The constant moment indicates uniformly accelerated motion. Therefore

$M = I_A \alpha_A$, so $\alpha_A = 5 / 1 = 5$ rad.s^{-2}. Since $\omega_A = \alpha_A t$, $\theta_A = \alpha_A t^2 / 2$ and $t = (2\theta_A / \alpha_A)^{1/2}$.

$t = 2 \times 3.142 / 5 = 1.26$ s.

The remainder of the body moves in the same time with an average angular velocity of 1.13 rad/s. Therefore $\theta_B = \omega_B t = 1.13 \times 1.26 = 1.42$ rad.

WE7. Work done W equals change of PE.

$$W = \int_{\theta = 0}^{\theta = \pi} (250 - 7\sin\theta)d\theta = [250\theta + 70\cos\theta]_0^{\pi} = 579.07 \text{ J.}$$

Initial hip joint height above the ground is obtained from the cosine rule:

$$h_I = (L_T^2 + L_L^2 - 2L_T L_L \cos\theta)^{1/2} = 0.22 \text{ m.}$$

Change of PE during the lift is:

$$(m_B + m_L)(h_F - h_I)g = (80 + m_L)[(0.4 + 0.46) - 0.22]9.81 = (502.72 + 6.28m_L) \text{ J.}$$
$$m_L = (579.07 - 502.72) / 6.28 = 12.16 \text{ kg.}$$

Can this be achieved? The answer is no. A limiting factor is whether the moment is sufficient to sustain the total load over the full range of joint motion. Readers can verify this graphically by drawing a diagram of the initial posture and calculating the moment due to gravity about the knee joint, then comparing this with the moment that can be produced from the equation. While there is sufficient work available from the muscles, the manner in which the activity can be performed will often determine success. Such a situation exists in many activities.

WE8. Since efficiency is work out divided by work in, or in other words work produced divided by total concentric metabolic cost of that work = 0.3, the latter is 579.07 / 0.3 = 1930 J. The moment needs to be applied through the same angle when the load is being lowered, but with an efficiency of 1.6 × .3, the total eccentric metabolic cost would be 1206 J.

WE9. By simple substitution:

$$(400 + 430)(\omega + 7.5) = (1700 + 430)7.5; \omega = 11.7 \text{ rad.s}^{-1}.$$

Power is $M\omega = 4699$ W.

WE10. At one-half of the prior angular velocity = 11.7 / 2 = 5.85, the maximal moment available is:

$$M = \{[(M_O + a)b] / (\omega + b)\} - a, \text{ which is } M = 767 \text{ N.m.}$$

This change is possible since it is greater than the required moment of 400 N.m.

At double the prior angular velocity = 11.7 × 2 = 23.4, the maximal moment available is:

$$M = \{[(M_O + a)b] / (\omega + b)\} - a, \text{ which is } M = 86.99 \text{ N.m.}$$

This change is not possible since it is less than the required moment of 400 N.m.

In the first case, the cyclist will need to use submaximal effort. In the second case, maximal effort would be insufficient to provide the moment required.

Glossary

abduction—Rotation of a body segment away from the center line of the body in the frontal plane about a sagittal axis.

abscissa—Horizontal axis of a two-dimensional graph in which the upward axis is the ordinate. Commonly called the X- and Y-axes, respectively.

absolute measurement—Kinematics and kinetics of one body segment with respect to some point or datum fixed in space.

acceleration—Rate of change of velocity with respect to time.

activation—Electrochemical process by which muscle is stimulated to contract.

adduction—Rotation of a body segment toward the center line of the body in the frontal plane about a sagittal axis.

angular acceleration—Rate of change of angular velocity with respect to time.

angular displacement—Change in angular position of the body.

angular impulse—Product of moment and time.

angular momentum—Quantity of rotational motion of a body, represented by the product of mass moment of inertia and angular velocity.

angular motion—Motion in a circular path around (or about) an axis of rotation.

angular velocity—Rate of change of angular displacement with respect to time.

articular surface—Smooth cartilage covering the ends of bones, allowing motion between connected segments.

articulation—General name for a joint and the motion of one segment relative to its connected adjacent segment.

axis of rotation—A straight line about which all parts of a body have the same angular velocity.

biomechanics—Application of mechanical principles to the study of biological systems.

center of mass (CM)—Theoretical point where the mass of an extended body can be considered concentrated; essentially the same as center of gravity.

center of pressure—The point on a body where a single force should be placed in order to have the same effect as a series of distributed forces.

centrifugal force—A pseudo-force acting away from the axis of rotation of a rotating body. See centripetal force.

centripetal force—Force acting toward the axis of rotation of a rotating body that provides acceleration to maintain a change in the direction of the velocity vector of rotating parts.

cocontraction—Simultaneous contraction of muscles on either side of a joint, for example flexors and extensors of the wrist.

coefficient of limiting friction—Ratio of forces parallel and perpendicular to a surface immediately prior to the slipping of a body on a surface.

coefficient of restitution—Ratio of velocities after and before bodies collide.

concentric contraction—Muscle contraction involving shortening.

concentric force—Force acting along a line through the center of mass of a body.

constraint force—Force that keeps a specific part of the body from moving.

contraction—Force production in a muscle.

couple—Turning effect of two or more noncollinear forces that have zero translational effect. See torque.

curvilinear motion—Motion in which all particles of a body travel in parallel curved lines. See also rectilinear motion.

deceleration—Negative value of acceleration indicating that a mass is slowing down.

degrees—Measure of angular (rotational) displacement that is commonly used, there being 360° in one rotation of displacement; 57.3° equals one radian, which is the unit used in mechanical calculations.

degrees of freedom (DOF)—Number of kinematic variables required to specify the motion of a segment completely.

dimensional analysis—Combination of mass, length, and time that describes any mechanical variable.

direct dynamics—Derivation of motion of a body from known forces and moments. Also known as forward dynamics.

displacement—Change of translational position in space.

distal—Away from the center of mass of the body.

dynamics—Mechanics of bodies in motion. See statics.

eccentric contraction—Muscle contraction involving lengthening.

eccentric force—A force acting along a line not passing through (or eccentric to) the center of mass of a body.

elasticity—Resistance to deformation of a body, which is related to the magnitude of the deformation. See viscosity.

energy—Capacity to do work. See kinetic energy and potential energy.

error—Amount by which a given variable departs from its true or intended value.

exponent—Mathematically the power to which a number is raised; an exponent of 3 means X times X times X = X^3.

force—That influence which accelerates a mass.

free body diagram—Diagram of a body free from any connections but including any external forces and moments.

free moment—Moment acting on a body without reference to an axis of rotation.

friction—Mutual force between a body and a surface opposing their relative motion parallel to the surface. Friction can be static or dynamic.

gravity—Mutual force of attraction between the earth and a body in newtons.

ground reaction force—Single force representing the combined effect of multiple forces between the ground and body parts touching the ground.

impulse—Integral of force with respect to time, which changes momentum of a body.

inertia—Resistance to acceleration by virtue of the mass of a body.

inertial force—Magnitude of the force opposing acceleration of a body.

internal force—Equal and opposite forces within a segmented body.

inverse dynamics—Derivation of forces and moments from knowledge of kinematics and inertia of body segments.

isometric contraction—Muscle contraction involving neither shortening nor lengthening.

isotonic contraction—Muscle contraction in which force is constant, a rare occurrence in normal human motion.

isovelocity contraction—Muscle contraction in which velocity is constant, a rare occurrence in normal human motion.

jerk—Rate of change of force with respect to time.

joint—Connection or articulation between adjacent segments.

joint capsule—Connective tissue encapsulating a joint and retaining synovial fluid as a lubricant for the articular surfaces.

joule—Unit of work and energy in kilogram.meters squared per second squared.

kilopond—Force equal to the attraction of the earth on a 1 kg mass; equal to 9.81 N.

kinematics—Study and description of motion without recourse to its causes.

kinetic energy—Capacity of a body to do work by virtue of its linear or angular motion in joules.

kinetics—Study of cause of motion such as force and work.

law of acceleration—Newton's second law relating force to acceleration of mass. $\sum F = ma$.

law of cosines—Resultant vector of two vectors acting in different directions.

law of inertia—Newton's first law, stating that a body continues in a state of rest or uniform motion in a straight line unless acted upon by an external force.

law of reaction—Newton's third law, stating that for every action there is an equal and opposite reaction.

length—Linear distance from one point to another on a body.

ligament—Strong elastic connective tissue, usually linking adjacent bones.

linear motion—Motion in a straight line. See also translation.

linear relationship—Straight-line relationship between two variables such that equal increments of change in one variable yield equal incremental changes in the other. It appears as a straight line on a graph relating the two variables.

local angular momentum—Product of angular velocity and moment of inertia with reference to an axis through the center of mass of a body.

mass—Resistance to acceleration by virtue of the amount of matter in the body in kilograms; otherwise known as inertia.

mechanics—Science of bodies in motion.

model—Physical or mathematical representation of a biomechanical system exhibiting the main properties of the latter.

moment—Abbreviated term for moment of force.

moment arm—Perpendicular distance between the line of action of a force and an axis of rotation.

moment of force—Product of a force and the perpendicular distance between the line of action of the force and an axis of rotation.

moment of inertia—Resistance to angular acceleration by virtue of the mass and distribution of a mass away from the axis of rotation in kilogram.meters squared.

momentum—Quantity of motion of a moving body represented by the product of mass and velocity of a body in kilogram.meters per second.

muscle—Organ producing force that is applied to bones. More properly known as "skeletal muscle."

muscular control—Determination of the motion of body parts by the timing of onset and magnitude of muscular force. Popularly known as "timing" in sport skills.

newton—Système International unit of force in kilogram.meters per second squared.

normal—Perpendicular to a surface or a straight line, or perpendicular to a tangent to a curved surface or curved line. Usually applied to the direction of a normal force.

ordinate—Upward axis of a two-dimensional graph in which the left-to-right axis is the abscissa. Commonly called the Y- and X-axes, respectively.

pendulum—Extended body, free to swing about a fixed axis under the influence of gravity only.

plane—Flat two-dimensional surface of zero thickness. "Flat" means having no irregularities; a plane can be in any orientation, including horizontal.

potential energy—Capacity to do work by virtue of position above some zero level.

pressure—Force applied to an area divided by the area; measured in pascals or newtons/meters squared.

principal axes—Mutually perpendicular longitudinal, transverse, and sagittal axes of a three-dimensional body that intersect at the center of mass. For many bodies, the principal moments of inertia about the various axes differ. A sphere differs from this rule.

pronation—Rotation of the forearm toward a palm-down position.

proximal—Toward the center of mass of the body.

radial—Direction along the radius of a circle.

radian—Angle subtended by the arc of a circle equal to the radius. It is unitless, equal to 57.3° approximately, and predominately used in mechanical calculations.

radius of gyration—Radius of a point mass from an axis of rotation that has the same moment of inertia as the original body.

rectilinear motion—Motion in which all particles of a body travel in parallel straight lines. See also translation.

reference frame—Three-dimensional mutually perpendicular set of axes relative to which motion is described; usually fixed to the earth.

relative measurement—Kinematics and kinetics of one body segment with respect to those of another.

remote angular momentum—Product of the instantaneous linear momentum of a body and the perpendicular distance to any arbitrary point. Also known as moment of momentum.

resultant—Vector sum of all vectors that have the same mechanical dimensions.

rigid body—Theoretical body in which all particles maintain a fixed relationship to each other, resulting in nondeformability and fixed inertial properties. All bodies are more or less rigid.

scalar—Quantity that has only magnitude (e.g., mass, time, and energy). See vector.

segment of body—Body parts such as forearm, thigh, and foot.

shoulder girdle—Clavicle (collarbone) and connected scapula (shoulder blade). Articulation is complex, involving motion between the sternum and clavicle and the clavicle and scapula.

shoulder joint—Articulation between the humerus (upper-arm bone) and scapula (shoulder blade).

single equivalent muscle (SEM)—A muscle that represents the combined action of a number of muscles contributing to a joint torque.

spatial—Relating to space. See temporal.

squared—In a mathematical context, multiplying a number by itself is squaring it. Multiplying by itself again is cubing it, and so on.

statics—Mechanics of stationary bodies. See dynamics.

strain—Deformation of a body divided by its initial length.

stress—Force applied to a body divided by its area perpendicular to the force.

supination—Rotation of the forearm toward a palm-up position.

temporal—Relating to time. See spatial.

tendon—Strong connective tissue with viscoelastic properties, usually connecting muscle to bone.

thermodynamics—Branch of mechanics that relates heat and energy.

torque—Turning effect of a force couple in newton.meters. The term torque strictly implies no net translational force.

translation—Motion of a body in which all particles travel in parallel straight lines.

vector—Mechanical variable that is quantified in terms of magnitude and direction (e.g., force, velocity, and momentum). See scalar.

velocity—Rate of change of position with respect to time.

viscoelasticity—Property of a body in which it resists deformation in proportion to both magnitude and rate of deformation.

viscosity—Resistance to deformation of a body that is related to the velocity of deformation. See elasticity.

waveform—The shape of the graph of the relationship between any continuously changing mechanical variable and time.

weight—Force of attraction of the earth on a body, equal to mass times gravitational acceleration, in newtons.

work—Product of force and displacement of the point of application of a force in the direction of the force, leading to a change in energy. Work can also be defined using the rotational equivalents of force and displacement.

Bibliography

Recommended readings are provided at the end of the chapter to which they apply. Following the bibliography, books are classified according to subject area rather than type of fundamental human movement. A list of some useful journals containing biomechanical information completes the bibliography. Readers should have little difficulty in adding to this list with the current electronic means of finding information.

Alexander, R. McN. (1968). *Animal mechanics.* London: Sidgwick and Jackson.

Alexander, R. McN. (1992). *The human machine.* New York: Columbia University Press.

Armstrong, C.G., and Mow, V.C. (1980). Friction, lubrication and wear of synovial joints. In *Scientific foundations of orthopaedics and traumatology,* ed. R. Owen, J. Goodfellow, and P. Bullough. London: William Heinemann Medical Books.

Bartel, D.L., Davy, D.T., and Keaveny, T.M. (2006). *Orthopaedic biomechanics: Mechanics and design in musculoskeletal systems.* Upper Saddle River, NJ: Pearson/Prentice Hall.

Batterman, C. (1974). *The techniques of springboard diving.* Cambridge, MA: MIT Press.

Beer, F., Johnston, E.R., Eisenberg, E., Clauser, W., Mazurek, D., and Cornwell, P. (2007). *Vector mechanics for engineers.* New York: McGraw-Hill Ryerson.

Caldwell, G.E., and Chapman, A.E. (1991). The general distribution problem: A physiological solution which includes antagonism. *Human Movement Science,* 10, 355-391.

Cavanagh, P.R. (1990). *Biomechanics of distance running.* Champaign, IL: Human Kinetics.

Chaffin, D.B., Andersson, G.B.J., and Martin, B.J. (2006). *Occupational biomechanics.* Hoboken, NJ: Wiley.

Chapman, A.E. (1980). The effect of a "wind-up" on forearm rotational velocity. *Canadian Journal of Applied Sport Sciences,* 5, 215-219.

Chapman, A.E. (1982). Hierarchy of changes induced by fatigue in sprinting. *Canadian Journal of Applied Sport Sciences,* 7, 116-122.

Chapman, A.E. (1985). The mechanical properties of human muscle. *Exercise and Sport Sciences Reviews,* 13, 443-501.

Chapman, A.E., and Caldwell, G.E. (1983a). Factors determining changes in lower limb energy during swing in treadmill running. *Journal of Biomechanics,* 16, 69-77.

Chapman, A.E., and Caldwell, G.E. (1983b). Kinetic limitations of maximal sprinting speed. *Journal of Biomechanics,* 16, 79-83.

Chapman, A.E., Caldwell, G.E., and Selbie, W.S. (1985). Mechanical output following muscle stretch in forearm supination against inertial loads. *Journal of Applied Physiology,* 59, 78-86.

Chapman, A.E., Leyland, A.J., Ross, S.M., and Ryall, M. (1991). Effect of floor conditions upon frictional characteristics of squash court shoes. *Journal of Sports Sciences,* 9, 33-41.

Chapman, A.E., Lonergan, R., and Caldwell, G.E. (1984). Kinetic sources of lower limb angular displacement in the recovery phase of sprinting. *Medicine and Science in Sports and Exercise,* 16, 382-388.

Chapman, A.E., and Medhurst, C.W. (1981). Cyclographic evidence of fatigue in sprinting. *Journal of Human Movement Studies,* 7, 255-272.

Currey, J.D. (1984). *The mechanical adaptations of bones.* Princeton, NJ: Princeton University Press.

Daniel, D.D., Akeson, W.H., and O'Connor, J.J. (1990). *Knee ligaments: Structure, function, injury and repair.* New York: Raven Press.

Dyson, G.H.G. (1973). *The mechanics of athletics.* 6th ed. London: University of London Press.

Edman, K.A.P., Elzinga, G., and Noble, M.I.M. (1978). Enhancement of mechanical performance by stretch during tetanic contraction of vertebrate skeletal muscle fibres. *Journal of Physiology,* 281, 139-155.

Edmonstone, M.A. (1993). Contribution of muscle to optimal performance in simulated overarm throwing. MSc thesis, Simon Fraser University, Burnaby, BC, Canada.

Fogiel, M. (1986). *The mechanics problem solver.* New York: Research and Education Association.

Freivalds, A. (2004). *Biomechanics of the upper limbs: Mechanics, modeling and musculoskeletal injuries.* Boca Raton, FL: CRC Press.

Gondin, W.R., and Sohmer, B. (1968). *Advanced algebra and calculus made simple.* London: Allen.

Gregor, R.J., and Abelew, T.A. (1994). Tendon force measurement and movement control: A review. *Medicine and Science in Sports and Exercise,* 26, 1350-1372.

Hay, J.G. (1978). *Biomechanics of sports techniques.* Englewood Cliffs, NJ: Prentice Hall.

Hay, J.G., and Reid, J.G. (1988). *Anatomy, mechanics and human motion.* Englewood Cliffs, NJ: Prentice Hall.

Herzog, W. (2000). *Skeletal muscle mechanics.* Chichester, New York: Wiley.

Hill, A.V. (1970). *First and last experiments in muscle mechanics.* Cambridge: Cambridge University Press.

Ivancevic, V.C., and Ivancevic, T.T. (2006). *Human-like biomechanics: A unified mathematical approach to human biomechanics and humanoid robotics.* Dordrecht (Great Britain): Springer.

Lanoue, F.R. (1940). Analysis of the basic factors involved in fancy diving. *Research Quarterly,* 11, 102-109.

Lonergan, R. (1988). Biomechanical rationale for preferred running style. MSc thesis, Simon Fraser University, Burnaby, BC, Canada.

MacIntosh, B.R., Gardiner, P.F., and McComas, A.J. (2006). *Skeletal muscle: Form and function.* Champaign, IL: Human Kinetics.

MacKenzie, C.L., and Iberall, T. (1994). *The grasping hand.* Amsterdam: North Holland.

McGill, S. (2002). *Low back disorders: Evidence-based prevention and rehabilitation.* Champaign, IL: Human Kinetics.

McGinnis, P.M. (2005). *Biomechanics of sport and exercise.* Champaign, IL: Human Kinetics.

McNeill-Alexander, R. (1992). *The human machine.* New York: Columbia University Press.

Nigg, B.M., and Herzog, W. (Eds.) (1994). *Biomechanics of the musculo-skeletal system.* New York: Wiley.

Nigg, B.M., MacIntosh, B.R., and Mester, J. (2000). *Biomechanics and biology of movement.* Champaign, IL: Human Kinetics.

Nordin, M., and Frankel, V.H. (2001). *Basic biomechanics of the musculoskeletal system.* Philadelphia: Lippincott, Williams & Wilkins.

Ozkaya, N., and Nordin, M. (1991). *Fundamentals of biomechanics: Equilibrium, motion, and deformation.* New York: Van Nostrand Reinhold.

Pandy, M.G., and Zajac, F.E. (1991). Optimal muscle coordination strategies for jumping. *Journal of Biomechanics,* 24, 1-10.

Pierrynowski, M.R. (1982). A physiological model for the solution of individual muscle forces. PhD dissertation, Simon Fraser University, Burnaby, BC, Canada.

Putnam, C.A. (1991). Interaction among segments during a kicking motion. In *Biomechanics VIII-B,* ed. H. Matsui and K. Kobayashi, pp. 688-694. Champaign, IL: Human Kinetics.

Reik, S., Chapman, A.E., and Milner, T. (1999). A simulation of muscle force and internal kinematics of extensor carpi radialis brevis during backhand tennis stroke: Implications for injury. *Clinical Biomechanics,* 14, 477-483.

Robertson, D.G.H., Caldwell, G.E., Hamill, J., Kamen, G., and Whittlesey, S.N. (2004). *Research methods in biomechanics*. Champaign, IL: Human Kinetics.

Rose, J., and Gamble, J.G. (2006). *Human walking*. Philadelphia: Lippincott, Williams & Wilkins.

Sagowski, S., and Piekarski, K. (1997). Biomechanics of the metacarpophalangeal joints. *Journal of Biomechanics, 10*, 205-209.

Salvendy, G. (2006). *Handbook of human factors and ergonomics*. Hoboken, NJ: Wiley.

Schmitt, K-U., Niederer, P.F., and Walz, F. (2004). *Trauma biomechanics: Introduction to accidental injury*. New York: Springer.

Thomson, D.B. (1983). The stretch response of human skeletal muscle in situ. MSc thesis, Simon Fraser University, Burnaby, BC, Canada.

Thomson, D.B., and Chapman, A.E. (1988). The mechanical response of active human muscle during and after stretch. *European Journal of Applied Physiology, 57*, 691-697.

Tichauer, E.R. (1978). *The biomechanical basis of ergonomics: Anatomy applied to the decision of work situations*. New York: Wiley.

Tubiana, R. (1981). *The hand*. Philadelphia: Saunders.

Valero-Cuevas, F.J., Zajak, F.E., and Burger, C.G. (1998). Subject independent patterns of muscular recruitment in large finger tip forces. *Journal of Biomechanics, 31*, 693-703.

Van Ingen Schenau, G.J., and Cavanagh, P.R. (1990). Power equations in endurance sports. *Journal of Biomechanics, 23*, 865-881.

Van Soest, A.J., Schwab, A.L., Bobbert, M.F., and van Ingen Schenau, G.J. (1993). The influence of the biarticularity of the gastrocnemius muscle on vertical-jumping achievement. *Journal of Biomechanics, 26*, 1-8.

Vogel, S. (2001). *Prime mover*. New York: Norton.

West, B.J., and Griffith, I. (2004). *Biodynamics: Why the wirewalker doesn't fall*. New York: Wiley-Liss.

White, A.A., and Panjabi, M.M. (1978). *Clinical biomechanics of the spine*. Philadelphia: Lippincott, Williams & Wilkins.

Whiting, W.C., and Zernicke, R.F. (1998). *Biomechanics of musculoskeletal injury*. Champaign, IL: Human Kinetics.

Yamaguchi, G.T. (2001). *Dynamic modeling of musculoskeletal motion*. Boston: Kluwer Academic.

Yeadon, M.R. (1984). The mechanics of twisting somersaults. Doctoral thesis, Loughborough University of Technology.

Books in Specific Areas

Anatomy

Hall, S.J. (1995). *Basic biomechanics*. St. Louis: Mosby-Year Book.

Hay, J.G., and Reid, J.G. (1988). *Anatomy, mechanics and human motion*. Englewood Cliffs, NJ: Prentice Hall.

Watkins, J. (1999). *Structure and function of the musculoskeletal system*. Champaign, IL: Human Kinetics.

Biomechanical Analysis

Allard, P. (1997). *Three-dimensional analysis of human locomotion*. Chichester, New York: Wiley.

Currey, J.D. (2002). *Bone: Structure and mechanics*. Princeton, NJ: Princeton University Press.

Latash, M.L., and Zatsiorsky, V.M. (2001). *Classics in movement science*. Champaign, IL: Human Kinetics.

Griffiths, I.W. (2006). *Principles of biomechanics and motion analysis*. Philadelphia: Lippincott, Williams & Wilkins.

McGinnis, P.M. (2005). *Biomechanics of sport and exercise.* Champaign, IL: Human Kinetics.

McNeill-Alexander, R. (1983). *Animal mechanics.* Oxford: Blackwell.

McNeill-Alexander, R. (1992). *The human machine.* New York: Columbia University Press.

Nigg, B.M., and Herzog, W. (Eds.) (1994). *Biomechanics of the musculo-skeletal system.* New York: Wiley.

Nigg, B.M., MacIntosh, B.R., and Mester, J. (2000). *Biomechanics and biology of movement.* Champaign, IL: Human Kinetics.

Ritter, A.B., Reisman, S.S., and Michniak, B.B. (2005). *Biomedical engineering principles.* Boca Raton, FL: Taylor and Francis.

Rose, J., and Gamble, J.G. (2006). *Human walking.* Philadelphia: Lippincott, Williams & Wilkins.

West, B.J., and Griffith, I. (2004). *Biodynamics: Why the wirewalker doesn't fall.* New York: Wiley-Liss.

Zatsiorsky, V.M. (1998). *Kinematics of human movement.* Champaign, IL: Human Kinetics.

Zatsiorsky, V.M. (2002). *Kinetics of human movement.* Champaign, IL: Human Kinetics.

Biomedical

Bartel, D.L., Davy, D.T., and Keaveny, T.M. (2006). *Orthopaedic biomechanics: Mechanics and design in musculoskeletal systems.* Upper Saddle River, NJ: Pearson/Prentice Hall.

Ritter, A.B., Reisman, S.S., and Michniak, B.B. (2005). *Biomedical engineering principles.* Boca Raton, FL: Taylor and Francis.

Injury, Safety, and Training

Bartel, D.L., Davy, D.T., and Keaveny, T.M. (2006). *Orthopaedic biomechanics: Mechanics and design in musculoskeletal systems.* Upper Saddle River, NJ: Pearson/Prentice Hall.

Freivalds, A. (2004). *Biomechanics of the upper limbs: Mechanics, modeling and musculoskeletal injuries.* Boca Raton, FL: CRC Press.

McGill, S. (2002). *Low back disorders: Evidence-based prevention and rehabilitation.* Champaign, IL: Human Kinetics.

Nordin, M., and Frankel, V.H. (2001). *Basic biomechanics of the musculoskeletal system.* Philadelphia: Lippincott, Williams & Wilkins.

Nyland, J. (2006). *Clinical decisions in therapeutic exercise: Planning and implementation,* ed. J. Nyland. Upper Saddle River, NJ: Pearson/Prentice Hall.

Radin, E.L. (1979). *Practical biomechanics for the orthopedic surgeon.* New York: Wiley.

Schmitt, K-U., Niederer, P.F., and Walz, F. (2004). *Trauma biomechanics: Introduction to accidental injury.* New York: Springer.

White, A.A., and Panjabi, M.M. (1978). *Clinical biomechanics of the spine.* Philadelphia: Lippincott, Williams & Wilkins.

Whiting, W.C., and Zernicke, R.F. (1998). *Biomechanics of musculoskeletal injury.* Champaign, IL: Human Kinetics.

Mathematics

Gondin, W.R., and Sohmer, B. (1968). *Advanced algebra and calculus made simple.* London: Allen.

Mechanics

Beer, F., Johnston, E.R., Eisenberg, E., Clauser, W., Mazurek, D., and Cornwell, P. (2007). *Vector mechanics for engineers.* New York: McGraw-Hill Ryerson.

Fogiel, M. (1986). *The mechanics problem solver.* New York: Research and Education Association.

Methods

Robertson, D.G.H., Caldwell, G.E., Hamill, J., Kamen, G., and Whittlesey, S.N. (2004). *Research methods in biomechanics.* Champaign, IL: Human Kinetics.

Muscle Mechanics

Chapman, A.E. (1985). The mechanical properties of human muscle. *Exercise and Sport Sciences Reviews,* 13, 443-501.

Herzog, W. (2000). *Skeletal muscle mechanics.* Chichester, New York: Wiley.

Hill, A.V. (1970). *First and last experiments in muscle mechanics.* Cambridge: Cambridge University Press.

MacIntosh, B.R., Gardiner, P.F., and McComas, A.J. (2006). *Skeletal muscle: Form and function.* Champaign, IL: Human Kinetics.

Vogel, S. (2001). *Prime mover.* New York: Norton.

Occupational

Chaffin, D.B., Andersson, G.B.J., and Martin, B.J. (2006). *Occupational biomechanics.* Hoboken, NJ: Wiley.

Salvendy, G. (2006). *Handbook of human factors and ergonomics.* Hoboken, NJ: Wiley.

Tichauer, E.R. (1978). *The biomechanical basis of ergonomics: Anatomy applied to the decision of work situations.* New York: Wiley.

Simulation and Modeling

Ivancevic, V.C., and Ivancevic, T.T. (2006). *Human-like biomechanics: A unified mathematical approach to human biomechanics and humanoid robotics.* Dordrecht (Great Britain): Springer.

Yamaguchi, G.T. (2001). *Dynamic modeling of musculoskeletal motion.* Boston: Kluwer Academic.

Sport

Hay, J.G. (1978). *Biomechanics of sports techniques.* Englewood Cliffs, NJ: Prentice Hall.

McGinnis, P.M. (2005). *Biomechanics of sport and exercise.* Champaign, IL: Human Kinetics.

Tissue Biomechanics

Currey, J.D. (2002). *Bone: Structure and mechanics.* Princeton, NJ: Princeton University Press.

Nordin, M., and Frankel, V.H. (2001). *Basic biomechanics of the musculoskeletal system.* Philadelphia: Lippincott, Williams & Wilkins.

Journals

Clinical Biomechanics

International Journal of Sport Biomechanics

Journal of Applied Biomechanics

Journal of Biomechanics

See also *Conference Proceedings of the International Society of Biomechanics* and the *World Congress of Biomechanics.*

Index

height of at takeoff of a jump 134-135

mechanics of jumping 134-135

movement of 96-97

overview 23-24

in a representation of muscle action 36-37

in standing 56-57, 56f

upward motion of 45-46, 46f

center of pressure (COP)

in walking 102-103

central nervous system 39

centripetal force 123

chemical energy 41-42

circular motion 20-21, 123, 260

climbing

aim of 218

biomechanics of 218-220

enhancement of 221-222

mechanics of 218

practical examples 222-224

variations of 220-221

coefficient of restitution (CR) 29

collagen fiber ligaments 9

collateral ligaments of the knee 7f

COM. *See* center of mass

conservation and transfer of energy 38-39

contractile component (CC) 40

contractions. *See* muscle contractions

contralateral motion for running 102

CR. *See* coefficient of restitution

crouch lift 174-175

cruciate ligaments 6-7

cruciate ligaments of the knee 7f

cycling

muscle power for 43

D

dimensional analysis 14-15

direct precession 248

displacement

angular 65f

of the CM 57-58

described 15

and time 17f, 18

from velocity, calculating 17f

distance running 117

diving 70, 143-144

dorsiflexion, ankle 99

dorsiflexor moment 99, 100f

double pendulum 100-101

downhill running 124

downhill walking 103-104, 104f

dynamic analysis, inverse 47-48

dynamic balancing 61-62

dynamic case 45-46

dynamic functioning

of muscles 39f, 40

E

eccentric muscular contraction 42, 104

elastic component of muscles 40

elasticity

of cartilage 6

in muscles 39-40

electrochemical energy to the muscles 112-113

electromyographic activity of muscles 49

endoskeleton 4-5

energy. *See also* mechanical energy

chemical 41-42

conservation and transfer of 38-39

cost in running 113, 120

electrochemical, to the muscles 112-113

exchanged from kinetic to potential forms 97f

kinetic 42

metabolic 79

relationship between force and extension 6f

rotational 38

rotational kinetic energy (RKE) 27-28

work-energy relationship 27

energy cost of distance running 117

energy dissipation 6, 86, 97

enhancement

airborne maneuvers 249-250

of carrying 187

of catching 212-213

of climbing 221-222

of gripping 159

of jumping 144-146

of lifting and lowering 180-181

of pulling and pushing 165-167

of running 124-125

and safety of falling and landing 87-91, 90ff

and safety of walking 108-109

of standing 61

of swinging 230-231

of throwing and striking 203-206

equations of uniform motion 17-18

equilibrium, static 56-57

externally applied force 45-46

F

fainting 90

falling. *See also* practical examples

arrested by the arms 92f

due to fainting 90

due to reduced floor friction 107

About the Author

Your author in his early years and recently with his model.

Arthur E. Chapman, PhD, is Professor Emeritus in the School of Kinesiology at Simon Fraser University in Burnaby, British Columbia, Canada, where he has taught and researched since 1970. Chapman has published more than 35 articles and presented more than 45 papers for refereed conferences, seminars, and workshops throughout the world. His past research interests included validation and modification of mechanical models of human muscle by means of direct observation in vivo, and the mechanical properties of squash balls, racquets, and shoes and their implications for manufacturing and strategy in the game. Current interest is in computer simulation of control and performance of sporting movements, kinematic and kinetic criteria of skills involving gross body movements, and the modeling of human bodily motion using external inputs of force and internal inputs of muscle force.

At Simon Fraser University, Chapman has served as a member of the University Ethics Committee, as chair of the Departmental Safety Committee, and chair of the Human Movement Stream for the Undergraduate Curriculum Committee. Chapman is a past member of the Canadian Association of Sports Sciences, and a founding member of both the Canadian Society of Biomechanics and the International Society of Biomechanics.

Between 1997 and 2000, Chapman served as an interviewer for the University of British Columbia Medical Admissions Board. As a biomechanist, he has served as an expert witness in numerous court cases throughout Canada providing human biomechanical analysis of automobile accidents, sports injuries, trips, and falls.

Chapman received his PhD in Biomechanics in 1975 from the University of London, England. A 1965 Fulbright Scholar, Chapman was also selected as the Rosenstadt Research Professor for the University of Toronto in 1992.